ENGINEERING ECONOMY

ENGINEERING ECONOMY

N V R Naidu
Professor
Department of Industrial Engineering & Management
M S Ramaiah Institute of Technology
Bangalore, Karnataka

K M Babu
Professor & Head
Department of Industrial Engineering & Management
BMS College of Engineering
Bangalore, Karnataka

G Rajendra
Assistant Professor
Department of Industrial Engineering & Management
Dr. Ambedkar Institute of Technology
Bangalore, Karnataka

NEW AGE INTERNATIONAL (P) LIMITED, PUBLISHERS
LONDON • NEW DELHI • NAIROBI
Bangalore • Chennai • Cochin • Guwahati • Hyderabad • Kolkata • Lucknow • Mumbai
Visit us at **www.newagepublishers.com**

Published by New Age International (P) Ltd., Publishers
First Edition: 2006
Reprint: 2018

GLOBAL OFFICES

- **New Delhi** **NEW AGE INTERNATIONAL (P) LIMITED, PUBLISHERS**
7/30 A, Daryaganj, New Delhi-110002, (INDIA)
Tel.: (011) 23253771, 23253472, **Telefax:** 23267437, 43551305
E-mail: contactus@newagepublishers.com • Visit us at www.newagepublishers.com
- **London** **NEW AGE INTERNATIONAL (UK) LTD.**
27 Old Gloucester Street, London, WC1N 3AX, UK
E-mail: info@newacademicscience.co.uk • Visit us at www.newacademicscience.co.uk
- **Nairobi** **NEW AGE GOLDEN (EAST AFRICA) LTD.**
Ground Floor, Westlands Arcade, Chiromo Road (Next to Naivas Supermarket)
Westlands, Nairobi, KENYA, **Tel.:** 00-254-713848772, 00-254-725700286
E-mail: kenya@newagepublishers.com

BRANCHES

- **Bangalore** 37/10, 8th Cross (Near Hanuman Temple), Azad Nagar, Chamarajpet, Bangalore- 560 018
Tel.: (080) 26756823, **Telefax:** 26756820, **E-mail: bangalore@newagepublishers.com**
- **Chennai** 26, Damodaran Street, T. Nagar, Chennai-600 017, **Tel.:** (044) 24353401, **Telefax:** 24351463
E-mail: chennai@newagepublishers.com
- **Cochin** CC-39/1016, Carrier Station Road, Ernakulam South, Cochin-682 016, **Tel.:** (0484) 2377303, **Telefax:** 4051304
E-mail: cochin@newagepublishers.com
- **Guwahati** Hemsen Complex, Mohd. Shah Road, Paltan Bazar, Near Starline Hotel, Guwahati-781 008,
Tel.: (0361) 2513881, **Telefax:** 2543669, **E-mail: guwahati@newagepublishers.com**
- **Hyderabad** 105, 1st Floor, Madhiray Kaveri Tower, 3-2-19, Azam Jahi Road, Near Kumar Theater, Nimboliadda Kachiguda, Hyderabad-500 027, **Tel.:** (040) 24652456, **Telefax:** 24652457
E-mail: hyderabad@newagepublishers.com
- **Kolkata** RDB Chambers (Formerly Lotus Cinema) 106A, 1st Floor, S N Banerjee Road, Kolkata-700 014
Tel.: (033) 22273773, **Telefax:** 22275247, **E-mail: kolkata@newagepublishers.com**
- **Lucknow** 16-A, Jopling Road, Lucknow-226 001, **Tel.:** (0522) 2209578, 4045297, **Telefax:** 2204098
E-mail: lucknow@newagepublishers.com
- **Mumbai** 142C, Victor House, Ground Floor, N.M. Joshi Marg, Lower Parel, Mumbai-400 013
Tel.: (022) 24927869, **Telefax:** 24915415, **E-mail: mumbai@newagepublishers.com**
- **New Delhi** 22, Golden House, Daryaganj, New Delhi-110 002, **Tel.:** (011) 23262368, 23262370, **Telefax:** 43551305
E-mail: sales@newagepublishers.com

ISBN: 978-81-224-1909-2

C-17-05-10450

Printed in India at Print'O'Pack, Delhi.
Typeset at Kalyani Computers, Delhi.

NEW AGE INTERNATIONAL (P) LIMITED, PUBLISHERS
7/30 A, Daryaganj, New Delhi-110002
Visit us at **www.newagepublishers.com**
(CIN: U74899DL1966PTC004618)

This book is dedicated
to Lotus Feet of
Lord Venkateshwara
and to our
Beloved Parents

PREFACE

Engineering Economy has now emerged as the most spectacular aspect of Economics and Financial Management. This book covers fundamental concepts of Engineering Economy and Financial Management. Practising professionals usually rate their Engineering Economy course as one of the most useful subjects studied in college.

Engineering Economics is a core in decision making based on comparisons of the worth of alternatives. Courses of action must be taken with respect to their costs. Decisions must be made at all organisational levels in both public and private sectors of the economy. Tools for decision making range from standardized work sheets for discounted cash flow evaluating to refinements necessary for sensitivity and risk analysis.

The book provides information regarding history of Engineering Economy, represents the information of traditional engineering concerns for operating economics to include financial consideration and decision analysis techniques.

It presents time–value calculations and comparisons of alternatives based on their equivalent annual worth, present worth and rate of return. Interest factors follow the suggested functional notations of the American Society for Engineering Education and interest calculations are displayed in cash flow diagrams.

It gives structure evaluations to determine investment alternatives or replacement policy and highlights of depreciation and tax concept of corporate income considerations and also the break-even analysis in examining current conditions.

This book contains a number of solved problems for each chapter and provides different situations and for ease of graphs and appended with a good number of exercise problems.

This book also contains the field of financial management which has undergone many changes. It is intended for undergraduate and also for post-graduate students in Engineering and Management. It also structures in decision making in the three interrelated financial areas: investment–long term as well as current assets, financial and divident policy and at the end of the book tables are provided.

The authors wish to express their sincere thanks to the principals and managements of their respective colleges. Further, they would like to thank Mr. Saumya Gupta, Managing Director and V.R. Babu, New Age International (P) Limited, Publishers, for their commitment and encouragement in bringing out this book in time with good quality.

Bangalore

Dr. N.V.R. Naidu
Dr. K.M. Babu
Sri. G. Rajendra

CONTENTS

1

INTRODUCTION

1.1 INTRODUCTION

Engineers are planners and builders. They are also problem solvers, managers, and decision makers. Engineering economics touches each of these activities. Plans and productions must be financed. Much of the management function is directed toward economic objectives and is monitored by economic measures. Engineering economics is closely aligned with conventional microeconomics. It is devoted to problem solving and decision making at the operations level. It can lead to suboptimization a condition in which a solution satisfies tactical objectives at the expense of strategic effectiveness but careful attention to the collection and analysis of data minimizes the danger. Evaluations rely mainly on mathematical models and cost data, but judgement and experience are pivotal inputs.

1.2 ENGINEERING DECISION MAKERS

The following general questions are representative of those that an engineer might encounter:

- Which one of several competing engineering designs should be selected?
- Should the machine now in use be replaced with a new one?
- With limited capital available, which investment alternative should be funded?
- Would it be preferable to pursue a safer conservative course of action or to follow a riskier one that offers higher potential returns?
- How many units of production have to be sold before profit can be made? This area is commonly called break-even analysis.
- Among several proposals for funding that yield substantially equivalent worth while results but have different cash flow patterns, which is preferable.
- Are the benefits expected from a public service project large enough to make its implementation costs acceptable?

Two characteristics of the questions above should be apparent. First, each deals with a choice among alternative; second, involve economic considerations. These considerations are embodied in the decision-making role of engineering economists to

1. Identify alternative uses for limited resources and obtain appropriate data.
2. Analyzes the data to determine the preferred alternative.

The breadth of problems, depth of analysis, and scope of application that a practicing engineer encounters very widely. In the following typical situations economic decisions are required.

Should a manufacturing plant in its own production facility, knowing that major investment will be needed in new equipment and that expensive training procedures will have to be implemented, or should the plant subcontract to an outside vendor? A university is planning a new foot ball stadium. Should the stadium be constructed now with a planned seating capacity of 80,000, or should it first be constructed with 65,000 seats with a planned end-zone enclosure to bring it to 80,000 seats in 5 years? Projected attendance revenues, expected increases in labour costs in 5 years, and potential stadium use problems during expansion are all factors that need to be considered. A manufacturing engineer is planning a high-speed production line that will use automated transfer mechanisms to move and position products from one automated work station to the next. More complex work stations will allow more operations to be completed at a work station at the expense of lower production rates per hour. However, such a situation could have the advantage of allowing fewer expensive transfer mechanisms. Given forecasts of products demand for the next 5 years should the engineer plan for a one-shift operation with a certain number of transfer mechanisms or for a two-shift operation with fewer transfer mechanisms? A decision is simply the selection from two or more course of action, whether it takes place in construction or production operations, services or manufacturing industries, private or public agencies. Most major decisions, even personal ones, have economic overtones. This consistent usage makes the subject of engineering economics especially challenging and rewarding.

1.3 ENGINEERING AND ECONOMICS

Engineers were mainly concerned with the design, construction and operation of machines, structures and processes. They gave less attention to the resources, human and physical that produced the final product. Many factors have since contributed to an expansion of engineering responsibilities and concerns. Engineers are now expected not only to generate novel technological solution but also to make skillful financial analyses of the effects of implementation. In today's close and tangled relations among industry, the public and government, cost and value analysis are expected to be more detailed and inclusive (*e.g.*, worker safety, environmental effects, consumer protection, resource conservation) than ever before. Without these analysis, an entire project can easily become more of a burden than a benefit. Most definitions of engineering suggest that the mission of engineers is to transform the resources of nature for the benefit of the human race. A growing awareness of the finite limits of the earth's resources has added a pressing dimension to engineering evaluations. The focus on scarce resources welds engineering to economics. Scientists are devoted to the discovery and explanation of nature's laws. Engineers work with scientists and translate the revelations to practical applications. The "laws" of economics are not as precise as those of physics, but their obvious applications to the production and utilization of scarce resources ensures increasing attention from engineers.

1.4 PRINCIPLES OF ENGINEERING ECONOMY

The development, study and application of any methodology must begin with a basic foundation. The foundation for engineering economy is a set of principles, on fundamental concepts, that provide a sound basics for development of the methodology. The principles must be adhered to and in most instances are also integral to decision making. The following are seven principles of Engineering Economics.

1. The choice (decision) is among alternatives. The feasible alternatives need to be identified and then defined for subsequent analysis.
2. Only the differences in expected future outcomes among the alternatives are relevant to their comparison and should be considered in the decision.
3. The prospective outcomes of the feasible alternatives, economic and other, should be consistently developed from a defined view point (perspective).
4. Using a common unit of measurement to enumerate as many of the prospective outcomes as possible will make easier the analysis and comparison of the feasible alternatives.
5. Selection of a preferred alternative (decision making) requires the use of a criterion (on criteria). The decision process should consider the outcomes enumerated in the monetary unit, and those expressed in some other unit of measurement on made explicit in a descriptive manner.
6. Uncertainty is inherent in projecting (on estimating) the future outcomes of the feasible alternatives and should be recognized in their analysis and comparison.
7. Improved decision making results from an adaptive process; to the extent practicable, initial projected outcomes of the selected alternative and actual results achieved should be subsequently compared.

1.5 PROBLEM SOLVING AND DECISION MAKING

An engineering economist draws upon the accumulated knowledge of engineering and economics to fashion and employ tools to identify a preferred course of action. There is still considerable debate about their theoretical bases and how they should be used. There are many aspects to consider and many ways to consider them.

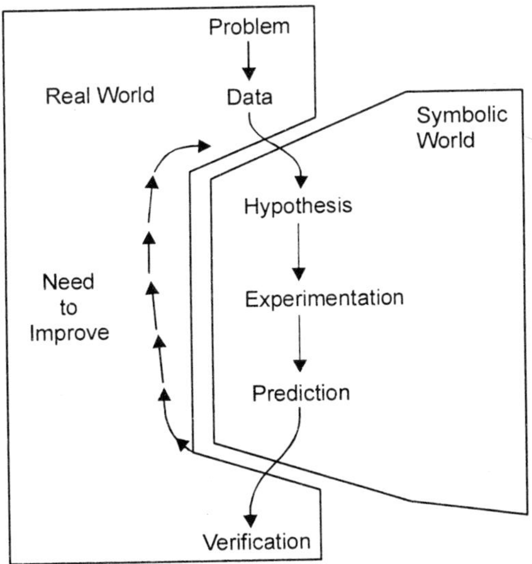

Fig. 1.1 Problem solving process.

The fundamental approach to economic problem solving is to elaborate on the time-honored scientific method. The method is anchored in two worlds: the real, everyday working world and the abstract, scientifically oriented world as shown in Fig. 1.1. Problems in engineering and managerial economy originate in the real world of economic planning, management, and control. The problem is confined and clarified by data from the real world. This information is combined with scientific principles supplied by the analyst to formulate a hypothesis is symbolic terms. By manipulating and experimenting with the abstractions of the real world, the analyst can simulate multiple configurations of reality that otherwise would be too costly on too inconvenient to investigate. From this activity a prediction usually emerges.

The predicted behaviour is converted back to reality for testing in the form of hardware designs, or commands. If it is valid, the problem is solved. If not, the cycle is repeated with the added information that the previous approach was unsuccessful.

1.6 INTUITION AND ANALYSIS

Engineers generally attack practical problems with solution dead-lines instead of engaging in esoteric issues for long-term enlightenment, their mission might appear relatively simple. Engineering economic evaluations could even seem mundane, since they usually rely on data from the market place and technology from the shelf-simply grab prices from a catalog, plug them into a handy formula, and grind out an answer. Occasionally, such a routine work.

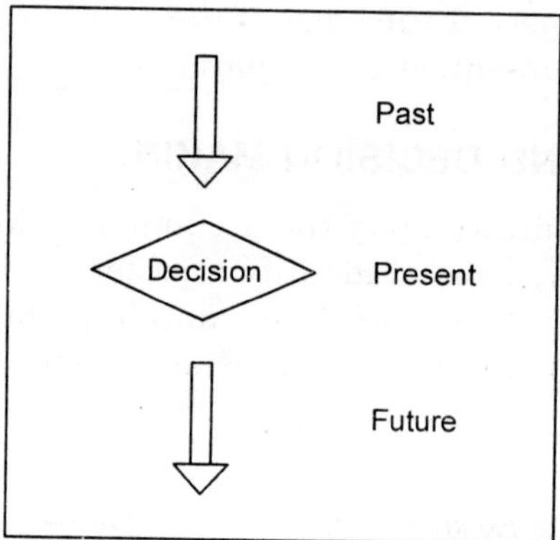

Fig. 1.2 Decision-making process.

As represented in Fig. 1.2, A decision made now is based an data from past performances and establishes a course of action that will result in some future outcomes. When the decision is shallow and the outcomes are not of much consequence, a reflex response based upon intuition is feasible. Instinctive judgements are often formalized by standard operations procedures (SOPs).

There SOP forms represent collective intuition derived from experience. They have a secure place in economic evaluations, but their use should be tempered by economic principles and a continuing audit to verify that previous judgements are appropriate for current decisions.

As the solution procedures progress, factors that are difficult to quantify often arise. These are called intangibles, they represent aspects of a problem that cannot be translated readily to monetary values. Judgement also enters the process in determining whether a solution is well enough founded to be accepted. Thus intuition and judgement complement analysis methods by contributing to better decisions.

1.7 TACTICS AND STRATEGY

Some problems are virtually handed to an analyst on a platter, complete with data trimmings. More commonly, a problem is ill-defined, and the analyst is forced to seek the intent of a solution before applying analytical tools. Recognizing the difference between tactical and strategic considerations may clarify the purpose.

Strategy and tactics historically are military terms associated with broad plans from the high command and specific schedules from lower echelons, respectively. Strategy sets ultimate objectives and the associated tactics define the multiple maneuvers required to achieve the objectives.

Strategic and tactical considerations have essentially the same meaning for economic studies. There are several strategies available to an organization. A stralegic decision ideally selects the overall plan that makes the best use of the organization's resources, in accordance with it's long-range objectives. A strategic industrial decision could be a choice from several different product designs to develop or products to promote. In government, strategic evaluations could take the form of benefit-cost analysis to select the preferred method of flood control or development of recreational sites.

A strategic plan can normally be implemented in a number of ways. For example, each industrial design or product has practical alternatives, which kind of machine to employ or materials to use, tactics for flood control might involve choices among dams, levees, dredging etc.

The relationship between strategies and tactics offers some constructive insights. The effectiveness of each strategy in initially estimated from the effect it will have on system objectives. It thus serves as a guide to the area in which tactics will produce the highest efficiency. The actual efficiency of each tactic is determined from a study of the activities required to conduct the tactical operation.

1.8 SENSITIVITY AND SUB-OPTIMIZATION

A sensitivity analysis can be conducted on any problem to explore the effects of deviations from the original problem conditions. Since most engineering economic problems extend over a period of years, future cash flows are necessarily estimated. These estimations may be quite reliable. It is often enlightening to observe how the attractiveness of alternatives varies as the initial estimates are altered.

In general, sub-optimization occurs when there is a larger problem than the analyst had visualized. It is always tempting to employ a classical text-book solution to a real-world problem, whether or not it truly fits the actual conditions. Another cause of sub-optimal solutions is the legitimate analysis technique of partitioning a large problem into a set of interdependant smaller problems during a preliminary investigation to avoid being bogged down in a deluge of details. Trouble enters when tentative solutions to the sub problems are not integrated. Advance in computer science and operations research may eventually allow analysis of an entire complex system in a single evaluation, but, it helps to be aware of the areas in which sub-optimization is most likely to occur. Three regularly encumbered perspectives that lead to sub-optimization are described below.

1. The cross-eyed view.
2. The short-sighted view.
3. Turn vision view point.

1.9 ENGINEERING ECONOMIC DECISION MAZE

The important decisions in engineering economics entail consideration of future events.

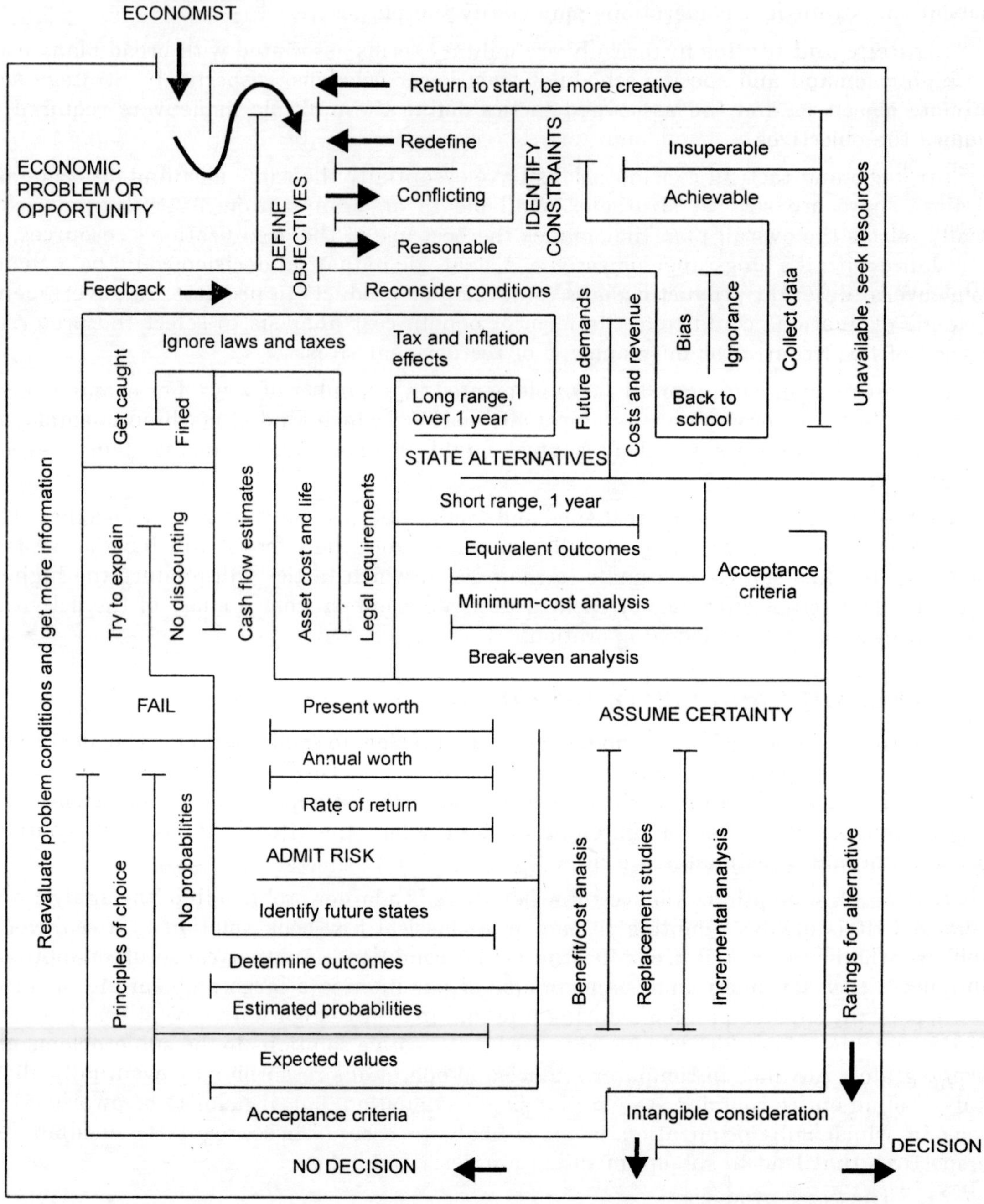

Fig. 1.3 Engineering economic decision maze.

A focus on the future has always had a special and irresistible appeal, it also encumbers the mission of engineering economists. It is must be search the past to understand the present and survey the present for hints about the future and consolidate the accumulated results into a pattern that a susceptible to analysis and then select a decision rule to yield a verdict. The complexities involved are similar to going through a maze as shown in Fig. 1.3.

It would take a much larger maze to portray are the pitfalls and challenges of economic analyses and information is included to expose the anatomy of engineering economics and to map the contents of economic analysis. There are many paths by which we could travel to get from a problem to a solution. Which path is utilized depends on the nature of the problem and the type of analysis that is most appropriate. Engineering economics is rich in application opportunities and offers rewarding challenges to its practitioners.

1.10 LAW OF SUPPLY AND DEMAND

An interesting aspect of the economy is that the demand and supply of a product are interdependent and they are sensitive with respect to the price of product. The interrelationships between them are shown in Fig. 1.4.

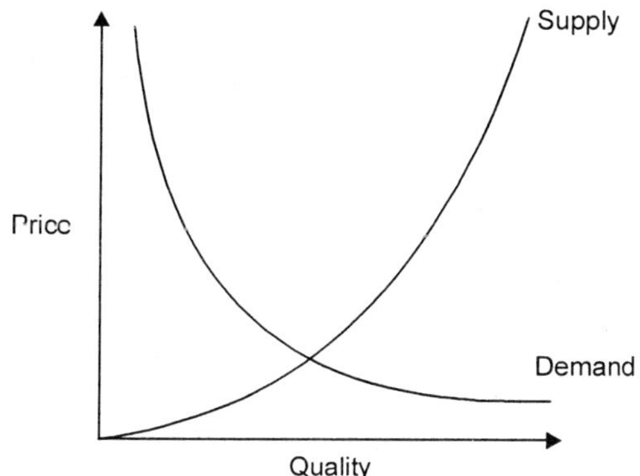

Fig. 1.4 Demand and supply curve.

It is clear that from Fig. 1.4, when there is a decrease in the price of a product, the demand for the product increases and its supply decreases. When lowering the price of the product makes the producers restrain from releasing more quantities of the product in the market. Hence, the supply of the product is decreased. The point of intersection of the supply curve and the demand curve is known as the equilibrium point. The quantity of supply is equal to the quantity of demand. This point is called equilibrium point.

Factors Influencing Demand

The shape of the demand curve is influenced by the following factors:

- Income of the people
- Prices of related goods
- Tastes of customers.

If the income level of the people increases, automatically their purchasing power will naturally improve. This would definitely shift the demand curve to the north-east direction as shown in Fig. 1.4. For example, the price of computer sets is lowered drastically its demand would naturally go up. As a result, the demand for its associated product, namely as shown in Fig. 1.4. For example, the price of computer sets is lowered drastically its demand would naturally go up. As a result, the demand for its associated product, namely printers, CD, floppy, CD writers would also increase. Hence, the prices of related goods influences the demand of a product.

Over a period of time, the preference of the people for a particular product may increase and also affect its demand.

Factors Influencing Supply

The shape of the supply curve is affected by the following factors:

- Costs of the inputs
- Technology
- Weather
- Prices of related goods

If the cost of inputs increases, naturally the cost of the products will go up. In this situation the product profit margin per unit will be less. The producer's will then reduce the production quality which in turn will affect the supply of the product. For example, if the prices of fertilizers and cost of labour are increased significantly, in agriculture, the profit margin per bag of paddy will be reduced. Then, the farmers will reduce the area of cultivation and hence, the quality of supply of paddy will be reduced at the prevailing prices of the paddy.

If there is an advancement in technology used in the manufacture of the product in the long run, there will be a reduction in the production cost per unit. Then it will enable greater profit margin per unit at the price of the product. Hence, the producer will be tempted to supply more products to the market. Weather also has a direct bearing on the supply of products. For example, during winter woollen products will have demand more. This means the prices of woollen goods will be increased in winter. So, naturally, the manufacturer will supply more volume of woollen goods during winter.

INTEREST AND INTEREST FACTORS

2.1 INTRODUCTION

Interest is the cost of using capital. In earliest times, before money was coined, capital was represented by wealth in the form of personal possessions and interest was paid in kind.

Example. A loan of feed to a neighbour before planting was returned after harvest with an additional increment.

The concept of interest has not changed much through the centuries but the modern but the modern credit structure differs markedly from that of antiquity. Today, there are many credit instruments and most people use them. Business and Government are the biggest borrowers. Business seek the use of capital goods to increase productivity. Governments borrow against future tax revenues to finance high ways, welfare programs and public services.

The majority of engineering economy studies involve commitment of capital for extended period of time, so the effect of time must be considered.

Example. A owner today is worth more than one or more years from now because of the interest (or profit) it can earn.

2.2 THE ORIGINS OF INTEREST

Like taxes, interest has existed from earliest recorded human history. Records reveal it's existence in Babylon in 2000 B.C.

History also reveals that the idea of Interest became so well established that a form of International bankers existed in 575 B.C. with home offices in Babylon. The Interest changed for a loan in the tense of 6 to 25% and some time legally sanctioned as high as 40%. Interest is nothing but the rent to be paid for the money borrowed. The period may be a week, monthly, quarterly, bi-annual or yearly.

There are two main rates of interest which are commonly practiced.

(*i*) Simple Interest

(*ii*) Compound Interest

(*a*) Nominal interest rates

(*b*) Effective interest rates

(*c*) Continuous compounding.

2.3 SIMPLE INTEREST

When a simple interest rate is quoted, the interest earned is directly proportional to the capital involved in the loan. The interest and the number of interest periods for which the principal is committed, the Interest and Interest rate are said to be simple.

Simple Interest is not used frequently in commercial practice in modern times.

The formula for simple interest.

$$I = P \times N \times i$$

I = Interest earned for particular time periods.

P = Principal amount lent or borrowed

N = No. of interest periods (eg. years)

i = Interest rate per interest period.

If P is a fixed value, the annual interest charged is constant. Therefore, the total amount a borrower obligates to pay a lender is.

$$F = P + I = P + PiN = P\,(1 + iN)$$

Example. The rental cost of money is a loan of Rs. 1000 for 2 months @ 10%. Use Simple Interest.

The amount to be repaid is

$$P = 1000$$

$$N = \frac{2}{12}\text{ years}$$

$$i = 10\%$$

$$F = P(1 + iN)$$

$$F = 1000\left(1+0.1\times\frac{2}{12}\right)$$

$$= 1016.67$$

Exact simple interest when the two months are January and February in a non-leap year. The future sum are

$$F = P\left[1+i\left[\frac{31+28}{365}\right]\right] = 1016.16$$

If the principal or amount borrowed P is a fixed value, the annual Interest charged is constant. Therefore, the total amount a borrower is obligated to pay a lender is

$$F = P + I$$

$$= P + PiN$$

$$= P\,(1 + iN)$$

When F is a future sum of money to be paid when N is not a full year. There are two ways to calculate the simple interest earned during the period of the loan.

(1) When ordinary Simple Interest is used the year is divided into twelve 30-days periods or a year is considered.

(2) In exact Simple Interest a year has precisely the calender number of days and N is the fraction of the number of days the loan is in effect that year.

An example of Simple Interest as the rental cost of money is a loan Rs. 2000 for 2 months at 10 percent. With ordinary simple interest, the amount to be repaid is

$$F = P(1 + iN)$$

$$N = \frac{2}{12}$$

$$F = 2000\left(1+\frac{2}{12}\right)$$

with exact simple interest for February in leap year and march the future sum is

$$F = P\left[1+i\left[\frac{28+31}{365}\right]\right]$$

$$F = 2000\left[1+0.1\left(\frac{59}{365}\right)\right]$$

$$F = 2032$$

SOLVED PROBLEMS

Problem 1. *A loan at Rs. 200 is made for a period of 13 months from January 1 to January 31 the following year at a simple interest rate of 10%, what future amount is due at the end of the loan period.*

Solution. Using ordinary simple interest,

The total amount repaid after 13 months is

$$F = P + PiN$$

$$= 200+200\times 0.10\times\left(1+\frac{1}{12}\right) = 221.67$$

Using exact simple interest

The future value is (it is not a leap year)

$$F = P + PiN$$

$$F = 200+200\times 0.10\left(1+\frac{31}{365}\right)$$

$$F = 221.70$$

Problem 2. *What sum must be loaned at 8% simple interest to earn Rs. 350 in 4 years.*

Solution.

$$P = ?$$

$$i = 8\%$$

$$I = 350$$

$$N = 4$$

$$I = PiN$$

$$\frac{I}{iN} = P$$

$$\frac{350}{0.08 \times 4} = 1093.75$$

$$P = 1093.75$$

Problem 3. *How long will it take Rs. 800 to yield Rs. 72 in simple interest at 4%*

Solution.

$$P = 800$$

$$I = 72$$

$$i = 4\%$$

$$N = 7$$

$$I = P \times i \times N$$

$$72 = 800 \times 0.04 \times N$$

$$\frac{72}{800 \times 0.04} = N$$

$$N = 2.25 \text{ years} = 2\tfrac{1}{4} \text{ years.}$$

Problem 4. *At what rate will 65.07 yield Rs. 8.75 in simple interest in 3 years 6 months?*

Solution.

$$P = 65.07$$

$$I = 8.75$$

$$i = ?$$

$$N = 3.6 \text{ years}$$

$$I = P \times i \times N$$

$$8.75 = 65.07 \times i \times 3.6$$

$$\frac{8.75}{65.07 \times 3.6} = i$$

$$i = 3.8\%.$$

Problem 5. *How long will it take any sum to triple itself at 5% simple interest rate?*

$$i = 5$$

$$P = 100$$

$$I = 300$$

$$N = ?$$

$$I = P \times i \times N$$

$$300 = 100 \times \frac{5}{100} \times N$$

$$\frac{300}{5} = N$$

$$600 = N$$

$$N = 60 \text{ years.}$$

Problem 6. *Determine the effective interest rate for a nominal annual rate of 6% that is compounded*

(i) *Semi-annually*

(ii) *Quarterly*

(iii) *Monthly*

(iv) *Daily.*

Solution. (1) Semi-annually

$$i_{eff} = \left(1+\frac{r}{m}\right)^n - 1$$

$$= \left(1+\frac{0.06}{2}\right)^2 - 1$$

$$= 0.0609 = 6.09\%.$$

(2) Quarterly

$$i_{eff} = \left(1+\frac{r}{m}\right)^m - 1$$

$$= \left(1+\frac{0.06}{4}\right)^4 - 1$$

$$= 0.0613 = 6.13\%.$$

(3) Monthly

$$i_{eff} = \left(1+\frac{r}{m}\right)^m - 1$$

$$= \left(1+\frac{0.06}{12}\right)^{12} - 1$$

$$= 0.0616 = 6.16\%.$$

(4) Daily

$$i_{eff} = \left(1+\frac{r}{m}\right)^m - 1$$

$$= \left(1+\frac{0.06}{365}\right)^{365} - 1$$

$$= 0.0618 = 6.18\%$$

Problem 7. *A Personal loan of Rs. 1000 is made for a period of 18 months at an interest rate of 1½ percent per month on the unpaid balance. If the entire amount owed is repaid in a lump sum at the end of that time, determine the effective annual interest rate.*

Solution.

$$P = 1000$$

$$N = 18 \text{ Months}$$

$$i = 1\tfrac{1}{2} \times 12 = 18\% \text{ P.A.}$$

$$i_{eff} = \left(1+\frac{r}{m}\right)^m - 1$$

$$= \left[\left(1+\frac{0.18}{18}\right)^{18} - 1\right]$$

$$= 0.1961$$

$$i_{eff} = 19.61\%.$$

Problem 8. *Find the compound amount of Rs. 100 for 4 years at 6% compounded annually.*

Solution.

$$P = 100$$
$$N = 4 \text{ years}$$
$$i = 6\%$$
$$F = ?$$
$$F = P(1 + i)^N$$
$$= 100(1 + 0.06)^4 = 100 \times 1.26248$$
$$F = 126.248$$

Problem 9. *A loan of Rs. 2000, if the interest rate is 10% per year. If interest had not been paid each year but, had been allowed to compound, how much interest would be due to the lender as a lump sum at the end of six years?*

Solution.

$$P = 2000$$
$$i = 10\%$$
$$N = 6$$
$$I = ?$$
$$F = P(1 + i)^n$$
$$= 2000(1 + 0.1)^6 = 2000(1.1)^6$$
$$= 2000(1.77156) = 3543.12$$
$$\text{Int.} = F - P = 3543.12 - 2000$$
$$\text{Interests} = 1543.12$$

Problem 10. *Accumulate a principle of Rs. 1000 for 5 years 9 months at a nominal rate of 12% compounded monthly. How much interest is earned?*

Solution.

$$P = 1000$$
$$N = 5\text{y. } 9 \text{ months} = 69 \text{ months or } 69 \text{ Interest-periods}$$
$$i = 12\% \approx 1 \text{ month is } 1\%$$
$$F_{69} = ?$$
$$F_{69} = P(1 + i)^n$$
$$= 1000(1 + 0.01)^{69}$$
$$F_{69} = 1986.9$$

2.4 NOTATION AND CASH FLOW DIAGRAMS

The following notation is used for Compound Interest Calculations:

i = effective interest rate per interest period

N = No. of compound periods.

P = Present sum of money; the equivalent worth of one or more cash flows at a reference point in time called the present.

F = Future sum of money; the equivalent worth of one or more cash flows at a reference point in time called the future.

A = End of period cash flows in a uniform series continuing for a specified number of periods, starting at the end of the first period.

* The use of cash flow (time) diagrams is strongly recommended for situations in which the analyst needs to clarify or visualize what is involved when flow of money occurs at various times.

* A cash flow is the difference between total cash inflows (receipts) and cash outflow (expenditures) for a specified period of time.

Cash flows are important in Engineering economy because they form the basis for evaluating alternatives.

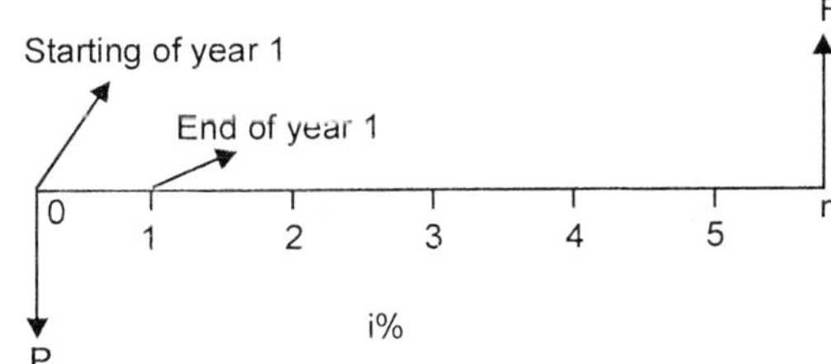

The cash flow diagram employs several convertions.

1. The horizontal line is a time scale with time. Progression of time moving left to right. The periods are marked to intervals of time 0, 1, 2, 3.......... rather than points on the time scale. For example the end of period 2 is coincident with the beginning of period 3.
2. The arrows signify cash flows. If a distinction needs to be made, downward arrows represent expenses (negative cash flows or cash outflows) and upward arrows represent receipts (positive cash flow or cash inflows).
3. The cash flow diagram is dependent on point of view. If the direction of all arrows had been reversed, the problem would be diagrammed from the borrower's view point.

Interest Formulas for Discrete Compounding and Cash Flows.

There are six most common discrete compound interest factors. These factors are derived and explained by example problems in the following sections and some are regular practiced in business.

The formulas also assume discrete (*i.e.*, lump sum) cash flows spaced at the end of equal time intervals on a cash flow diagram.

Interest Formulas

For Single Cash Flows

1. Single Payment Compound Amount
2. Single Payment Present Worth

For Uniform Series (annuities)

1. Uniform Series Compound Amount
2. Uniform Series Present Worth
3. Equal Payment Series Sinking Fund
4. Equal Payment Series Annual Equivalent Amount
5. Arithmetic Gradient Conversion Factor (to uniform series).

2.5 INTEREST FORMULAS FOR DISCRETE COMPOUNDING AND DISCRETE CASH FLOWS

Table 1 : Discrete Compounding Interest Factors and Symbols

To find	*Given*	*Factors which to multiply "Given"*	*Factor Name*	*Factor functional symbol*
For single cash flows				
F	P	$(i+i)^N$	Single Payment compound amount	(F/P, i%N)
P	F	$\frac{1}{(1+i)^N}$	Single Payment present worth	(P/F, i%N)
For uniform series (annuities)				
F	A	$\frac{(1+i)^N - 1}{i}$	Uniform series compound amount	(F/A, i%N)
P	A	$\frac{(1+i)^N - 1}{i(1+i)^N}$	Uniform series present worth	(P/A, i%N)
A	F	$\frac{i}{(1+i)^N - 1}$	Sinking fund	(A/F, i%N)
A	P	$\frac{i(1+i)^N}{(1+i)^N - 1}$	Capital recovery	(A/P, i%N)

1. Compound Amount Factor (Single Payment)

Use

To find F, given P

Symbols : (F/P, i%N)

Formula :

$$F = P(F/P, i\%N)$$

$$F = P(1 + i)^N$$

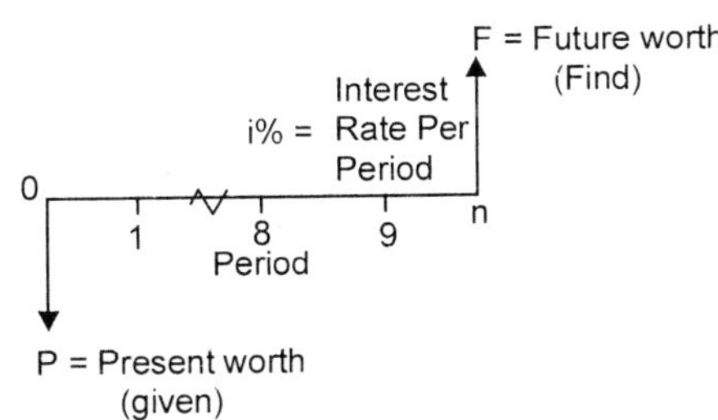

PROBLEMS

Problem 1. *A Future amount F, is equivalent to Rs. 1500. Now when eight years separates the amounts and the annual interest is 12%. What is the value of F?*

Solution.

$$P = 1500$$
$$F = ?$$
$$N = 8$$
$$i = 12\%$$
$$F = P(1 + i)^N$$
$$= 1500\ (1 + 0.12)^8$$
$$F = 3713.9$$

Problem 2. *A person deposits a sum of Rs. 10,000/- in a bank at a nominal rate of Interest of 12% for 10 years. Find the maturity amount of the deposit after 10 years. If the compounding is done quarterly.*

Solution.

$$P = \text{Rs. } 10{,}000$$
$$i = 12\%\ \text{(compounded quarterly)}$$
$$N = 10\ \text{years}$$
$$F = ?$$

When compounding is done for any period other than annual, either of the following two methods can be utilised.

Method 1

In this method, rate of interest 'i' is broken into many interest periods for which compounding is to take place in a year. 'N' becomes the total number of interest periods.

No. of Interest Periods per year = 4

No. of Interest Periods for 10 years = 10 × 4 = 40

Revised no. of periods (Quarters) n = 40

Rate of interest per quarter i = 12/4 = 3% = 0.03.

$$\therefore \quad F = 10{,}000\ (1 + 0.03)^{40}$$
$$F = 32620.37$$

Method 2

Find effective rate of interest

An effective rate of Interest (R) is found using the formula.

$$R = \left(1+\frac{i}{N}\right)^{N} - 1 \qquad i = \text{Annual rate of Interest (given)}$$

$$R = \left(1+\frac{0.12}{4}\right)^{4} - 1 \qquad N = \text{No. of interest periods year.}$$

R = 0.1255 or 12.55%

Hence, R replaces 'i' in the formula

$$F = P(1 + i)^N$$
$$= 1000(1 + 0.1255)^{10}$$
$$F = 32{,}620.$$

Problem 3. *Suppose you borrow Rs. 8000 now, with the promise to repay the loan principal plus accumulated interest in four years at i = 10% per year. How much would you owe at the end of four years?*

Solution.

$$i = 10\%$$
$$N = 4 \text{ years.}$$
$$P = 8000$$
$$F = ?$$
$$F = P(1 + i)^N$$
$$= 8000(1 + 0.10)^4$$
$$= 11{,}713.00$$

Cash Flow Diagram

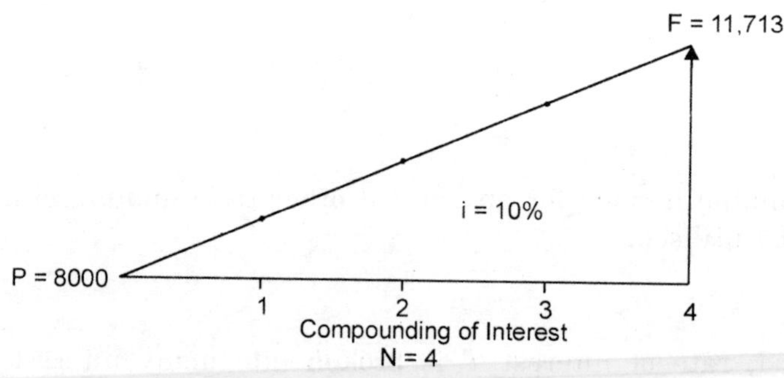

Problem 4. *Suppose that 10,000 is borrowed now at 15% interest per annum. A partial repayment of 3000 is made four years from now. The amount that will be remain to be paid then is most nearly (a) 7000 (b) 8050 (c) 8500 (d) 13000 (e) 14490.*

3000
0 1 2 3 4
10000 17490

$$F_4 = P(1 + i)^4$$
$$= 10{,}000(1 + 0.15)^4$$
$$= 10{,}000(1.15)^4$$
$$= 17490$$

The amount to be paid is

$$17490 - 3000 = 14490$$

The choice is (e).

2. Present Worth Factor (Single Payment)

Use : To find P, given F

Symbols : (P/F, i%N)

Formula :

$$P = F(1/(1 + i)^N)$$
$$= F(P/F, i\%N)$$

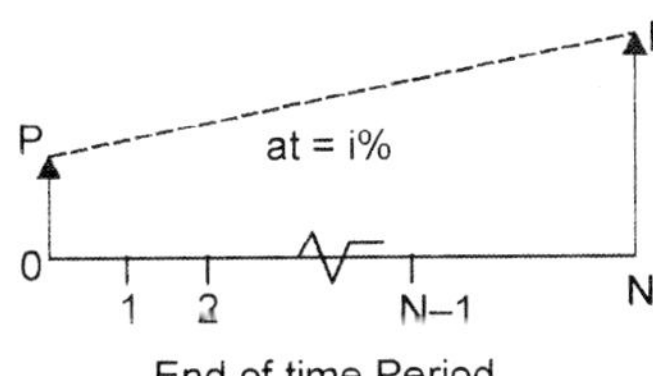

Problem 1. *An investor has an option to purchase a tract of land that will be worth Rs. 10000 in six years. If the value of the land increases at 8% each year, how much should the investor be willing to pay now for this property.*

Solution.

$$F = 10{,}000$$
$$N = 6 \text{ years}$$
$$i = 8\%$$
$$P = ?$$
$$P = F(P/F, i\%N)$$
$$= 10{,}000(P/F, 8\%,6)$$
$$= \frac{10{,}000}{(1+i)^N} = \frac{10{,}000}{(1+0.08)^6}$$

$$\boxed{P = 6302.00.}$$

Problem 2. *A person wishes to have a future sum of Rs. 10 lakhs for his daughter's engineering education in 15 years from now, what is the single payment that he should deposit now. So, that he gets the desired amount after 10 years? The banks gives 12% rate of interest compounded annually.*

Solution.

$$F = 10{,}00000$$
$$i = 12\%$$
$$N = 10 \text{ years}$$
$$P = ?$$
$$P = F(P/F, i\%N)$$
$$= 10{,}00000(P/F, 12\%,10)$$

$$= \frac{10,00000}{(1+0.12)^{10}}$$

$$\boxed{P = 321973.3}$$

Problem 3. *If the same person wishes to have choice of investing in a private bank which pays a rate of interest of 11% but compounded quarterly, should he go for it? His desire is to receive 10,00000, 10 years from now.*

Solution. F = 10 lakhs

i = 11%

No. of interest periods per year is four = 4

Total no. of interest periods for 10 years. = 4 × 10 = 40

Interest rate for one year = 11% = 0.11

$$\text{Interest rate quarterly} = \frac{0.11}{4}$$

$$P = F(P/F, i\%N)$$
$$= 10,00000\ (P/F, 0.11/4, 40)$$

$$P = \frac{F}{(1+i)^n}$$

$$= \frac{10,00000}{(1+0.11/4)^{40}}$$

$$\boxed{P = 337852.2}$$

In private bank at rate of interest 11% compounded quarterly in order to realize Rs. 10 lakhs after 10 years

Clearly the earlier option is better, so, he should not deposit in private bank.

3. Series Compound-Amount Factor (uniform series)

Use : To Find F, given A.

Symbols : (F/A, i%N)

Formula :

$$F = A\frac{(1+i)^N - 1}{i}$$

$$F = A\ (F/A, i\%N)$$

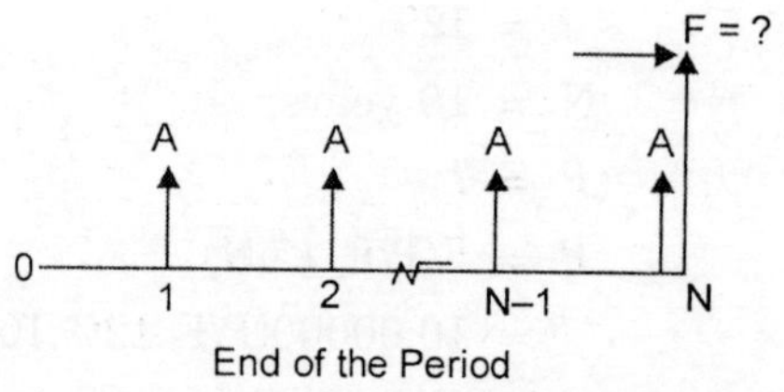

Problem 1. *A 45 year old person is planning for his retired life. He plans to divert Rs. 30,000/- from his bonus as investment every year for the next 15 years. The banks gives 10% interest rate compounded annually. Find the maturity value of his account when he is 60 years old.*

Solution.

$$A = 30,000$$
$$N = 15 \text{ years}$$
$$i = 10\%$$
$$F = ?$$
$$F = A(F/A, i\%N)$$
$$= 30,000\ (F/A,\ 10\%,\ 15)$$
$$= 30,000\left[\frac{(1+0.1)^{15}-1}{0.1}\right]$$
$$F = 9,53,174/-$$

The person receives the amount at the age 60 years. if he invests Rs. 30,000/- every year for the next 15 years. cash flow diagram

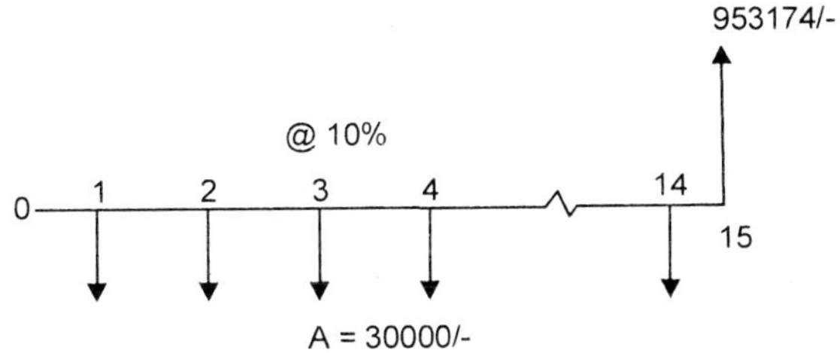

Problem 2. *A woman desires to have Rs. 100,000 in her retirement savings plan after working for 25 years. She will accomplish this by depositing A Rs. each year in a savings account that earns 6% per year. How much must she save each year.*

Solution.

$$F = 100,000$$
$$i = 6\%$$
$$N = 25 \text{ years.}$$
$$A = ?$$
$$A = F(A/F, i\%N)$$
$$= 100,000\ (A/F,\ 6\%,\ 25).$$
$$= 100,000(0.0182)$$
$$= 1820$$ From table discrete compound periods.

4. Sinking Fund Factor (Uniform Series)

Use : To find A, given F.

Symbols : (A/F, i%N)

Formula : $A = F\left(\frac{i}{(1+i)^N - 1}\right)$

$A = F(A/F, i\%N)$

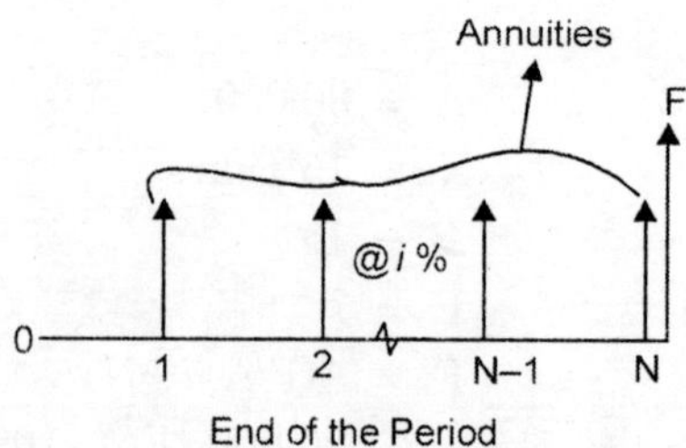

Problem 1. *A person estimates an expenditure of Rs. 5 lakh for his daughter's wedding about 10 years. from now. He plans to deposit an equal amount at the end of every year for the next 10 years at a rate of interest of 10% compounded annually. Find the equivalent amount that must be deposited at the end of every year for the next 8 years.*

Solution.

$F = 5$ lakhs.

$N = 10$ years.

$A = ?$

$i = 10\%$

$A = F(A/F, i\%N)$

$= 500000\ (A/F, 10\%, 10)$

$= 500000 \times \frac{i}{(1+i)^N - 1}$

$= 500000 \times \frac{0.1}{(1.1)^{10} - 1}$

$A = 31372.697$

The annual amount to be paid is 31372.697.

5. Series Present Worth Factor (Uniform Series)

Use : To find P, given A

Symbols : (P/A, i%N)

Formula : $P = A\left[\frac{(1+i)^N - 1}{i(1+i)^N}\right]$

$P = A(P/A, i\%N)$

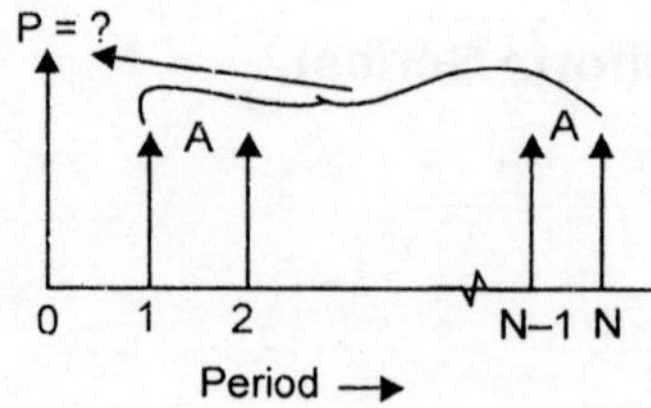

Problem 1. *It is estimated that a certain piece of equipment car save Rs. 6000 per year in labour and materials costs. The equipment has an expected life to five years and no salvage value. If the company must earn a 20% rate of return on such investments, how much could be justified now for the purchase of this piece of equipment? Draw a cash flow diagram.*

Solution.

$$\begin{aligned} A &= 6000 \\ N &= 5 \text{ years.} \\ Sr &= \text{No} \\ i &= 20\% \\ P &= ? \\ P &= A(P/A, i\%N) \\ &= 6000\left[\frac{(1+i)^N - 1}{i(1+i)^N}\right] \\ &= 6000\left[\frac{(1.2)^5 - 1}{0.2(1.2)^5}\right] \\ P &= 17943.7 \end{aligned}$$

6000 6000 6000 6000 6000

1 2 3 4 5

i = 20%

P = 17943.7 Period →

Problem 2. *Suppose that installation of low-loss thermal windows in your area is expected to save Rs. 150 a year on your home heating bill for the next 18 years. If you can earn 8% a year on other investments, how much could you afford to spend now for these windows?*

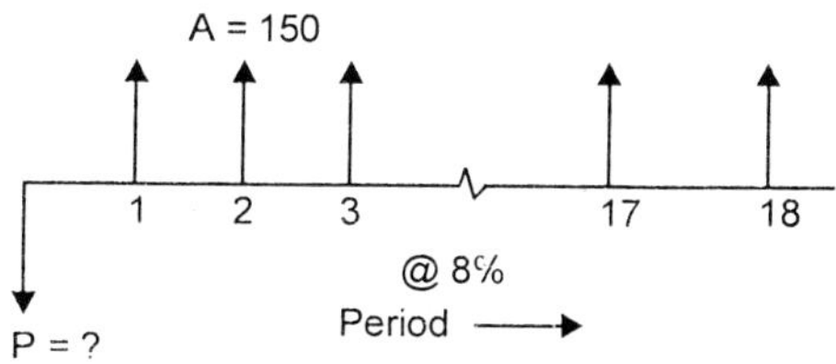

Solution.

$$\begin{aligned} A &= \text{Rs. } 150 \\ N &= 18 \text{ years.} \\ i &= 8\% \\ P &= ? \\ P &= A(P/A, i\%N) \\ &= A(P/A, 8\%, 18) \end{aligned}$$

$$P = 150\frac{(1+0.08)^{18}-1}{0.08(1+0.08)^{18}}$$

$$P = 1405.8$$

6. Capital Recovery Factor (Uniform Series)

Use : To find A, given P

Symbols : (A/P, i%N)

Formula :

$$A = P\left[\frac{i(1+i)^N}{(1+i)^N-1}\right]$$

$$A = P(A/P, i\%N)$$

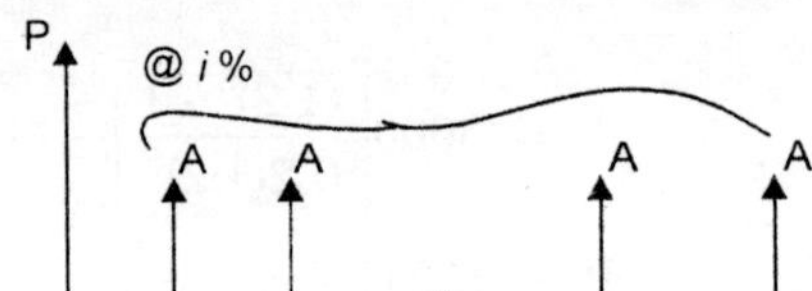

Problem 1. *A proposed product modification to avoid production difficulties will require an immediate expenditure of Rs. 14,000 to modify certain dies. What annual savings must be realized to recover this expenditure in four years with interest at 10%?*

Solution.

$$P = 14000$$

$$A = ?$$

$$N = 4$$

$$i = 10\%$$

$$A = P(A/P, i\%N)$$

$$A = P(A/P, 10\%, 4)$$

$$A = 14000\left[\frac{0.1(1+0.1)^4}{(1+0.1)^4-1}\right]$$

$$\boxed{A = 4416.6}$$

Problem 2. *If Rs. 25000 is deposited now into a savings account that earns 12% per year, what uniform annual amount could be withdrawn at the end of each year for 10 years so that nothing would be left in the account after the tenth withdraw?*

Solution.

$$P = 25000$$

$$i = 12\%$$

$$A = ?$$

$$N = 10$$

$$A = P(A/P, i\%N)$$
$$= P(A/P, 12\%, 10)$$
$$= 25000\left[\frac{0.12(1+0.12)^{10}}{(1+0.12)^{10}-1}\right]$$
$$\boxed{A = 4424.6}$$

7. Arithmetic Gradient Conversion Factor (Uniform Series)

Use : To find A given G.

Symbols : (A/G, i%N)

Formula :

$$A = G\left[\frac{1}{i} - \frac{N}{(1+i)^{N-1}}\right]$$

$$A = G(A/G, i, N)$$

G = Gradient increase in the cash flow receipts or disbursements.

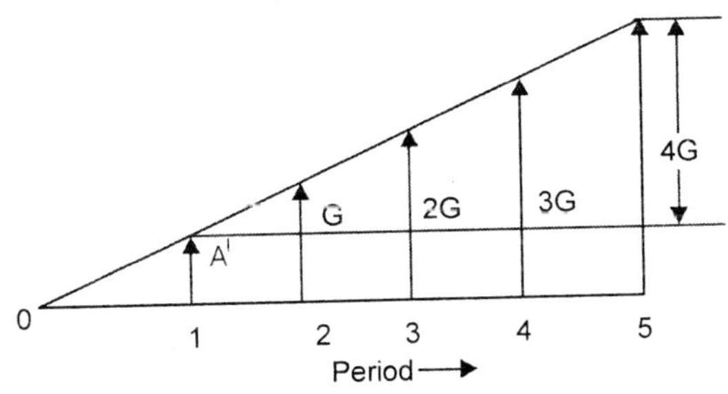

Problem 1. *Assume that an endowment was originally set up to provide a Rs. 10000. First payment with payments decreasing by 1000 each year during the 10 year endowment life. What constant annual payment for 10 years would be equivalent to the original endowment plan if i = 8%?*

Solution.

$$A' = 10000$$
$$G = 1000$$
$$N = 10 \text{ years}$$
$$i = 8\%$$
$$A = A' - G(A/G, i\%, N)$$
$$= 10000 - 1000\left[\frac{1}{0.08} - \frac{10}{(1+0.08)^{10}-1}\right]$$
$$\boxed{A = 6128.69}$$

Problem 2. *A film star is at the height of his carrier. He wants to invest Rs. 10 lakhs from the end of this year and follow it up with 9 lakhs, 8 lakhs and so on for the next five years, when his income would go on diminishing. Find the maturity amount 6 years later if a film producer agrees to pay him 15% rate of interest, compounded annually.*

Solution.

$A' = 10$ lakhs

$G = 1$ lakhs

$N = 6$ years

$F = ?$

$A = ?$

$i = 15\%$

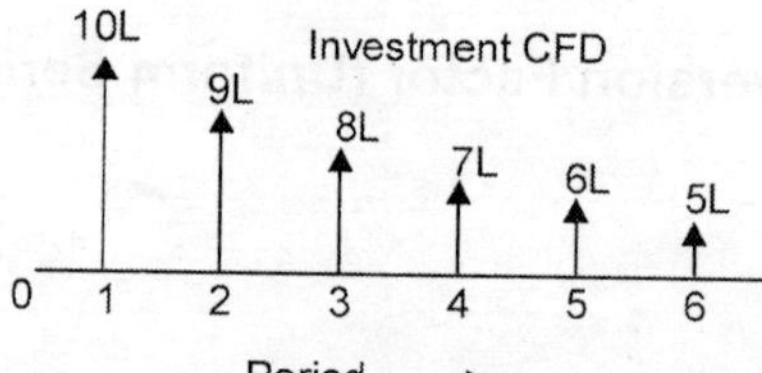

$$A = A^1 - G\,(A/G,\ i\%,\ N)$$

$= 1000000 - 100000$ (A/G, 15% 6) from the table of Discrete Series.

$= 1000000 - 100000[2.09719]$

$$\boxed{A = 7{,}90{,}280}$$

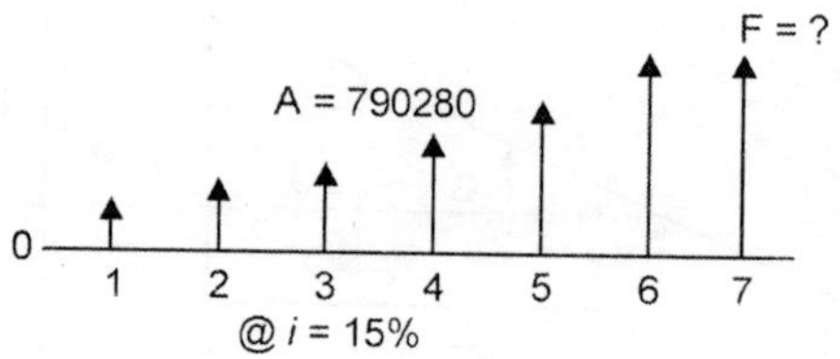

$$F = A(F/A,\ i\%N)$$

$= 790280$ (F/A, 15% 6) from table of Discrete Series.

$= 790280\ (8.75374) = 69{,}17{,}905 = 65$

SOLVED PROBLEMS

Problem 1. *If an investor invests a sum of Rs. 100 in a fixed deposit for five years with an interest rate of 15% compounded annually, the accumulated amount at the end of every year will be as shown in table 1.*

Solution.

Table 1 : Compound Amounts

Year End	*Interest (Rs.)*	*Compound Amount (Rs.)*
0	–	100.0
1	15.00	115.00
2	17.25	132.25
3	19.84	152.09
4	22.81	174.90
5	26.24	201.14

The find the compound amount as shown in table 1 is

$$F = P \times (1 + i)^N.$$

Problem 2. *A person deposits a sum of Rs. 100000 in a bank for his son's education who will be admitted to a professional course after 6 years. The bank pays 15% interest rate, compounded annually. Find the future amount of the deposited money at the time of admitting his son in the professional course.*

Solution.

$$P = 100000$$
$$N = 6 \text{ years}$$
$$i = 15\%$$
$$F = ?$$
$$F = P(F/P, i\%N)$$
$$F = P(1 + i)^N$$
$$F = 100000(1 + 15/100)^6$$
$$\boxed{F = 2{,}31{,}306.1}$$

Problem 3. *A person needs a sum of Rs. 200,000 for his daughter's marriage which will take place 15 years from now. Find the amount of money that he should deposit now in a bank if the bank gives 18% interest, compounded annually.*

Solution.

$$F = 200{,}000$$
$$N = 15 \text{ years}$$
$$P = ?$$
$$i = 18\%$$
$$P = F(P/F, i\%N)$$
$$= 200{,}000 \text{ (P/F, 18\% 15)}$$
$$= 200{,}000 \frac{1}{(1+i)^N}$$
$$= 200{,}000 \frac{1}{(1+18/100)^{15}}$$
$$\boxed{P = 16703.2}$$

Problem 4. *A financial institution introduces a plan to pay a sum of Rs. 15,00,000 after 10 years at the rate of 18% compounded annually. Find the annual equivalent amount that a person should invest at the end of every year for the next 10 years to receive Rs. 15,00,000 after 10 years from the institution.*

Solution.

$$F = 15{,}00{,}000$$
$$N = 10 \text{ years}$$
$$i = 18\%$$
$$A = ?$$
$$A = F(A/F, i\%N)$$
$$= 15{,}00{,}000 \text{ (A/F, 18\% 10)}$$
$$= 15{,}00{,}000 \left[\frac{18\%}{(1+18\%)^{10} - 1}\right]$$

$$= 15{,}00{,}000\left[\frac{0.18}{(1.18)^{10}-1}\right]$$

$$\boxed{A = 63771.9}$$

Problem 5. *A company 3 years ago borrowed 40,000 to pay for a new machine tool, agreeing to repay the loan in 100 monthly payments at an annual nominal interest rate of 12% compounded monthly. The company now wants to pay off the loan. How much would this payment be, assuming no penalty costs for early payout?*

Solution.

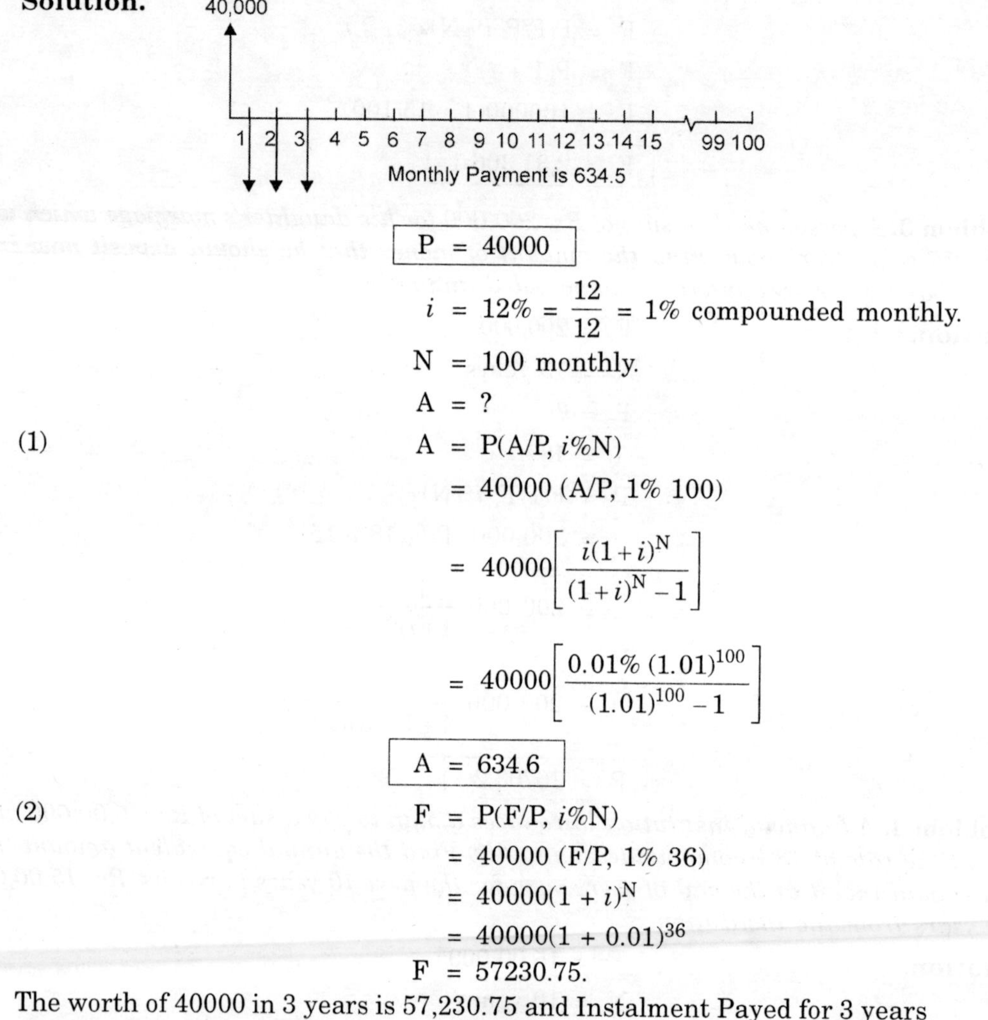

$$\boxed{P = 40000}$$

$$i = 12\% = \frac{12}{12} = 1\% \text{ compounded monthly.}$$

$$N = 100 \text{ monthly.}$$

$$A = ?$$

(1) $$A = P(A/P, i\%N)$$

$$= 40000\ (A/P, 1\%\ 100)$$

$$= 40000\left[\frac{i(1+i)^N}{(1+i)^N-1}\right]$$

$$= 40000\left[\frac{0.01\%\ (1.01)^{100}}{(1.01)^{100}-1}\right]$$

$$\boxed{A = 634.6}$$

(2) $$F = P(F/P, i\%N)$$

$$= 40000\ (F/P, 1\%\ 36)$$

$$= 40000(1 + i)^N$$

$$= 40000(1 + 0.01)^{36}$$

$$F = 57230.75.$$

The worth of 40000 in 3 years is 57,230.75 and Instalment Payed for 3 years

$$= 634.6 \times 36$$

$$= 22845.6$$

The pay off amount @ the end of 3 years.

$$= 57230.75 - 22845.6$$

$$\boxed{\text{Total Amount} = 34385.2}$$

Problem 6. *An automobile company recently advertised its car for a down payment of Rs. 1,50,000. Alternatively, the car can be taken home by customers without making any payment, but they have to pay an equal yearly amount of Rs. 25,000 for 15 years at an interest rate of 18% compounded annually. Suggest the best alternative to the customers.*

Solution. (1)

$$\text{CAR Value} = 150000$$
$$A = 25000$$
$$N = 15 \text{ years}$$
$$i = 18\%$$
$$F = ?$$
$$F = A(F/A, i\%N)$$
$$= 25000\ (F/A, 18\%\ 15)$$
$$= 25000\left[\frac{(1+i)^n - 1}{i}\right]$$
$$= 25000\left[\frac{(1+0.18)^{15} - 1}{0.18}\right]$$
$$\boxed{F = 1524131.6}$$

Problem 7. *A person is planning for his retired life. He has 10 more years of service. He would like to deposit 20% of his salary, which is Rs. 10,000 at the end of the first year and there after he wishes to deposit the same amount (Rs. 10,000) with an annual increase at Rs. 2000 for the next 9 years with an interest rate of 20%. Find the total amount at the end of the 10th year of the above series.*

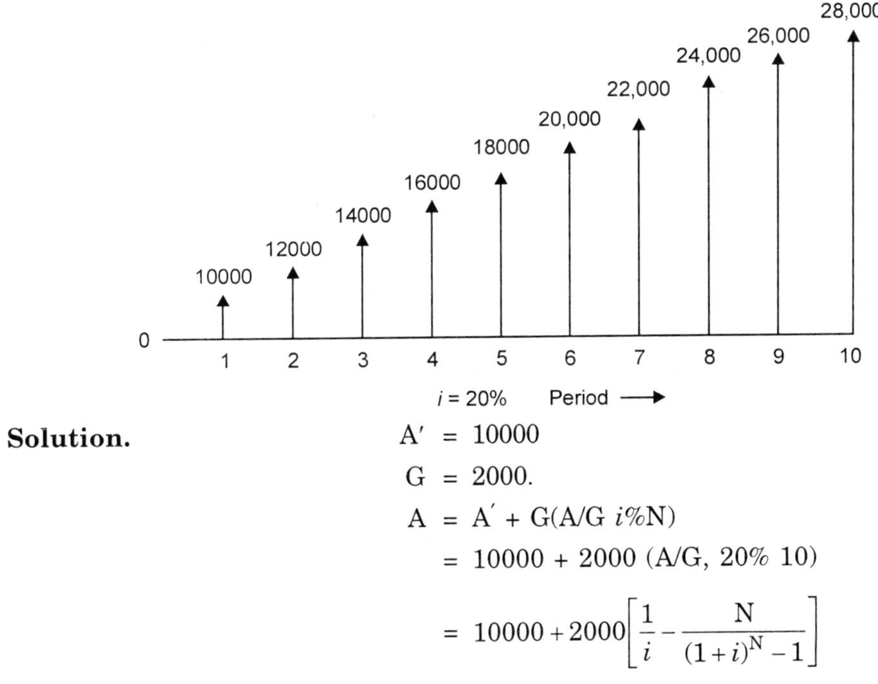

Solution.

$$A' = 10000$$
$$G = 2000.$$
$$A = A' + G(A/G\ i\%N)$$
$$= 10000 + 2000\ (A/G, 20\%\ 10)$$
$$= 10000 + 2000\left[\frac{1}{i} - \frac{N}{(1+i)^N - 1}\right]$$

$$= 10000 + 2000\left[\frac{1}{0.2} - \frac{10}{(1.2)^{10} - 1}\right]$$

$$A = 16147.7$$

$$F = A(F/A, i\%N)$$

$$= 16147.7 \text{ (F/A, 20\% 10)}$$

$$= 16147.7\left[\frac{(1+i)^N - 1}{i}\right]$$

$$= 16147.7\left[\frac{(1.2)^{10} - 1}{0.2}\right]$$

$$\boxed{F = 419173.01}$$

Problem 8. *A company is planning to buy an inspection device (coordinate - measuring machine) for Rs. 45,000. The expected life of the device in 5 years and the expected annual operating costs and taxes are Rs. 600 for the first year with an added increase per year Rs. 100 for years 2 through 5. Maintenance costs win be zero in the first 2 years because of the warranty but are expected to be Rs. 1000 in year 3, Rs. 1500 in year 4 & Rs. 2000 in year 5. What is the minimum desired annual economic benefit of the device, assuming that these benefits will just offset the annual costs? The company uses an interest rate of 10%. For economic evaluations.*

Solution:

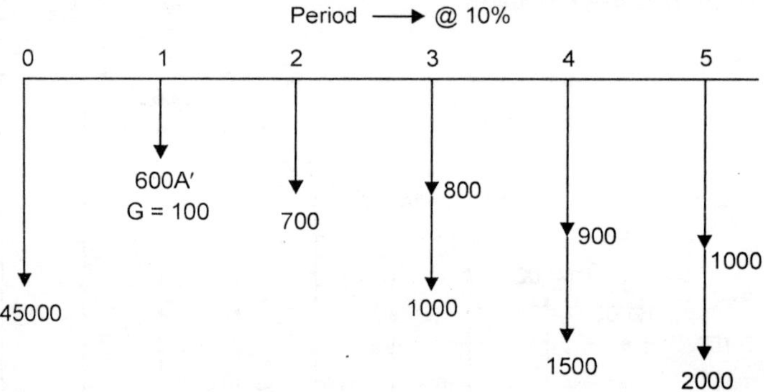

AW = 45000 (A/P, i%N) + [600 + 100 (A/G, i%N)] + [1000 (P/F, i%3) + 1500 (P/F, i%4) + 2000 (P/F, i%5)] (A/P, i%5)

= 4500 (A/P, 10, 5) + [600 + 100 (A/G, 10% 5] + [1000 (P/F, 10% 3) + 1500 (P/F, 10% 4) 2000 (P/F, 10% 5)] (A/P, 10% 5)

$$= 4500\left[\frac{0.1(1+0.1)^5}{(1+0.1)^5 - 1}\right] + \left[600 + 100\left[\frac{1}{(0.1)} - \frac{5}{(1+0.1)^5 - 1}\right]\right]$$

$$+\left[1000\frac{1}{(1+0.1)^3} + 1500 + \frac{1}{(1+0.1)^4} 2000\frac{1}{(1+0.1)^5}\right]\left[\frac{0.1(1+0.1)^5}{(1+0.1)^5 - 1}\right]$$

$= 4500\ (0.2638) + [600 + 100\ (1.8101)] + [1000(0.7513) + 1500(0.6830)$
$+ 2000(0.6209)]0.2638$

$= 11871.0 + 781.01 + 796.1$

$$\boxed{A = 13448.11}$$

Problem 9. *A person invests a sum of Rs. 50,000 in a bank at a nominal interest rate of 18% for 15 years. The compounding is monthly. Find the maturity amount of the deposit after 15 years.*

Solution.

$P = 50000$

Nominal int. $= 18\%$

$i = 18/12 = 1.5.$

$N = 15 \text{ years} = 15 \times 12 = 180$

Compounding months.

$F = ?$

$F = P(F/P, i\%N)$

$= 50000\ (F/P, 1.5\%, 180]$

$= 50000\ (1 + i)^N$

$$= 50000\left[1+\frac{1.5}{100}\right]^{180}$$

$$\boxed{F = 7{,}29{,}218.4}$$

EXERCISE PROBLEMS

1. A person who is just 30 years old is planning for his retired life. He plans to invest an equal sum of Rs. 10,000 at the end of every year for the next 30 years starting from the end of next year. The bank gives 15% interest rate, compounded annually. Find the maturity value of his account when he is 60 years old.
2. A company is planning to expand it's business after 5 years from now. The expected money required for the expansion programme is Rs. 5,00,00,000. The company can invest Rs. 50,00,000 at the end of every year for the next five years. If the assumed rate of return of investment is 18% for the company, check whether the accumulated sum in the account would be sufficient to meet the fund for the expansion programme. If not, find the difference in amounts for which the company should make some other arrangement after 5 years.
3. The amount of Rs. 1200 per year is to be paid into an account over each of the next 5 years. Using a nominal interest rate of 12 percent per year, determine the total amount that the account will contain at the end of the fifth year under the following conditions.
   1. Deposits made at the first of each year with simple interest.
   2. Deposits made at the end of each year, with interest compounded annually.
   3. Deposits made at the end of each month with interest compounded annually.
   4. Deposits made at the end of each year with interest compounded monthly.
4. A building site for a new gasoline station was purchased 10 years ago for Rs. 50,000. The site has recently been sold for 120,000. Disregarding any taxes. Determine the rate of interest obtained on the initial investment.

5. An inventor has been offered Rs. 12,000 per year for the next 5 years and Rs. 6000 annually for the following 7 years for the exclusive rights to an invention. At what price could the inventor offered to sell the rights to earn 10 percent, disregarding taxes?

6. Consider the following cash flow diagram. Find the total amount at the end of the 10th year at an interest rate of 12% compounded annually.

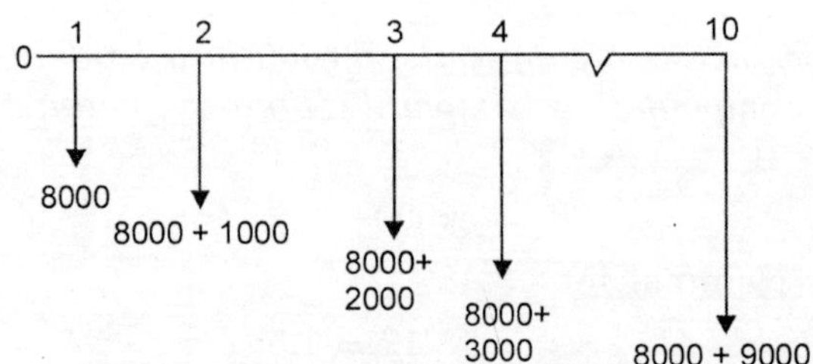

7. A working woman is planning for her retired life. She has 20 more years of service. She would like to deposit 10% of her salary which is Rs. 5000 at the end of the first year and there after she wishes to deposit the same amount (Rs. 5000) with an annual increase of Rs. 1000 for the next 14 years with an interest rate of 18%. Find the total amount at the end of the 15th year of the above series.

8. A financial institution introduces a plan to pay a sum of Rs. 15,00,000 after 10 years at the rate of 18% compounded annually. Find the annual equivalent amount that a person should invest at the end of every year for the next 10 years to receive Rs. 15,00,000 after 10 years from the institution.

9. A new piece of material handling equipment costs of Rs. 20,000/- and is expected to save Rs. 7500/- the first year of operation. Maintenance and operating cost increases are expected to reduce the net savings by Rs. 500 per year for each additional year of operation until the equipment is worn out at the end of 8 years. Determine the net persent worth of the equipment at an interest rate of 12 percent.

10. A certain government agency utilizes mid period cash flow convention in its engineering economy studies. Project R-127 has this estimated cash flow pattern.

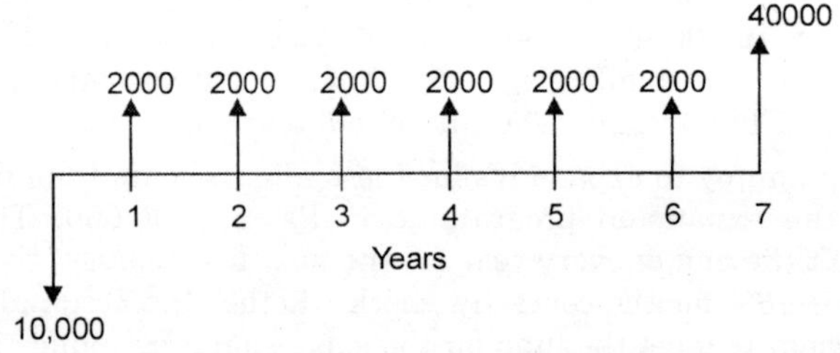

If i = 10% per year, answer the following questions.

(*i*) What is the present equivalent value of this project at the beginning of year?

(*ii*) What is the annual equivalent value of the project over the seven-year period.

(*iii*) What is the future equivalent value at the end of year 7?

PRESENT WORTH COMPARISONS

3.1 INTRODUCTION

All Engineering Economy Studies of capital projects should consider the return that a given project will or should produce. A basic question is whether a proposed capital investment and its associated expenditures can be recoverd by revenues or savings. Over time on the capital that is sufficiently attractive in view of risks involved and potential alternative uses.

The pattern of capital investment, revenue or savings cash flows and cost cash flows can be different in various projects there is no single method for performing engineering economic analysis.

The engineering economic decisions based on the executed present worth of all incomes and expenditure associated with those decisions regardless of when those activities occurred. Present worth (PW) comparisons are made only between "coterminated proposals" and "common multiple" of assets lives or by a study period that ends with the disposal of all assets. There are five basic methods to have results of selected alternative or to be analysed. they are:

Equivalent worth:

(*i*) Present Worth (PW)

(*ii*) Annual Worth (AW)

(*iii*) Future Worth (FW)

Rate of Returns

(*iv*) Internal Rate of Return (IRR)

(*v*) External Rate of Return (ERR)

The first three methods convert all cash flows into equivalent worths at some point or points in time using an interest rate before taxes equal to the minimum attractive rate of return (MARR). Establishing the MARR, Management resolved the policy and number of considerations. The considerations are (1) the number of potential projects, their purpose and their financial attractiveness. (2) the availability, source and cost of capital funds and (3) the perceived risks associated with the investment opportunities. Also, the type of organization involved will affect the selection. Determination of the MARR is a critical policy issue that affects the strategic welfare of any enterprise.

The last two methods listed are different ways to calculate an annual rate of profit or savings resulting from an investment so that a rate of return can in turn be compared against the MARR.

Many Economists prefer a present-worth analysis because it reveals the sum in today's Rs. that is equivalent to a future cash flow stream and PW models are less subject to misinterpretation.

3.2 CONDITIONS FOR PRESENT WORTH COMPARISONS

The present worth of a cash flow stream is denoted by PW - the PW of receipts and disbursements associated with a particular course of action.

The ingredients of a present-worth comparison are the amount and timing (N) of cash flow and interest rate i at which the flow is discounted.

The following conditions are for present worth, comparisons.

1. Cash flows are known: The accuracy of cash flow estimates is always suspect because future developments cannot be anticipated completely. Transactions that occur now at time 0, should be accurate.

2. Cash flows are in constant value Rs.

The buying power of money is assumed to remain unchanged during the study period. This is such an important assumption and the concept of constant Rs. is paramount to the evaluation of alternatives.

3. The interest rate is known:

Different interest rates have a significant effect on the magnitude of the calculated present worth as shown in Fig. 3.1.

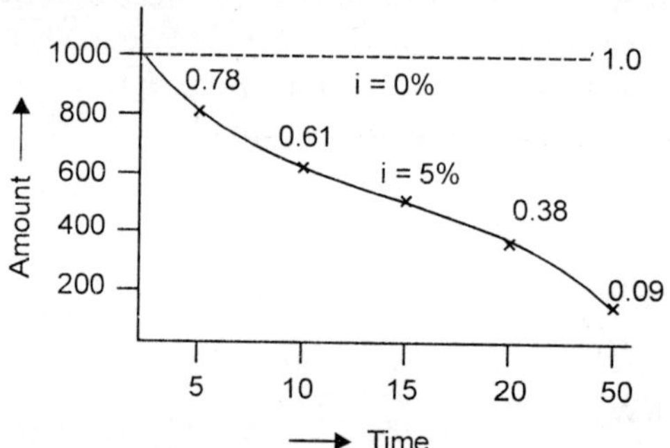

Fig. 3.1 Different Interest rate

The rate of return i required by an organization is a function of its cost of capital, attitude towards risk and investment policy. Alternate course of action for the same proposal are normally compared by using the same interest rate.

4. Comparisons are made with before-tax cash flows. Inclusion of income taxes greatly expands the calculation effect for a comparison and correspondingly increases reality.

5. Comparisons do not include intangible considerations: Intangibles are difficult to quantify factors that pertain to a certain situations. The impression created by a design is an intangible factor in evaluating that design and an important one for marketing. It would not be included in a present-worth comparison unless its economic consequences could be reasonably estimated.

6. Comparisons do not include consideration of the availability of funds to implement alternatives.

It is explicitly assumed that funds will be found to finance a course of action if the benefits are large enough. For instance, an old, inefficient machine could be kept in operation because appears to be insufficient capital available in the organizations to afford a replacement. An engineering economic analysis might-point out that the savings from replacing the outdated machine would be so great that the organization could not afford to not find funds for a replacement.

3.3 BASIC PRESENT WORTH COMPARISONS

The present worth of a cash flow overtime is its value today is represented as time 'o' in a cash flow diagram. Two general patterns are apparent in present-worth calculations: present-worth equivalence and net present worths. Both types of analysis will give the same results and interpretation.

3.4 PRESENT WORTH EQUIVALENCE

One pattern determines the present-worth equivalence of a series of future transactions. The purpose is to secure one figure that represents all the transactions for instance, a series of expenses that will occur in the future can be discounted to obtain its PW and decision can be made about whether an investment of the PW amount should be made now to avoid the expenses.

Example: *The lease on a warehouse amounts to Rs. 5000 per month for 5 years. If payments are made on the first of each month, what is the present worth of the agreement at a nominal annual interest rate of 12% compounded monthly.*

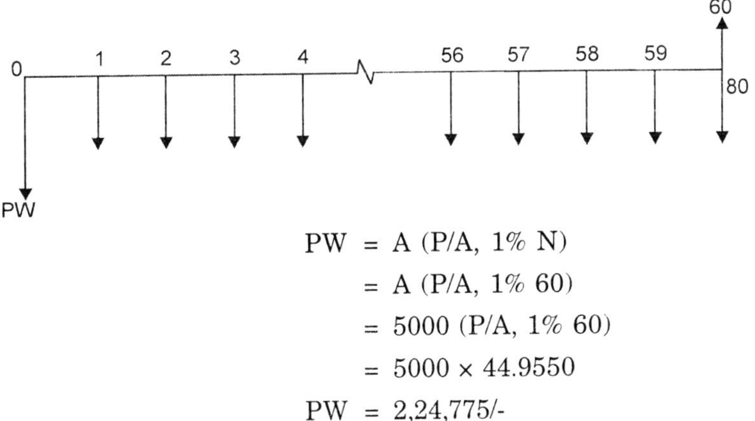

$$\begin{aligned} PW &= A\ (P/A,\ 1\%\ N) \\ &= A\ (P/A,\ 1\%\ 60) \\ &= 5000\ (P/A,\ 1\%\ 60) \\ &= 5000 \times 44.9550 \\ PW &= 2{,}24{,}775/- \end{aligned}$$

3.5 NET PRESENT WORTH

The second general pattern for PW calculations has an initial outlay at time o followed by a series of receipts & disbursements.

Net present-worth = PW (benefits) – PW (costs)

The criterion for choosing between mutually exclusive. Alternatives is select the one that max-net present worth or simply the one that yields the layer positive PW. A negative PW means that the alternatives does not satisfy the rate of return requirement.

Example: *A piece of new equipments was proposed by engineers to increase the productivity of a certain manual welding operation. The investment is Rs. 25000/- and the equipment will have salvage value of Rs.5000 at the end of 5 years. Increased productivity will gain Rs.8000/- per year after extra operating costs have been subtracted from the additional production. Draw a cash flow diagram. If the firm's minimum attractive rate of return is 20% per year, is this proposal a sound one? Use the present worth method.*

Solutions: Total PW = PW of cash receipts – PW of cash outlays

= 8000 (P/A, 20% 5) + 5000 (P/F, 20% 5) – 25000

= 8000 (2.9906) + 500 (0.4019) – 25000 using table the value is

PW = 934.3

Solved Problems:

1. A company borrowed 1,00,000 to finance a new product the loan was for 20 years at a nominal interest rate of 8 percent compounded semiannually. It was to be repaid in 40 equal payments. After one half the payments were made, the company decided to pay the remaining balance in one final payment at the end of the 10th year how much was owed?

N = 20 Years. = 20 × 2 = 40 Instalment

i = 8% Compounded semi-annually i = 4%

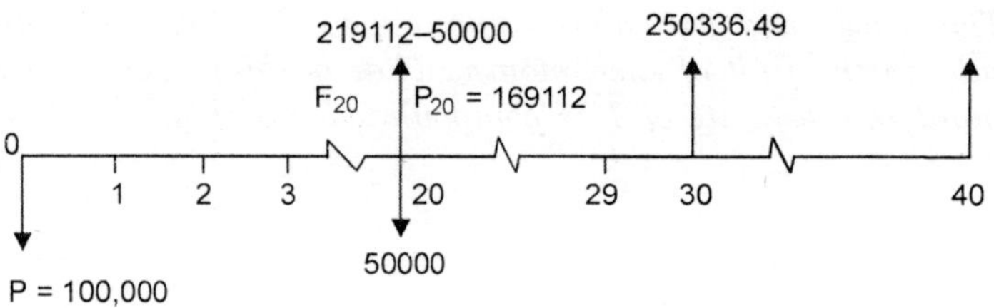

(1) F_{20} = P (F/P, i% N)

= P (F/P 4% 20) using table

= 100,000 (2.19112)

= 219112

(2) F_{30} = P (F/P, i% N)

= 169112 (F/P, 4%, 10) using table

= 169112 (1.4803)

= 250336.49

The company owed to finance is 2,50,336.49.

2. A proposed improvement in an assembly line will have an initial purchase and installation cost of Rs. 175,000. The annual maintenance cost will be Rs.6000. Periodic overhauls once every 3 years excluding the last year of use, will cost Rs. 11,500 each. The improvement will have a useful life of 9 years at which time it will have no salvage value, what is the present worth of the 9 year costs of the improvement at i = 8%?

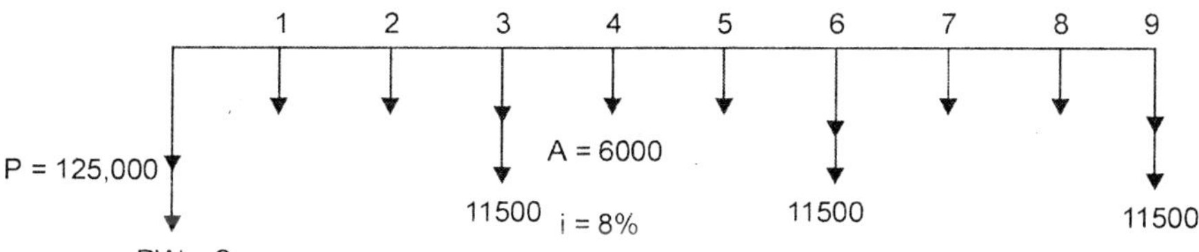

$$
\begin{aligned}
PW &= A\,(P/A, i\%N) + F(P/F, i\%N) + F(P/F, i\%N) + F(P/F, i\%N) \\
&= 6000\ (P/A;\ 8\%\ 9) + 11500\ (P/F,\ 8\%\ 3) + 11500\ (P/F,\ 8\%\ 6) + 11500\ (P/F,\ 8\%\ 9) \\
&= 6000\ (6.2489) + 11500\ (0.7938) + 11500\ (0.63017) + 11500\ (0.5003)\ \text{ using table} \\
&= 37493.4 + 9128.7 + 7246.9 + 5753.5 \\
&= 59622.50 + \text{Initial Investment} \\
&= 59622.50 + 175000 \\
&= 234622.50
\end{aligned}
$$

3. A bakery is thinking of purchasing a small delivery truck that has a first cost of Rs. 18,000 and is to be kept in service for 6 year. At what time the salvage value is expected to be 2500. Maintenance and operating costs are estimated at Rs. 2500 the first year and will increase at a rate of Rs.200/year. Determine the PW of this vehicle using interest rate of 12%.

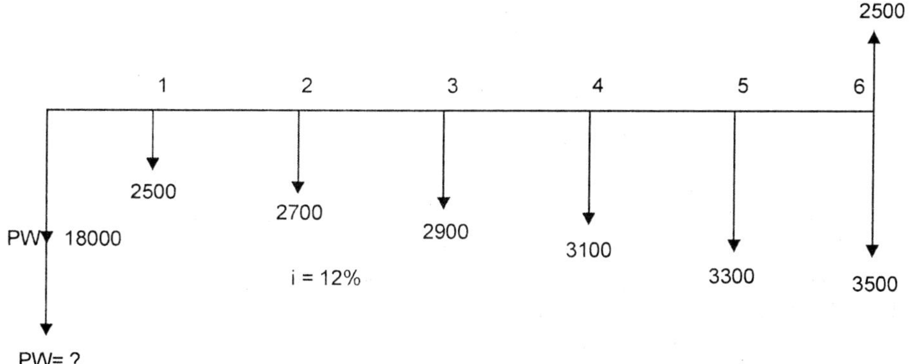

$$
\begin{aligned}
A &= A' + G\ (A/G,\ i\%\ N) \\
&= 2500 + 200\ (A/G,\ 12\%\ 6) \\
&= 2500 + 200 \left[\frac{1}{0.12} - \frac{6}{(1+0.12)^6 - 1}\right] \\
&= 2500 + 434.4 \\
&= 2934.4
\end{aligned}
$$

$$
\begin{aligned}
PW &= A\ (P/A,\ i\%\ N) - F(P/F,\ i\%\ N) \\
&= 2934.4\ (P/A,\ 12\%,\ 6) - 2500\ (P/F,\ 12\%\ 6) \\
&= 2934.4\ \frac{(1+0.12)^6 - 1}{0.12(1+0.12)^6} - 200\frac{1}{(1+i)^N}
\end{aligned}
$$

$$= 12064.5 - 2500\frac{1}{(1+0.12)^6}$$
$$= 12064.5 - 1266.6$$
$$= 10797.9$$

4. A small dam and an irrigation system are exclusive to cost Rs. 300,000. Annual maintenance and operating costs are executed to be Rs.40000 the first year and will increase at a rate of 10% per year. Determine the equivalent present worth of building and operating the system with interest of 10% over a 30 year life.

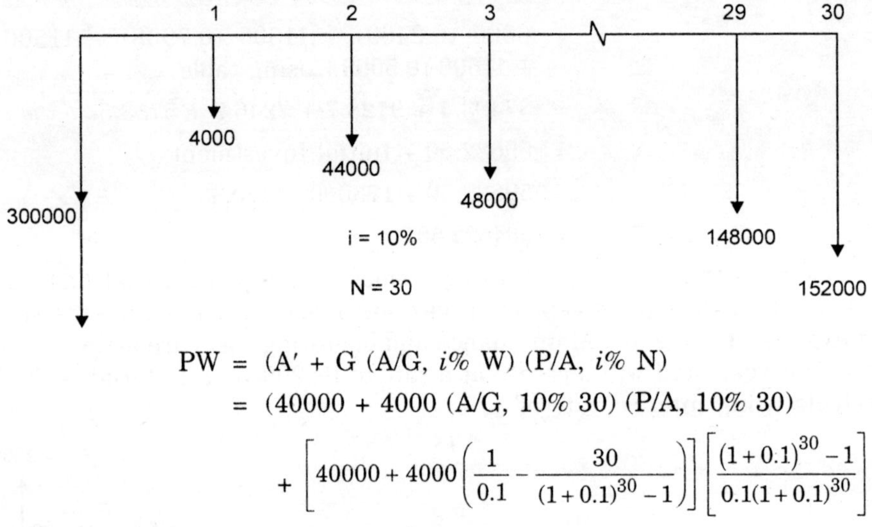

$$\text{PW} = (A' + G\ (A/G,\ i\%\ W)\ (P/A,\ i\%\ N)$$
$$= (40000 + 4000\ (A/G,\ 10\%\ 30)\ (P/A,\ 10\%\ 30)$$
$$+ \left[40000 + 4000\left(\frac{1}{0.1} - \frac{30}{(1+0.1)^{30}-1}\right)\right]\left[\frac{(1+0.1)^{30}-1}{0.1(1+0.1)^{30}}\right]$$
$$= (72704.9)\ (9.427)$$
$$= 685389.1$$

5. A newly developed electric car will cost 21000 to purchase. Operating and maintenance costs, including home charging of the batteries, are estimated to be Rs.350 for the first year with annual increase there after of Rs. 50 per year. Salvage value after 5 years is estimated to be Rs.6500. A new gasoline runabout will cost Rs.16000/- and will average 30 miles per gallon. Gasoline costs Rs.1.26 per gallon is expected to increase at a rate of Rs. 0.05 per year each of the next 4 years maintenance costs are estimated to be Rs. 300 per year including warranty coverage. Salvage value is estimated to be Rs. 1500 after 5 years of service. If the vehicles are expected to be driven for 20,000 miles per year, determine which option will have the lower cost over 5 years. Use PW analysis with a 10% rate of interest.

Solution:

Electric Car.

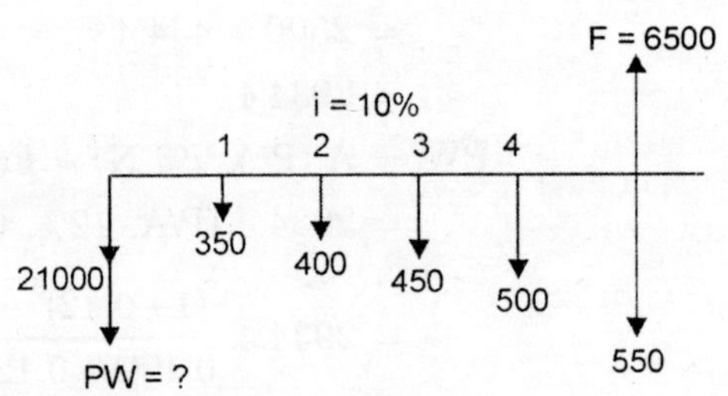

PW = –21000 –[350 + 50(A/G, 10%, 5) (P/A, 10%5)] + 6500 (P/F, 10% 5) using table

$$PW = -18633.40$$

(*b*) *New gasoline run about*

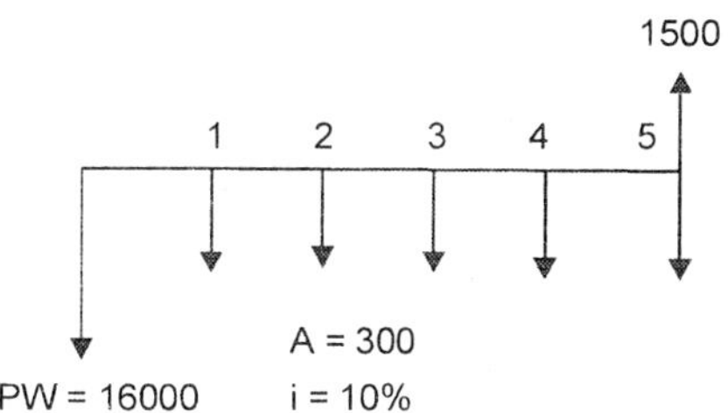

Gasoline cost/year

Expected to run

20000 miles per year

30 miles per gallon

∴ No. of gallons $= \frac{20000}{30}$

$= 666.67$

∴ Cost of gasoline per year is $= 666.67 \times 1.26$

$= 840$

It is expected to increae by Rs. 0.05 per litre/year

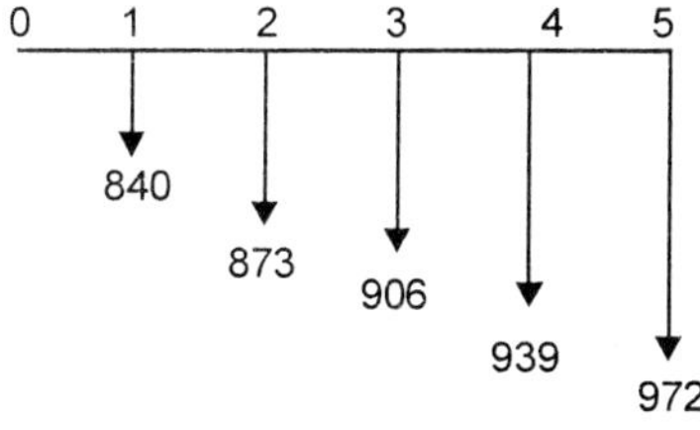

PW gasoline = [A + G(A/G, *i*% N] (P/A *i*% N)

= [840 + 33(A/G, 10%5)] (P/A, 10%5)

= [840 + 33 × (1.810)] (3.791)

= 899.7 × 3.791

PW gasoline = 3410.9

PW = –16000 – 300(P/A, 10%5) – 3410.9 + 1500(P/F, 10%5)

= –16000 – 300(3.791) – 3410.9 + 931.38

= –19616.82

6. A certain Indian treasury bond matures in 8 years has a face value of Rs. 10000. It means bond holder receives Rs.10000 cash when bond's maturity date is reached. The interest rate is 8% per year. Interest payments are made every 3 months and amount to 2% of the face value.

A prospective buyer of this bond would like to earn 10% nominal interest on his investment. How much should this buyer be willing to pay for the bond?

Solution: A = 200/qtr. (2% of face value)

$$i = \frac{10\%}{4} = 2.5\%$$

N = 8 × 4 = 32 qtr. of bond's life

PW = A(P/A, *i*%N) + F(P/F, *i*%N)

= 200(P/A, 2.5/qtr, 32/qtrs) + 10000 (P/F, 2.5% qtr, 32 qtr)

$$= 200\left[\frac{(1+i)^N - 1}{i(1+i)^N}\right] + 10000\left[\frac{1}{(1+i)^N}\right]$$

$$= 200\left[\frac{(1+0.025)^{32} - 1}{0.025\,(1+0.025)^{32}}\right] + 10000\left[\frac{1}{(1+0.025)^{32}}\right]$$

= 4369.8 + 4537.7

PW = 8907.5

7. Determine the equivalent present worth of the following series of year end cash flows. Extending over 8 years using annual interest rate of 20%. The amounts are Rs. 100 for 1st year, Rs. 200 for 2nd year, Rs. 500 for 3rd year Rs. 400 for 4th, 5th, 6th, 7th and 8th. Draw CFD for above series of payments.

Solution: N = 8 yeas

i = 20%

PW = ?

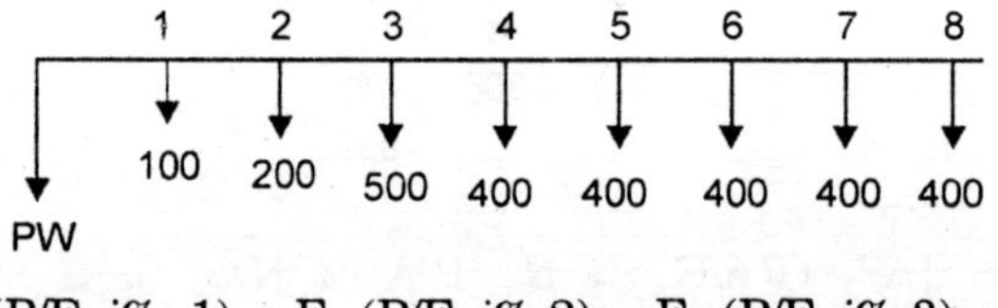

1 2 3 4 5 6 7 8

100 200 500 400 400 400 400 400

PW

PW = F_1 (P/F, *i*%. 1) + F_2 (P/F, *i*% 2) + F_3 (P/F, *i*% 3) + [A(P/A, *i*%5)] (P/F, *i*%3)

= 100 (P/F, 20%1) + 200(P/F, 20%, 2) + 500 (P/F, 20% 3) +

[(400 (P/A, 20% 5) (P/F, 20% 3) using table

= 100 (0.8333) + 200 (0.895) + 500 (0.5787) + [400(2.9906)] (0.5787)]

= 83.3 + 179 + 289.4 + 692.26

= 1243.96

8. A company is considering constructing a plant to manufacture a proposed new product. The land costs 300,000 the building costs 600,000. The equipment costs 250000 and 100000 working capital is required. It is expected that the product will result in sales of 750000 per year for 10 years at which time the land can be sold for 400,000 the building for 350000 the equipment for 50,000 and all of the working capital recovered. The annual out-of-pocket expenses for labour, materials and all

other items are estimated to total 475000. If the company requires a minimum return of 25% on projects of comparable risk, determine if it should invert in the new product line use prevent worth methods.

Solution:

Investment = 300000 + 600000 + 250000 + 100000
= 1250000
Salary = 750000/year for 10 years
Salvage value = 400000 + 350000 + 50000 + 100000 = 900000
Annual expinse = 475000/Year for 10 years
i = 25%
N = 10 Years.

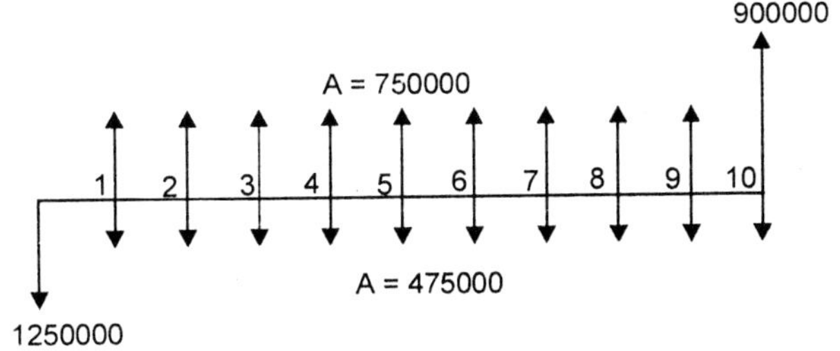

PW = –1250000+ A(P/A, i%N) – A(P/A, i% 10) + F (P/F, i% 10)

PW = –1250000 + 750000 (P/A, 25%, 10) – 475000 (P/A, 25% 10) + 900000(P/F, 25% 10)

PW = –1250000 + 750000 (3.571) – 475000 (3.571) + 900000(0.108) using table value

PW = –170775

It should invest in the new product line.

9. Determine the PW of the following proposal when the MARR is 15%.

	Proposal A
First cost	10000
Expected lift	5 yers.
Salvage value	–1000
Annual receipts	8000
Annual expenses	4000

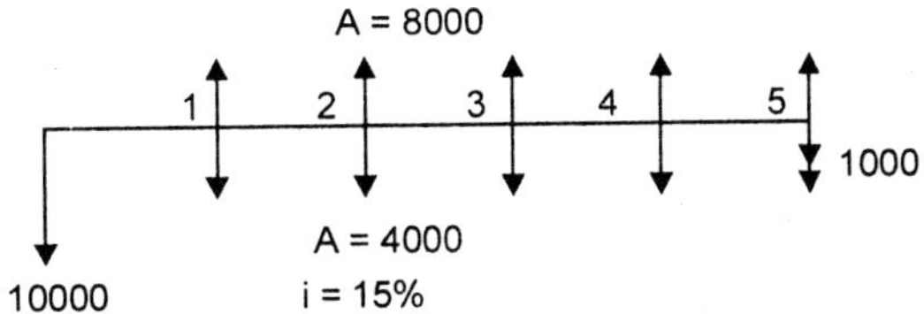

PW = –10000 – A(P.A, i%N) + A(P/A, i%N) – F(P/F, i%N)
= –10000 – 4000(P/A, 15% 5) + 8000 (P/A, 15%, 5) –1000 (P/F, 15%, 5)
= –10000 – 4000 (3.362) + 8000 (3.362) – 1000 (0.497)

PW = –10000 – 13448 + 26896 – 497

PW = 2951

10. Evaluate machine XYZ on the basis of the present worth method when the MARR is 12% pertinent cost data are as follows :

	Machine XYZ
First cost	13000
Useful life	15 years
Salvage value	3000
Annual operating cost	100
Overhaul end of 5th year	200
Overhaul end of 10th year	550

3000

1 2 3 4 5 6 7 8 9 10 11 12 13 14 15

A = 100 200 550

13000

PW = –13000 – A(P/A, *i*%N) – F(P/F, *i*% N) – F(P/F, *i*%N) + F(P/F, *i*%N)

= –13000 – 100(P/A, 12%15) – 200(P/F, 12%5) – 50(P/F, 12%10)s + 3000(P/F, 12%15)

using table

= –13000 – 100(6.811) –200(0.568) – 550(0.322) + 3000(0.183)

PW = –13000 – 681.1 – 113.6 – 177.1 + 549

= – 13971.8 + 549

PW = –13422.8

3.6 ASSETS WITH UNEQUAL LIVES

Unequal lives among feasible alternatives somewhat complicate their analysis and comparison. To make engineering economy studies in such cases, we adopt some feasible procedure in saving the alternatives on a comparable basis.

Two types of assumptions regarding the analysis period are employed :

(1) the repeatability assumptions

(2) The co-terminated assumption.

The repeatability assumption or common multiple method involver two main conditions.

1. The analysis period over which the alternatives are being compared is either indefinitely long or equal to a common multiple of lives of the alternatives.

Example: If assets had lives of 2, 3, 4 and 6 years, the least common multiples is 12 years, which means that the asset with a life of 2 years would be replaced 6 times during the analysis period the assets 3, 4, and 6 years lives would be replaced 4, 3 and 2 times respectively.

2. The economic consequences that are estimated to happen in an alternatives initial life span will happen in all succeeding life spans (replacements)

The repeatability assumption has limited use in engineering practice because actual situations seldom meet both conditions.

3.7 THE CO-TERMINATED ASSUMPTION OR STUDY-PERIOD METHOD

A more justifiable analysis is based on a specified duration that corresponds to the length of a project or the period of time the assets are expected to be in service. This planning horizon may be the period of needed service or any selected length of time such as,

1. The life of the shorter-lived alternative
2. The life of the longer-lived alternative
3. Less than the shortest alternative life
4. Greater than the longest alternative
5. In between the shorter and longest lives

The key point is that the selected analysis period is approximate for the decision situation under investigation.

SOLVED PROBLEMS

1. The following data presents for two feasible alternatives A & B for which revenues and costs are known and which have different lives. If the minimum alternative rate of return is 10%, show which feasible alternative is more desirable by using PW.

	A	*B*
Investment (first) cost	3500	5000
Annual revenue	1900	2500
Annual cost	645	1383
Useful life	4	8
Salvage value at end of useful lift	0	0

Solution:

PW Method

Particulars	*Alt. A*	*Alt. B*
Annual revenue		
1900 (P/A, 10% 8)	10136	
2500 (P/A, 10% 8)		13337
Total revenue (PW)	10136	13337
Annual cost		
645 (P/A, 10% 8)	3441	

(Contd...)

1383 (P/A 10% 8)		7378
Original Investment	3500	5000
First replacement		–
3500 (P/F, 10% 4)	2390	
Total PW costs	9331	12378
PW of revenue	805	959
PW of costs		

2. Machine A has a first cost of Rs. 9000, no salvage value at the end of its 6-year useful life, and annual operating costs of Rs. 5000. Machine B costs Rs. 16,000 now and has an expected resale value of Rs. 4000 at the end of its 9-year economic life. Operating costs for machine B are Rs. 4000 per year. Compare the two alternatives on the basis of their PW, using the repeated projects assumption at 10% annual interest.

	M/C A	*M/C B*
First cost	9000	16000.00
Salvage value	0	14000.00
Life	6 years	9 years
Annual cost	5000	4000.00

PW Method: *Present worth*		*M/C A*
Original Investment		–9000.00
First replacement		–5080.00
9000 (P/F, 10% 6)		
Second replacement		–2867.00
9000 (P/F, 10% 12)		
Annual Expenses		–41007.00
5000 (P/A, 10% 18)		–57954.00

Present Worth of M/C B		*M/C B*
Original Investment		–16000.00
First replacement		
(16000-4000) (P/F, 10%9)		–5088.00
Annual Expenses		–32805.00
4000 (P/A, 10%18)		–53893.60

As per the analysis by present worth method Machine B as lower cost compared to machine A using repeatability assumption.

3. Consider the following two matually exclusive alternatives related to an improvement project & recommend which one should be implemented. Use the present worth method?

	Machine *A*	*B*
Investment (first) cost	2000	30000
Salvage value	4000	0
Annual receipts	10000	14000
Annual costs	4400	8600
Useful lise (years)	5	10
Minimum attractive rate of return = 15%		

Solution: Analysis period = 10 years assuming repeatability.

Present Worth of Machine A

Original Investment	–20000 = 00
First replacement	–7955 = 20
(20000 – 4000)(P/F, 15% 5)	
Annual costs	
4400 (P/A, 15% 10)	–22082 = 72
Total PW costs	–50037.92
Total receipts are	
10,000 (P/A, 15% 10)	50188.00
	150.08

Present Machine B

First cost	–30000 = 00
Annual cost	
8600 (P/A, 15% 10)	–43161 = 68
Total PW cost	73161.68
Total receipts	
14000 (P/A, 15% 10)	70263.20
	–2898.48

Machine A is better than machine B. Machine A recommended for implementations.

4. A construction company is going to purchase several light duty trucks. Its 'MARR' before taxes is 18%. It is considering two makes and the following relevant data are available.

	Godrej	*Reliance*
Investment (first) cost	10000	15000
Salvage value at the end of life	2000	3000
Annual out-of pocket cost	4000	3000
Useful life.	3 Years	5 Years

(*a*) Which type of truck should be selected when the repeatability assumption is appropriate?

(*b*) Which type of truck would you recommend if the selected analysis period is 3 years (co-terminated assumption) and it is estimated that a Reliance truck will have a salvage value of Rs.5600 at that time?

Solution: Present Worth Using Repeatability Assumption:

	godrej
Original Investment	–10000
First replacement (10000 – 2000) (P/F, 18% 3)	–4869.1
Second replacement (10000 – 2000) (P/F, 18% 6)	– 2963.5
Third replacement (10000 – 2000) (P/F, 18% 9)	–1803.7
fourth replacement (10000 – 2000) (P/F, 18% 12)	–1097.8

Annual cost.

4000 (P/A, 18% 15)

$$4000\left[\frac{(1+0.18)^5-1}{0.18(1.18)^{15}}\right] = -20366.3$$

Total PW cost = –41100.4

Present Worth Reliance

Original Investment	– 15000 = 00
First replacement (15000 – 3000) (P/F, 18% 5)	–5245 = 30
Second replacement (15000 – 3000) (P/F, 18% 10)	–2292 = 77
	–22538.07

Annual cost

3000 (P/A, 18% 15)

$$3000\left[\frac{(1+0.18)^{15}-1}{0.18(1+0.18)^{15}}\right] \qquad -15274 = 70$$

– 37812 = 80

Reliance truck is best choice compared to Godrej.

(*b*) *Co-terminated Analysis*

Data

	Godrej	*Reliance*
Investment	10000	15000
Salvage value	2000	5600
Annual out of pocket costs	4000	3000
Life	3 years	3 years.

PW Godrej $= -10000 + 2000$ (P/F, 18% 3) – 4000 (P/A, 18% 3)

$$= -10000 + 2000\left[\frac{1}{1+0.18^3}\right] - 4000\left[\frac{(1+0.18)^3-1}{0.18(1+0.18)^3}\right]$$

$$PW_G = -17479.8$$

PW Reliance

= –15000 + 5600 (P/F, 18% 3) – 3000 (P/A, 18% 3)

$$= -15000 + 5600\left[\frac{1}{(1+0.18)^3}\right] - 3000\left[\frac{(1+0.18)^3-1}{0.18(1+0.18)^3}\right]$$

$$= -15000 + 3408.3 - 6522.8$$

$$\boxed{PW_R = -18114.5}$$

As per coteriminated analysis Godrej is better than Reliance.

5. Certain service can be performed satisfactorly either by process R or process. Process S has a first cost of Rs. 8000, an estimated service life of 10 yeras no salvage value and annual net receipts (revenue-costs) of Rs. 2400. The corresponding figures for process S are 18000. After 20 years salvage value equal to 20% of first cost and Rs. 4000. Assuming a minimum attractive rate of return of 15% before income taxes. Find the future of each process and specify which you would recommend. Use the repeatability assumption.

	Process R	*Process S*
First cost	8000	18000
Life	107 year	20 yrs
Salvage value	0	3600
Annual net receipts	2400	4000
i	15%	15%
F	?	?

Future Worth Method

Procers R

FW_R = 8000 (F/P, 15% 20) - 8000 (F/P, 15% 10) + 2400 (F/A, 15% 20)

$$= -8000\,[(1+0.15)^{20}] - 8000(1+0.15)^{10} + 2400\left[\frac{(1+0.15)^{20}-1}{0.15}\right]$$

$$= -130932.3 - 32364.5 + 245864.6$$

$$\boxed{FW_R = 82567.8}$$

Procers S

FW_S = –18000 (F/P, 15% 20) + 3600 + 4000 (F/A, 15% 20)

=–294597.7 + 3600 + 409774.3

$$\boxed{FW_S = 118776.6}$$

Process S is worth in choosing.

6. The following alternatives are available to accompany an objective of years duration.

	Plan A	*Plan B*	*Plan C*
Life cycle	6 years	3 years	4 years
First cost	Rs. 2000	Rs. 8000	Rs. 10000
Annual cost	Rs. 3200	Rs. 700	Rs. 7500

Compare the present worth of the alternatives using an interest rate of 7 percent?

Solution:

Plan A **PW**	
First cost	2000.00
First replacement 2000 (P/F, 7%6)	1332.80
Annual cost 3200 (P/A, 7%12)	25416.00
	28749.50

Plan B **PW**	
First cost	8000.00
First replacement 8000 (P/F, 7%3)	6530.40
2nd replacement 8000 (P/F, 7%6)	5331.20
3rd replacement (8000(P/F, 7%9)	4351.20
Annual cost 700 (P/A, 7%12)	5559.89
	29772.69

Plan C **PW**	
First cost	10000.00
First replacement 10000 (P/F, 7% 4)	7629.00
2nd replacement 10000 (P/F, 7% 8)	5820.00
Annual cost 500 (P/A, 7% 12)	3971.00
	27420.35

Plan C is better option compared to Plan A and Plan B.

7. Autocon company is evaluating three robots for possible use in its assembly operations (only one robot will be purchased). Data associated with there robots are as follows :

	Robot A	*Robot B*	*Robot C*
First costs Rs.	55000	58000	53000
Operating & maintenance Costs Rs.	3000/year	4500/year	4000/year
Expected incomes	40000/year	44000/year	38000/year
Estimated salvage value Rs.	4000	6000	4000

Assuming a technological life of 3 years and a desired interest rate of 12 percent, which robot seems to be preferable assuming all other factors are equal? Use a net present worth evaluation.

Solution: *Robot A*

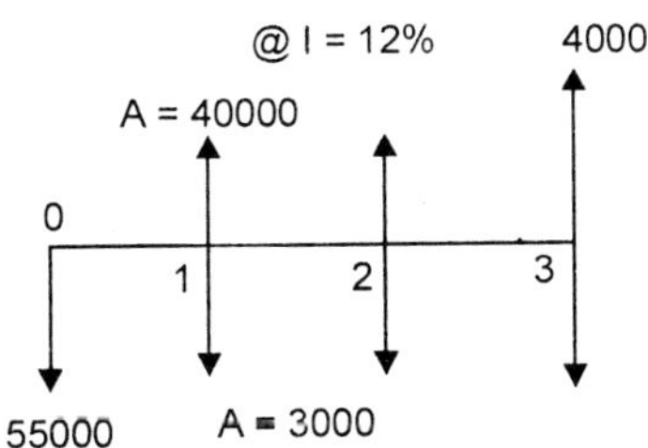

$$PW_A = -55000 - 3000\,(P/A, 12\%\ 3) + 40000\,(P/A, 12\%\ 3) + 4000\,(P/F, 12\%\ 3)$$

$$= -55000 - 7206 + 96080 + 2848$$

$$PW_A = 36722$$

Robot B

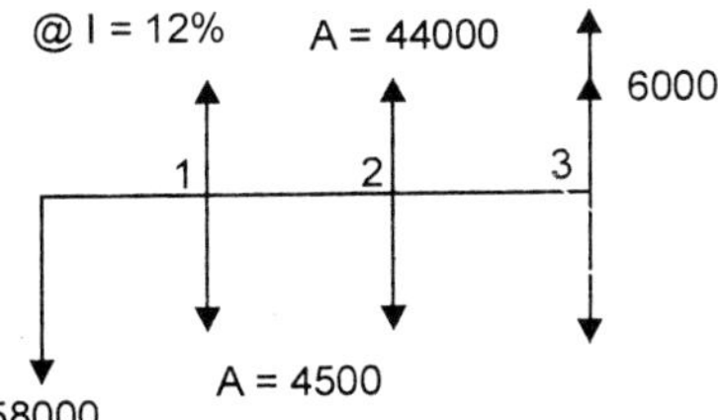

$$PW_B = -58000 - 4500\,(P/A, 12\%\ 3) + 44000\,(P/A, 12\%3) + 6000\,(P/F, 12\%3)$$

$$= -58000 - 10804 = 50 + 105644 + 4270.8$$

$$\boxed{PW_B = 41110.3}$$

Robot C

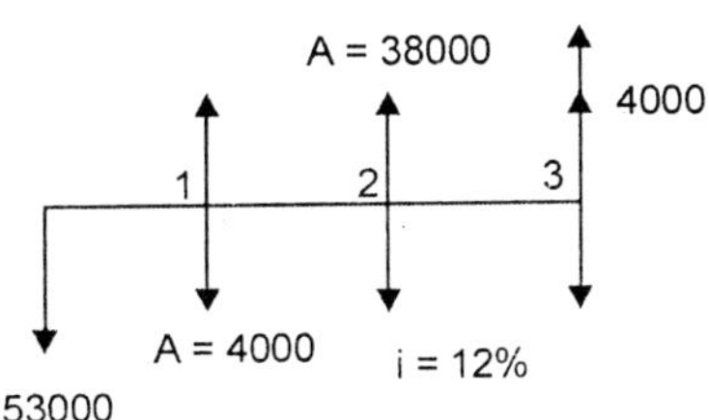

PW_C = –53000 – 4000 (P/A, 12%3) + 38000 (P/A, 12%3) + 4000 (P/F, 12%3)
= 53000 – 9604 + 91238 + 2847.2

PW_C = 31481.2

8. Two feasible Alternatives A&B, for which revenues and costs one known and which have different lives. If MARR is 10%, show which feasible alternative is more desirable.

	A	*B*
First cost	3500	5000
Annual revenue	1900	2500
Annual cost	645	1383
Useful life (years)	4	8
Salvage value @ end of useful life	0	0

If the expected period of required service from A or B is only 4 years. Because, the company has contract to produce a manufactured good for exactly 4 years. A choice must be made between A and B in view of co-teriminated life of 4 years and a MARR = 10%, which feasible alternative is more desirable?

Solution: As per the data given the life of Alternative 'B' is 8 years and the salvage value is 'O'. So, it is desirable to find the salvage value at the end of 4th years of the (Alternative B)

Salvage value of Alternative 'B' whose useful life has been = S_4 truncted.

S_4 = Present worth of remaining capital recovery cost amounts plus present worth of salvage value at the end of year 8.

Since, the salvage value at the end of year 8 is zero,

∴ S_4 = [(5000 (A/P, 10% 8)] (P/A, 10% 4)
= (937.22) (3.1699)

S_4 = 2971

Using Annual Worth Method

	A	*B*
Annual revenue	1900	2500
Annual cost Expenses	645	1383
Capital recovery: cost		
3500 (A/P, 10% 4)	1104	
5000 (A/P, 10% 4) 2971(A/F, 10% 4) –		937
Total Annual cost	1749	2320
(AW of revenue – AW of costs)	151	180

Alternative 'B' is the better choice.

9. Maruti Udyog has out come with an offer of taking home Zen at a down payment of Rs. 175000/-Alternatively the car can be taken home by giving an initial amount or Rs. 20,000/- and an equal yearly amount of Rs.15000 for 15 years at a rate of interest of 5% compounded annually. Suggest the best alternative for a customer to buy the car by PW method.

Solution:

(A) down payment = 175000/-

(B)

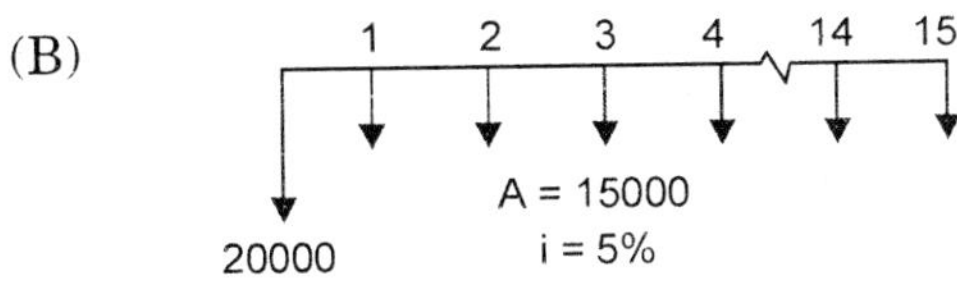

PW = +20000 + 15000 (P/A, i% N)

PW = +20000 + 15000 (P/A, 5% 15) using table

PW = +20000 + 15000 (10.3799)

PW = +175698.50

Alt B is best compared to 'A' because in case 'A' Alternative lumpsun amount of Rs.175000 where as in case 'B' Alternative initial amount is Rs.20000/- only.

3.8 COMPARISON OF ASSETS HAVING INFINITE LIVES

The special variation of the present worth method involves the determination of the worth of all receipts and expenses over an infinite length of time. This is known as the capitalized worth method. If only expenses are considered the output obtained by this method can be expressed as capitalized cost.

This method is convenient for convenient mutually exclusive alternatives when the period is indefinitely long or when the common multiple of the lives is very long and the repeatability assumption is applicable.

Capitalized cost is calculated in the same way as in a present-worth comparison, where N equals infinity.

$$(P/A,\ i\%N\ ) = \frac{1}{i}\frac{(1+i)^N - 1}{(1+i)^N}$$

The limit of (P/A, i%, N) as N approaches infinity is

$$(P/A,\ i\%,\ \infty) = \frac{1}{i}$$

If P represents the first cost.

Capitalized worth = P + A(P/A, i, ∞)

= P + A (1/i)

= P + A/i

Where A is the uniform difference between annual receipts and disbursements. When there is no revenue,

$$\text{Capitalized cost} = P + \frac{\text{disbursements}}{i}$$

Problem 1. *A proposed mill in an isolated area can be furnished with power and water by a gravity feed system. A stream high above the mill will be tapped to provide flow for water needs and power requirements by connecting it to the mill with a ditch & tunnel system or with a wood and concrete flume that winds its way down from the plateau. Either alternative will meet current and future needs, and both will utilize the same power-generating equipments.*

The ditch and funnel system will cost 500,000 with an annual maintenance cost of Rs. 2000. The flume has an initial cost of Rs. 200,000 and yearly maintenance cost of Rs. 12,000. In addition the wooden portion of the flume will have to be replaced every 10 years at a cost of Rs. 100,000.

Compare the alternatives on the basis of capitalized costs with an interest rate of 6 percent.

Solution:

1. The ditch and funnel system

$$P = \text{Rs. } 500{,}000$$

$$\text{M.C.} = \text{Rs. } 2000$$

$$i = 6\%$$

$$C = \text{Rs. } 500{,}000 + \frac{\text{Rs. } 2000}{0.06}$$

$$= \text{Rs. } 533{,}333.33$$

2. The flume

$$P = \text{M } 200{,}000$$

$$\text{MC} = \text{Rs. } 12000$$

Additional maintenance cost = Rs. 100,000 every 10 years.

$$\text{Capitalized} = P + \frac{\text{disbursement}}{i}$$

$$= 200000 + \frac{12000 + 100{,}000(A/F, 6\% 10)}{0.06}$$

$$= 526{,}500/-$$

Problem 2. *The Government of Karnataka has sanctioned Rs. 50000000 to a community trust for the construction of library block. Annual maintenance for the block estimated to be Rs. 10 lakhs. In addition, Rs. 8 lakhs will be required every 10 years for cleaning and major repairs.*

If the budget granted has to take care of perpetual maintenance, how much of the amount can be used for initial constructions costs? Deposited fund can cash 9% interest, compound annually. Assume the taxes and inflation do not come into picture.

Solution:

Capitalized cost = 5 crore

Annual maintenance cost = 10 lakhs.

additional maintenance cost every 10 years = 8 lakhs.

$$i = 9\%$$

$$\text{Capitalized cost} = \text{First Cost} + \frac{\text{disbursement}}{i}$$

$$\text{First cost} = \text{capitalized cost} - \frac{\text{Total annual costs}}{\text{Interest rate}}$$

$$= 50000000 - \frac{100000 + 800000(A/F, 9\%10)}{0.09}$$

$$\text{First cost} = 3,83,03,822.22$$

3.9 FUTURE WORTH COMPARISON

The equivalence concept that a present worth can be translated to a future worth at any given time at a given interest rate. Because a primary objective of all time value of money is to maximize the future wealth of the owners of a firm. The future worth criterion has become increasingly popular in recent years. The future worth of an alternative can be calculated in view of the MARR and compared with do-nothing option. If FW > 0 the alternative would be recommended.

Attention to secure cash flows may yield more accurate estimates of receipts and distersements because thinking tender to be in current Rs. Because the purchasing power of a unit of currency is eroded by inflation. Future-worth calculations are frequently utilized in escalation analyses, for evaluating the effects of inflation. The future worth at any time can be calculated by

$$FW = PW\ (F/P,\ i\ N)$$

$$FW = PW\ (1 + i\ N)$$

The FW ranking of alternatives has to be the same as the ranking based on a PW analysis.

Problems

1. A small company purchased now for 23000 will lose. 1200 each year the first 4 years. An additional 8000 invested in the company during the fourth year will result in a profit 5500 each year from the 5th year through the 15th year at the end of 15 years. The company can be sold for Rs. 33,000/- MARR = 12%.

Solution:

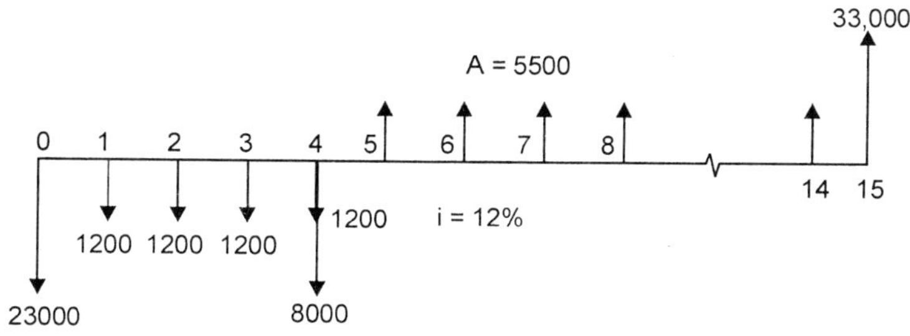

FW = –23000 (F/P, 12%15) – [(8000 (P/F, 12%4)] (F/P, 12%15) – [(1200 (P/A, 12%4)] (F/P, 12% 15) + (5500 (CF/A, 12% 11) + 33000

$= -125892.8 - 27827.8 - 19950.9 + 113600.9 + 33000$

$FW = -27070.6$

2. Monthly amounts of Rs. 200, each are deposited into an account that earns 12% nominal interest, compounded quarterly. After 48 deposits of Rs. 200 each, what is the future equivalent worth of the account? State your assumptions.

Solution:

Monthly deposited = Rs. 200

i = 12% = monthly 1%

N = 48 months.

$\therefore$ FW = A (F/A, i% N)

= 200 (F/A, 1%, 48)

$$= 200\left[\frac{(1+0.01)^{48}-1}{0.01}\right] = 200 \times 61.22$$

FW = 12244.5

3. Find the future worth in each of these situations.

(*a*)

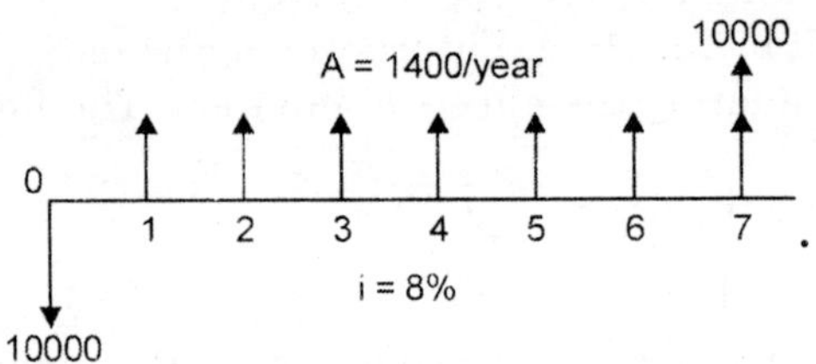

FW = –10000 (F/P, 8% 7) + 1400 (F/A, 8% 7) + 10000

= –10000 × 1.7138 + 1400 × 8.9228 + 10000

= –17138 + 12491.9 + 10000

FW = 5353.9

(*b*)

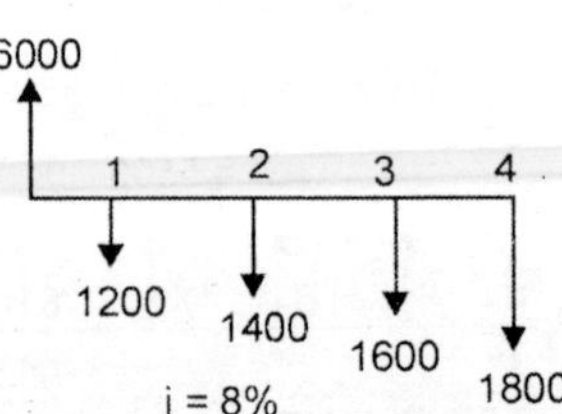

FW = 6000 (F/P, 8% 4) – 1200 (F/P, 8%4) – 1400 (F/P, 8% 3) – 1600 (F/P, 8%2) – 1800 using table

= 6000 × 1.3605 - 1200 × 1.3605 – 1400 × 1.2597 – 1600 × 1.1864 – 1800

= 8163 – 1632.6 – 1763.6 – 1898.3 – 1800

FW = 1068.5

(*c*)

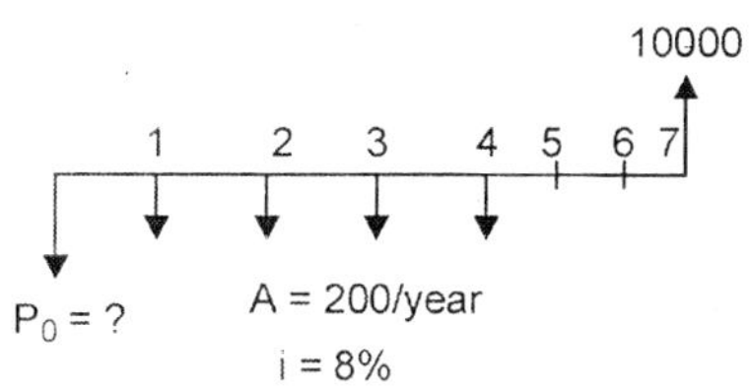

$$P_0 = A(P/A, i\%, N)$$
$$= 200(P/A, 8\%, 4)$$
$$= 200(3.3121)$$
$$P_0 = 662.42$$
$$FW = P_0(F/P, i\%, N)$$
$$= 662.42\ (F/P, 8\%, 7) + 10000$$
$$= 662.4 \times 1.7138 + 10000$$
$$\boxed{FW = 8864.8}$$

4. You purchased a building five years ago for 100,000. Its annual maintenance cost has been Rs. 5000 per year. At the end of 3 years, you have spent 9000 on roof repairs. At the end of five years, you sell the building for 120,000. During the period of ownership, you rented the building for 10000 per year paid at the beginning of each year use the future worth method to evaluate this investment when your MARR is 12% per year.

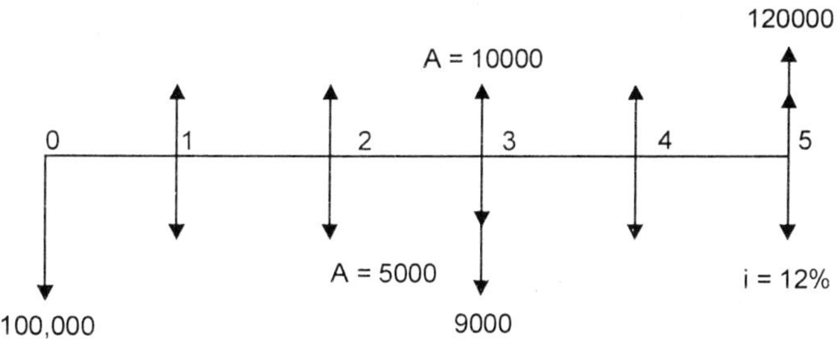

$$FW = -100000\ (F/P, 12\%\ 5) - 5000(F/A, 12\%5) - 9000\ (F/P, 12\%2) + 10000\ (F/A, 12\%5) + 120000$$
$$= -100000\ (1.7624) - 5000\ (1.7624) - 9000\ (1.2544) + 10000\ (6.3529) + 120000$$
$$= -176240 - 8812 - 11289.6 + 63524 + 120000$$
$$\boxed{FW = -12817.6}$$

5. A refining company entered into a contract for raw materials with an agreement to pay 600,000 now and 150000 per year beginning at the end of the fifth year. The contract was made for 10 years. At the end of the third year because of unexpected profits. The company requested that it be allowed to make a lumpsum payment in advance for the rest of the contract. Both, parties agreed that 7 percent compounded annually was a fair interest rate. What was the amount of the lumpsum?

Solution:

Contract deal now = 600000

remaining payments = 150000

Beginning from 5th year

N = 10 years.

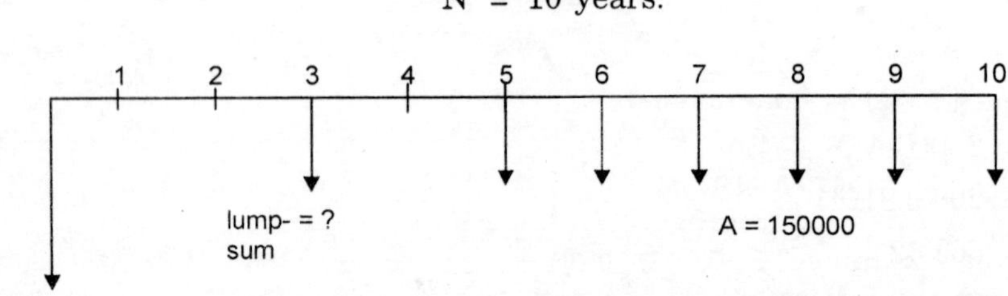

Lump sum amount = [A(P/A, *i*% N)] [(P/F, *i*%N)]

= [150000 × (P/A, 7%6)] (P/F, 7%1)

= (150000 × 4.7666) (0.9346)

Lumpsum amount = 668,229.7

6. In order to enter the market to produce a new toy for children, a manufacturer will have to make an immediate investment of Rs. 60,000 and additional investments of Rs. 5000 at the end of 1 year and 3000 more at the end of 2 years. Competing toys now are being produced by two large manufacturers, from a fairly extensive. study of the market it is believed that sufficient sales can be achieved to produce year-end before tax, net-cash flows as follows:

Year	1	2	3	4	5	6	7	8	9	10
Cash flow	–10000	5000	5000	20000	21000	26000	26000	20000	15600	7000

In addition, while it is believed that after 10 yeas the demand for the toy will no longer be sufficient to justify. Production, it is estimated that the physical assets would have a scrap value of about 8000. If capital is worth not less than 12% before taxes would you recommend undertaking the project? Make a recommendation with future method.

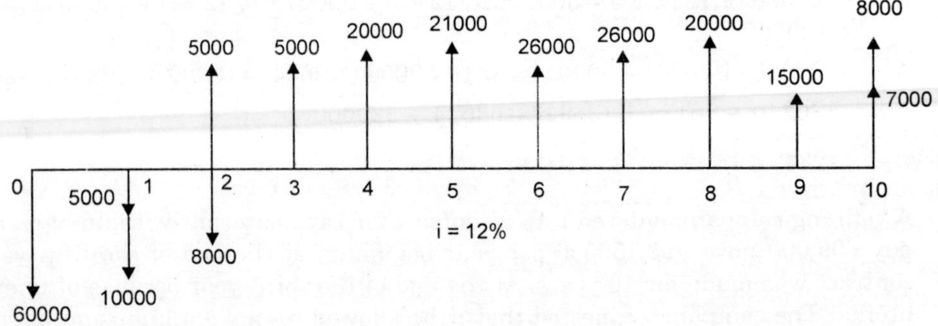

$$\begin{aligned} FW &= -60000\,(F/P,\ 12\%10) - 10000\,(F/P,\ 12\%9) - 5000\,(F/P,\ 12\%9) - 8000\,(F/P,\ 12\%8) \\ &\quad + 5000\,(F/P,\ 12\%8) + 5000\,(F/P,\ 12\%7) + 20000\,(F/P,\ 12\%\ 6) + 21000\,(F/P,\ 12\%5) \\ &\quad + 26000\,(F/P,\ 12\%\ 4)\ 26000\,(F/P,\ 12\%3) + 20000\,(F/P,\ 12\%\ 2) + 15000 \\ &\quad (F/P,\ 12\%1) + 7000 + 8000 \\ &= -186354 - 27731 - 13865.5 - 19802.9 + 12377 + 11053.5 + 39476.4 + 37010.4 \\ &\quad + 40913.6 + 36527.4 + 25213.5 + 16800 + 15000 \\ &= -13511.4 \end{aligned}$$

The project is not recommended.

3.10 PAY BACK COMPARISON

The pay back method is an extremely simple method used to obtain a rough estimate of the time that an investment will take to pay for itself.

All methods used so far shows profitability of a proposed alternative for a study period of N. The pay back method, mainly indicates a projects liquidity rather than its profitability. The pay back method has been used as a measure of a project's riskness since liquidity deals with how fast an investment can be recovered. Pay back method calculates the number of years required for positive cash flows to equal the total investment.

The formula for obtaining a rough measure of the time an investment takes to pay for itself is simple to use and understand.

$$\text{Pay back period} = \frac{\text{Required investment}}{\text{Annual receipts} - \text{annual disbursements}} = \frac{\text{First cost}}{\text{Net annual savings}}$$

In actual practice, the simple pay back formula is sometimes modified to recognize capital recovery through depreciation changes and to include without elaborations.

For example:

An investment of Rs. 1000 promises to return Rs.250 per year during its economic life of 5 years and yields.

$$\text{Pay back period} = \frac{1000}{250} = 4 \text{ years}$$

If the results of pay back calculations are questionable, wey are they used? There are at least two apparent reasons. First, they are simple calculations. Here, the depreciation and interest effects are ignored and the calculations are quick and easy and the results are intuitively logical.

Second, the other reason stems from a Bi-occupation with the flexibility of capital. If the money spent on an improvement is recovered rapidly the funds can be allocated again to other desired projects. Since, the pay back period criterion has serious weakneses, it should never be applied alone, it should only be applied as an aid in decision making.

Pay back method can also be calculated as follows : In this method it tells the no of years for an investment to pay for itself.

$$\text{Pay Back} = \frac{\text{Investment} - \text{salvage}}{\text{Operating Advantage / year}} = \frac{I - S}{OA / \text{year}}$$

The OA reflects the improvements in the flow either from increased income or decreased expenses or both but it does not yet *have a depreciation expenses—deducted* from it. Pay back measures how quickly the saving will recoup the investment amount. In its simplest form for the pay *back does not consider salvage values and taxes.*

However, when salvage and tax considerations apply, they should be included in the analysis. The investment should be reduced by the value of any salvage expected and the operating advantage should be after the tax advantages.

OA/Year after taxes = O/A/year – taxes/year

Problem 1. *The lake city bank is considering a purchase of a data processing storage unit which will cost Rs. 20,000 and will last 20 years and then have a guaranteed salvage value of Rs. 2000. It will generate savings of Rs. 4000/year before depreciation but necessioctes that Rs. 1000 of the savings must be paid in taxes. If management insists on a 5 year pay off period. Does this investment qualify.*

Solution: Given Investment I = Rs. 20000

Expected life = 20 years

Salvage life 'S' = Rs. 2000

O.A/Years = Rs. 4000/years

Taxes paid = Rs. 1000/years

O/A/years. after taxes = OA/year – taxes = 4000 – 1000 = Rs. 3000

Pay off period = 5 yrs.

$$\text{W.K.T. Pay back period} = \frac{I - S}{OA/\text{year}} = \frac{20000 - 2000}{3000}$$

$$\text{P.B.} = \frac{18000}{3000} = 6 \text{ years}$$

Conclusion: The investment does not meet the management criteria as the pay back period is more than the required one. (*i.e.,*) 5 years

Problem 2. *The Manager of a manufacturing for a large m/c shop is considering an investment with the following flow:*

Year	1	2	3	4	5	6	7	8	9	10	11	12
The Flows	0	25	100	200	500	500	500	500	400	300	200	1000

The project will cost Rs. 450, the pay back ltd for the firm is 3 years. Should be accept or reject the proposal. What shortcomings in the pay back method does this problem illustrate.

Solution: Investment 'I' = Rs. 450

In this problem, we have to see in how many years we will get the capital invested back.

First year cash flow = 0

2nd year cash flow = Rs.25

3rd year cash flow = Rs.100

4th year cash flow = Rs.200

Total Rs. 325

But we need our capital Rs.450 back which is still short of Rs.125 in 4 years. So now we will see in how much time we will get back the remaining amount.

After 4th year, we are getting Rs. 500 in 5th year.

So Rs. 500 we get in 1 year

Re. 1 we will get in $\frac{1}{500}$

Rs. 125 we will get in $\frac{1}{500} \times 125 = 0.25$ years.

That means we will get our Rs. 450 back within 4.25 years. which is greater than the required pay back.

The operating advantage or cash flow after the pay back period are more than the pay back period. It is the major disadvantage.

Problem 3. *A Rs. 40,000 extrusion machine is expected to be obsolete after 10 years with no salvage value during its life time, it should generate on Rs. 8000/yr. of which Rs. 3000 must be paid in taxes. What is the pay off period.*

Solution:

Given I = Rs. 40,000

S = 0

Expected life = 10 years

O.A/year = Rs. 8000

Taxes/year = 3000 Rs.

O.A/year after taxes = 8000 – 3000 = Rs. 5000

$$\text{Pay back period} = \frac{I - S}{O.A/\text{year}} = \frac{40,000 - 0}{5,000} = 8$$

∴ Pay back period = 8 years.

Problem 4. *A new Rs.16,400 automatic m/c will have operating cost of Rs. 0.30/unit produced whereas the existing m/c costs are Rs. 0.70/unit. The existing m/c has a market value of Rs. 8700 now and has another 5 years. of life. It would cost Rs. 500 to remove the existing machine and install the new one. If the firm requires 3 years pay back period, how many units must be produced annually to justify new m/c disregard taxes. Given Data.*

Solution:

New m/c	*Old m/c*
Cost of new m/c = Rs.16,400	Cost of old m/c = Rs.8700
Operating costs = 0.30/unit	Operating costs = 0.70/unit
	Life = 5 years

Removing and Installation charges of old m/c = Rs. 500

Pay back period = 3 years.

Units produced annually = ?

To Justify new m/c, means to go in for the new m/c

cost of the new m/c = Rs. 16,400

Market value of old m/c = Rs. 8,700

Investment = Rs. 7,700

But, Removing & installation of old & new m/c

Cost = Rs. 500

∴ Net investment 'I = Rs. 8,200 (7700 + 500)

Net operating advantage/unit = (0.7–0.3) = 0.4/unit

If there are N units, total operating advantage = 0.4 × N

$$\text{Now, Pay back} = \frac{I - S}{OA} = \frac{8200 - 0}{0.4 \times N}$$

$$3 = \frac{8200}{0.4\ N}$$

∴ N = 6,833 units/year are no. of units produced annually.

Advantages of Pay Back Method

1. Simple and quick to calculate.
2. Easy to understand.
3. A measure of time required, to return on original investment.

Disadvantages

1. If does not consider the economic life of investment.
2. Does not consider the total return on the investment.
3. Simple pay back does not consider the time value of money.

3.11 EXERCISE PROBLEMS

1. A project involves an initial outlay of Rs. 30,00,000 and with the following transactions for the next five years. The salvage value at the end of the project after five years is Rs.2,00,000. Draw a cash flow diagram of the project and find its present worth by as using i = 15% compounded annually.

End of years	*Maintenance & operating expense Rs.*	*Revenue Rs*
1	2,00,000	9,00,000
2	2,50,000	10,00,000
3	3,00,000	12,00,000
4	3,00,000	13,00,000
5	4,00,000	12,00,000

2. Find the present worth of the following cash flow series. Assume i = 15% Compounded annually

End of	0	1	2	3	4	5
Cash flow Rs.	–10,000	30,000	30,000	30000	30000	30000

3. The details of the feasibility report of a project are as shown below. Check the feasibility of the project based on present worth method using i = 20%.

Initial outlay = Rs. 50,00,000

Life of the project = 20 years

Annual equivalent revenue = 15,00,000

Modernizing cost at the end of the 10th year = Rs. 20,00,000

Salvage value at the end project life = 5,00,000

4. What is the maximum amount that you could afford to bid for a bond with a face value of Rs. 5000 and a coupon rate of 8 percent payable semiannually, if your minimum attractive rate of return is 10 percent? The bond matures in 6 years.

5. A wealthy industrial economist dies & her will specifies the 5 million Rupees of her estate will go to Bangalore University (BU) to fund a small Engineering economy building as well as 20 graduate scholarships per year over the next 20 years.

   The scholarships are to have a value of 12000 per year for the first year and should increase at a rate of Rs. 1500 per year over the following 19 years. BU requries that Rs. 15,000 starting with the third year of the bequest, be reserved for building maintenance and operating costs. Those costs are to have a linear increase of Rs. 2000 per year starting with year 4. Assuming that a 10 percent interest rate is used for such analysis, determine how much will be available for building first costs.

6. Consider the following cash flow diagram. Find the present worth using an interest rate of 15% compounded annual.

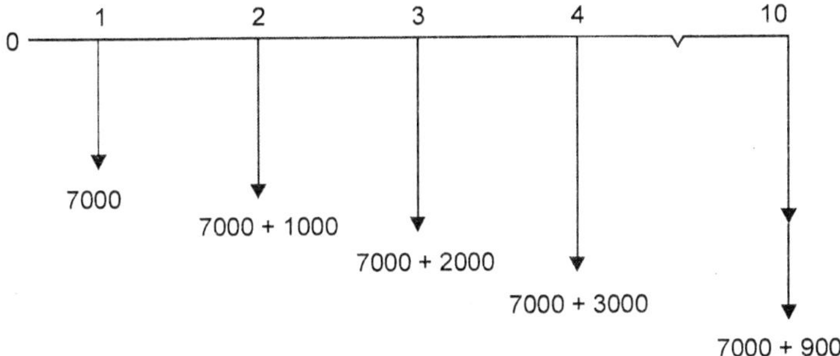

7. An automobile company recently advertised it's car for down payment of Rs. 1,50,000. Alternatively, the car can be taken home by customers without making any payment, but they have to pay an equal yearly amount of Rs 25,000 for 15 years at an interest rate of 18% compounded annually. You are asked to advise the best alternative for the customers based on the present worth method of comparison.

8. A marina has two alternative plans for constructions a small-boat landing on a lake behind the sales building, one is a wooden dock and the other is a metal and concrete wharf. Data for the two plants are as shown.

	Wood	*Metal and concrete*
First-cost	35,000	55,000
Period before replacement	10 yrs.	15 yrs.
Salvage value	5,000	0
Annual maintenance	6,000	3,200

   Using a MARR of 10 percent, compare the present worth of the two plans. Assume that both will provide adequate service and that replacement costs will be the same as the original costs.

9. Evaluate machine XYZ on the basis of the present worth method when the minimum alternative rate of return is 12%. Pertinent cost data are as follows:

	Machine XYZ
First-cost	13000
Useful life	15 years
Salvage value	3000
Annual operation costs	100
Overhaul—end of 5th year	200
Overhaul—end of tenth year	550

10. Data for 3 alternative plans are as follows.

Alternative	*Investments*	*Salvage value*	*Life*	*Annual net cash flow*
X	6000	0	3	2600
Y	12000	3000	6	2500
Z	18000	0	6	4000

If MARR is 10% which plan is to be selected based on present worth method ?

EQUIVALENT ANNUAL WORTH COMPARISONS

4.1 INTRODUCTION

Engineering Projects and business ventures can be accomplished by more than one feasible method or alternative. The feasible alternatives being considered require the investment of different amounts of capital and their annual revenues and costs may vary, five basic methods of analyzing cash flows were discussed in chapter 3 and applicable to analyses in this chapter, and their method provide a basis for economic comparison of feasible alternatives for an engineering project or other business undertaking. These methods result in the correct selection of a preferred alternative from a set of mutually exclusive feasible alternatives.

The integration of all the functions within a manufacturing enterprises leads to a better way of producing a product. Their function include communications design, production, assembly, marketing and so on. The recent trend in manufacturing products is done by computer controlled machining centers, commonly called direct numerical control (DNC) machines may cost hundreds of thousands of dollars. Evaluations of these machines against older, more conventional processes will have to consider quality, productivity, equipment reliability, possible scheduling simplification, work in process reduction and other potential benefits. Many economic decisions can be assisted by determining costs, expenditures and net worth on the basis of annual or periodic timing.

Manufacturing and other engineering alternative evaluations often just make more sense when viewed on an annual basis.

4.2 EQUIVALENT ANNUAL WORTH COMPARISON METHOD

In an annual-worth method all the receipts and disbursements occurring over a period are converted to an equivalent uniform yearly amount. Cost accounting procedures, depreciation expenses, tax calculations and other summary reports are annual.

Equivalent annual-worth comparisons produce results compatible with present worth and rate of return comparisons for a set of common assumptions; a preference for an alternative exhibited by one method will be mirrored by the other two. Annual worth calculations are frequently part of the computations required to develop. Present-worth and rate-of-return, values and parallel computations by different method due useful for complementary comparisons that improve the clarity of an analysis. The six conditions listed in Chapter 3 for basic present-worth comparisons also apply to basic annual-worth comparisons cash flows and interest rates are known. Cash flows are before taxes and in constant Rs. and comparisons include neither intangible considerations nor limits due to availability of financing.

There are two general situations.

(1) All the feasible alternatives under consideration have identical useful lives.

(2) The alternatives have different useful lives.

The analyst must determine the analysis (study) period that best reflects actual conditions particular to the capital investment situation being evaluated and provides a sound basis for decision making.

4.2.1 Structure of a Capital Recovery Annuity

The main structure of annual-worth calculations is the capital recovery factor, which converts a lumpsum to an equivalent annuity. This annuity usually represents an investment in an asset that is expected to generate a positive future cash flow; and the duration of the annuity is therefore the life of the asset. The cost of asset is uniform negative payments. This negative cash flow is offset by the positive revenue produced by the asset in establishing the net equivalent annual worth of the investment.

The capital recovery factor (A/P, i, N) accounts for both the repayments of invested capital P and the interest earned on the unrecovered portion of the investment. The structure of an annuity in which varying amounts from the equal payments are allocated to capital recovery and interest is understood by a simple application.

Example 1. An asset is purchased for Rs. 40,000/. Expected life is 4 years and no salvage value at the end of its life. The purchaser intends to recover the 40,000/- investment over 4 years plus the interest of the 40,000/- would have earned if it had been invested elsewhere. Interest rate is 10%. The series of equal payments that will return the capital plus interest is computed as

Solution:

$$\text{Equivalent annual payment A} = \text{P(A/P, 10\%, 4)}$$

$$= 40{,}000\left[\frac{i(1+i)^N}{(1+i)^N - 1}\right]$$

$$= 40{,}000\left[\frac{0.1(1+0.1)^4}{(1+0.1)^4 - 1}\right]$$

$$= 12618.8$$

$$\boxed{A \approx 12619}$$

Example 2. *A large gasoline station is required by the city to install vapor containment equipment on it's gasoline pump nozzles and storage tank vents the immediate conversion cost will be 180,000 with an estimated Rs. 600 per year for maintenance. It will be necessary to update the equipment every 3 years at a cost of 3500. The station pumps an average of 1 million gallons of gasoline per month. On an annual basis, what would be the price increase per gallon necessary to pay for the conversion over a 6-year period? Include the sixth year's update cost in your analysis and assume an interest rate of 14 percent.*

Solution.

$$P = 180{,}000$$

$$A = 600/\text{year}$$

Once 3 years. For update equipment = Rs. 3500

$$i = 14\%$$

$$N = 6 \text{ years}$$

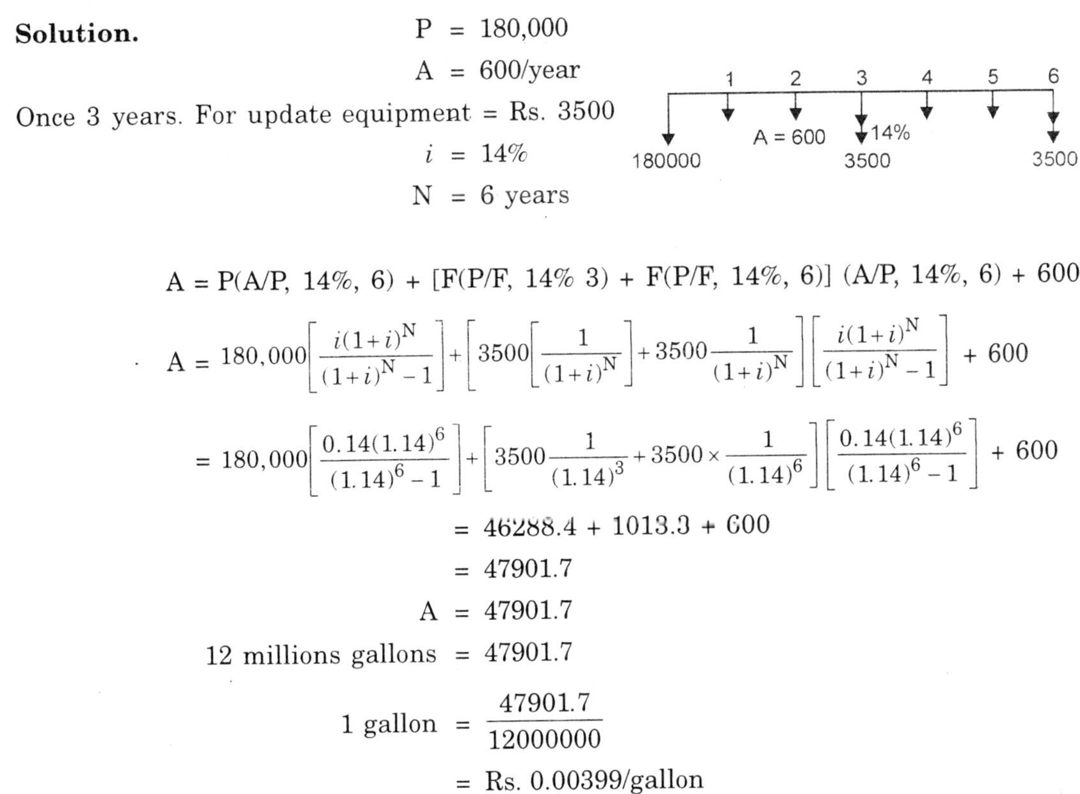

$$A = P(A/P,\ 14\%,\ 6) + [F(P/F,\ 14\%\ 3) + F(P/F,\ 14\%,\ 6)]\ (A/P,\ 14\%,\ 6) + 600$$

$$A = 180{,}000\left[\frac{i(1+i)^N}{(1+i)^N - 1}\right] + \left[3500\left[\frac{1}{(1+i)^N}\right] + 3500\frac{1}{(1+i)^N}\right]\left[\frac{i(1+i)^N}{(1+i)^N - 1}\right] + 600$$

$$= 180{,}000\left[\frac{0.14(1.14)^6}{(1.14)^6 - 1}\right] + \left[3500\frac{1}{(1.14)^3} + 3500 \times \frac{1}{(1.14)^6}\right]\left[\frac{0.14(1.14)^6}{(1.14)^6 - 1}\right] + 600$$

$$= 46288.4 + 1013.3 + 600$$

$$= 47901.7$$

$$A = 47901.7$$

$$12 \text{ millions gallons} = 47901.7$$

$$1 \text{ gallon} = \frac{47901.7}{12000000}$$

$$= \text{Rs. } 0.00399/\text{gallon}$$

$$= 0.4 \text{ Paise/gallon}$$

4.3 SITUATIONS FOR EQUIVALENT ANNUAL WORTH COMPARISONS

The term annual worth suggests a positive value. The calculations will produce a negative value. A negative annual worth indicates the equivalent value of negative cash flow for disbursements is greater than the corresponding positive flow of receipts. Then the objective is to identify the alternative with the least equivalent annual cost (negative cash flow) we use the term equivalent annual cost term equivalent annual cost (EAC) to designate only costs and we will use the term equivalent annual worth (EAW) when costs and incomes (benefits) are both present. It is often very difficult to not worth of required study time, for example the income produced by a copying machine is troublesome to derive exactly its output is utilized by many people, many departments working on many projects.

When any costs are involved, it is convenient to ignore the minus sign convention and both when cost and revenue are involved.

4.3.1 Consolidation of Cash Flows

In any project the general question will be what's it worth. It is difficult to expect from the proposal until the receipts and disbursements associated with its conduct are collectively analyzed. Improvements programs are prime examples.

Organizations regularly engage in programs to improve productivity, reduce accidents, raise quality etc. Each is a worth while goal, expected to have positive rewards, but each has cost too. Consolidating the costs and potential rewards may take the form of a net annuity.

Example 1. *A consultant firm proposes to provide "self-inspections" training for clerks who work with finance section. The program lasts 1 year, cost Rs. 2000 per month and professes to improve quality while reduce clerical time. The use of this program savings in the first months should amount to Rs. 800 and should increase by Rs. 400/month for the rest of the year. However, operational difficult and work interference are expected to boost clerical costs by Rs. 1200 the first month, but this amount decline in equal increments at the rate of Rs. 100 per month. If the required return on money is 12% compounded monthly and the program must pay for itself with in 1 year. Should its consultants be hired?*

Solution.

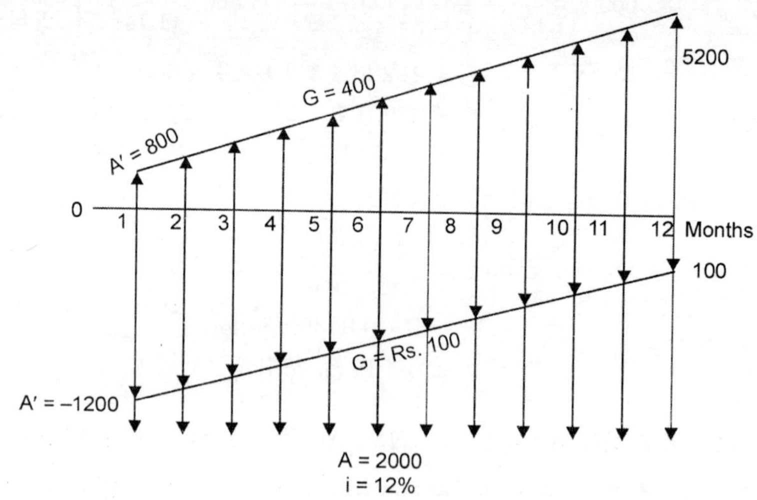

$$i = \frac{r}{N} = \frac{0.12}{12} = 0.01 \text{ per period.}$$

$$N = 12 \text{ months.}$$

Equivalent monthly worth of savings

$$= 800 + 400\,(A/G, 1, 12)$$
$$= 800 + 400\,(5.3681)$$
$$= \text{Rs. } 2947$$

Equivalent monthly worth of costs

$$= -2000 - [1200 - 100\,(A/G, 1, 12)]$$
$$= -2000 - [1200 - 100(5.3681)$$
$$= -2000 - [1200 - 536.8]$$
$$= -2000 - 663.2$$
$$= -2663.2$$

Equivalent net monthly cash flow $= 2947 - 2663$

$= \textbf{Rs. 284.}$

4.3.2 Recovery of Inverted Capital

The investers want to know the recovery of inverted capital plus the desired rate of return. It is spread over the life of the investment, it is convenient to convert capital recovery costs to the same annual pattern. The consequential result of combining uniform cost of revenue flows is a positive, zero or negative series of payments respectively, categorizes the investment as gratifying, adequate or insufficient.

Example 1. *The purchase of a new bus will reduce labour cost for maintenance by Rs. 15000 per year. The price of the bus is Rs. 93000 and it's operating costs will exceed those of present equipment by Rs. 250 per month. The salvage value is expected to be 18000 in 8 years. Should the bus be purchased when the current available interest rate is 7%?*

Solution.

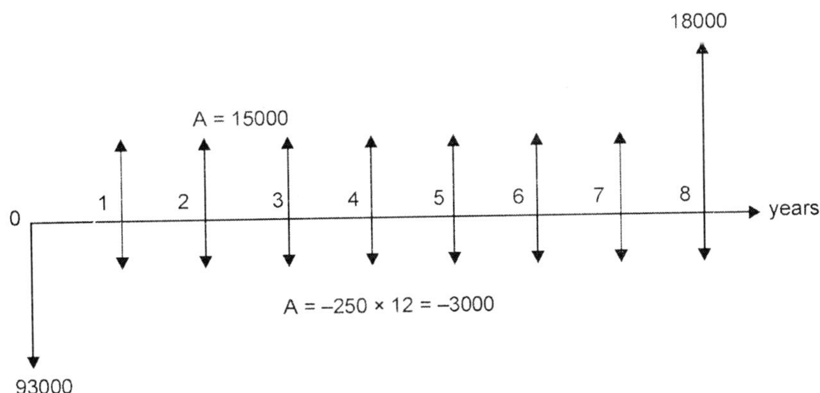

$$\text{EAW} = 93000\,(A/P, 7, 8) + 18000\,(A/F, 7, 8) - 3000 \text{ using table value}$$

$$= 93000\,(0.16747) + 18000\,(0.09747) + 15000$$

$$\text{EAW} = \text{Rs. } -1820$$

Equivalent Annual Worth (EAW) calculations shows that the purchase and use of the bus will cause a loss equivalent of Rs. 1820 per year for 8 years.

The capital recovery factor leads to the same solution. When capital recovery costs are registered negatively and the net annual savings are positive.

$$\text{EAW} = \text{annual savings} - \text{capital recovery costs}$$

$$= 15000 - 3000 - [(P - S)\,(A/P, 7, 8) + Si]$$

$$= 12000 - [(93000 - 18000)\,(0.16747) + 18000 \times 0.07]$$

$$= 12000 - (12500 + 1260)$$

$$= \mathbf{-1820}$$

4.3.2 Net Cash Flow Comparison

Next question arises, which one is better?

If worth is measured by revenues, the criterion is strictly economic, highest net worth is preferred. But, when alternatives have only costs and no income, a low EAC is preferred.

Example. 1. *The machine A and B have same service life of 5 years. The other expenses are given below. If the money is worth 10% P.A. which machine is more economical?*

Cash flows	*M/CA*	*M/CB*
First cost	250000	150000
Uniform end of year maintenance	2000	4000
Over-haul at the end of 3rd year	–	3000
Salvage value	10,000	–
Yearly savings on account of less inspection.	1000	–

M/CA:

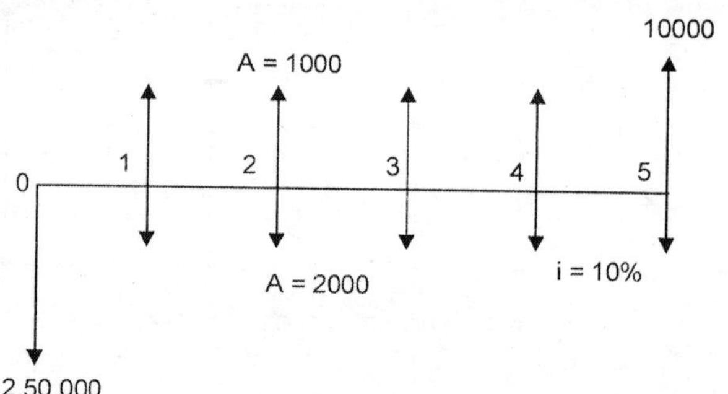

AW = –250000 (A/P, *i*, N) + [10000(P/F, *i*% N)] (A/P, *i*% N) +1000 – 2000

= –250000 (A/P, 10% 5) + [10000 (P/F 10% 5)] (A/P, 10% 5) –1000 + 2000

= –65950 + 1637.9 + 1000 – 2000

AW_A = –65312.1

M/CB :

1 2 3 4 5

A = 4000

3000

150000

AW = –150,000 (A/P, *i*%N) – [(3000 (P/F, *i*%N) (A/P, *i*% N)] – 4000

= –150,000 (A/P, 10% 5) – (3000(P/F, 10% 3) (A/P, 10% 5) – 4000

= –39570 – 618.3 –4000

AW = –44188.3

Many economic decisions are based only on costs rather than net equivalent annual cost study.

4.4 CONSIDERATION OF ASSET LIFE

Translating cash flows to equivalent annuities is a mechanical process that becomes almost automatic with practice. Understanding the meaning of an economic comparison and being able to explain it's significance to others are the critical skills.

4.4.1 Definitions of Assets Life

The time value analysis, N takes on a special meaning to represent the life of an asset that loses value as a function of use or time. The frequently used term to describe the life of an asset are listed as follows:

Ownership Life or Service Life is the period of time an asset is kept in service by the owner's. It implies useful service from the time of purchase until disposal. A machine can have a physical life longer than its service life the machine is still physically sound, but there is no useful function for it to perform.

Accounting Life is a life expectancy based primarily on book keeping and tax considerations. It may or may not correspond to the period of usefulness and economic desirability.

Economic Life is the period of time extending from date of installation to date of retirement (by demotion or disposal) from the primary intended service. The need for retirement is signalled in an engineering economy study. When the equivalent cost of a new asset (challenger) is less than the equivalent cost of keeping asset (defender) for an additional period of time.

4.4.2 Comparison of Assets with Equal and Unequal Lives

The lease or buy questions is raised with increasing regularity that corresponds to the rapid growth of leasing companies. It is now possible to lease almost any type of production equipment that is not custom-designed for narrowly specialized service. Important tax and inflation considerations involved in the lease or buy choice.

Example 1. *Alternative with Equal Annual Costs.*

A machine needed for 3 years can be purchased for 70,000 and sold at the end of the period for about 20,000. A comparable machine can be leased for 27737/-per year. If a firm expects a return of 20% on investments should it buy or lease the machine when end-of-year payments are expected.

Solution. Equivalent annual cost to buy

$$\text{EAC} = (70{,}000 - 20{,}000)\ (\text{A/P},\ i\%\ \text{N}) + 20{,}000 \times i$$

$$= 50{,}000\ (\text{A/P},\ 20\%,\ 3) + 20{,}000 \times \frac{20}{100}$$

$$= 50{,}000 \times 0.47473 + 4000$$

$$= 27736.5$$

$$\text{EAC} \approx 27737/-$$

$$\text{Annual cost of lease} = \mathbf{27737/-}$$

Since, the results are the same for both leasing and buying the decision will likely affected by the existence or absence of other projects in need of finding.

The two alternatives are compared on the basis of their costs because the income resulting from their contributions is not available and it is believed that both are capable of producing that contribution.

Comparison of assets with unequal lives

Example 2: *Two machines performs the same function. First machine has a cost of Rs. 9500/- relatively high operating costs of Rs. 1900 per year more than those of the second machine and short life is 4 years. The second machine Rs. 25100 and can be kept in service economically for 8 years. The scrap value from either machine at the end of its life will barely cover it's removal cost which is preferred when the minimum alternative rate of return is 8 percent?*

Solution.

First Machine

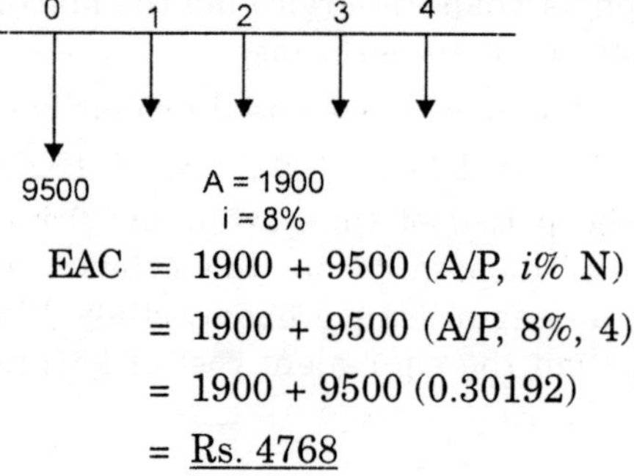

$$\begin{aligned} \text{EAC} &= 1900 + 9500\ (\text{A/P}, i\%\ \text{N}) \\ &= 1900 + 9500\ (\text{A/P}, 8\%, 4) \\ &= 1900 + 9500\ (0.30192) \\ &= \underline{\text{Rs. } 4768} \end{aligned}$$

Second Machine

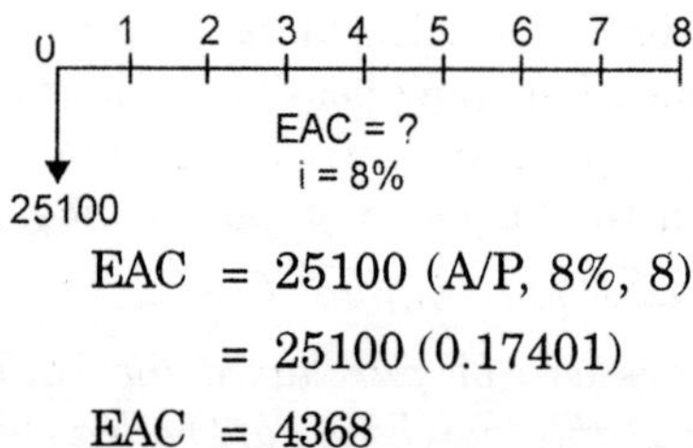

$$\begin{aligned} \text{EAC} &= 25100\ (\text{A/P}, 8\%, 8) \\ &= 25100\ (0.17401) \\ \text{EAC} &= \underline{4368} \end{aligned}$$

The second machine has a lower annual cost for service during next 8 years and is preferred.

4.4.3 Perpetual Life

Sometimes asset is treated as if it will last forever. The assumption of infinite life in terms of capital recovery is more reasonable than the physical interpretation. Examples Like Egyptian Pyramids or the Great Wall of China, lasts forever, but the difference between infinity and 100 years in the numerical value of the capital recovery factor is quite small.

$$(\text{A/P}, i, \text{N}) = \frac{i(1+i)^{\text{N}}}{(1+i)^{\text{N}} - 1}$$

as N gets very large

$$\left[(1+i)^{\text{N}} - 1\right] \rightarrow (1+i)^{\text{N}}$$

The limit of the capital recovery factor as N approaches infinity is $(A/P, i, N)_{N \to \infty} = i$.

In an economic comparison involving as asset with an infinite life **Ex.** as land. The interest rate replaces the capital recovery factor.

The human-made assets most closely approaching perpetual life are dams, tunnels, canals and monuments. The nature of very-long-lived assets relegates them mostly to public projects and study period is 50 years or recognition of changing public needs and technological advances that generate new ways to fulfil the needs.

The similarity between the perpetual-life assumption for calculating equivalent annual worth and the capitalized cost method associated with present worth models should be apparent.

4.5 USE OF A SINKING FUND

The sinking fund factor is applied to compute the annuity required to accumulate a certain future amount. Organizations are sometimes obligated by legislated or contractual agreements to establish a fund, separate from their internal operations to accumulate a specified amount by a specified time. This accumulation is called a sinking fund. Provision for sinking fund requires set aside a portion of the income derived from sales,or taxes each year in order to retire a bond issue.

SOLVED PROBLEMS

Problem 1. *A company has three proposals for expanding it's business operations. The details are as follows:*

Alternative	*Initial cost Rs.*	*Annual revenue Rs.*	*Life yrs.*
A_1	*25,00,000*	*8,00,000*	*10*
A_2	*20,00,000*	*6,00,000*	*10*
A_3	*30,00,000*	*10,00,000*	*10*

Each alternative has in significant salvage value at the end of it's life. Assuming an interest rate of 15% compounded annually. Find the best alternative for expanding the business operations of the company using the annual equivalent method.

Solution.

Alternative 1

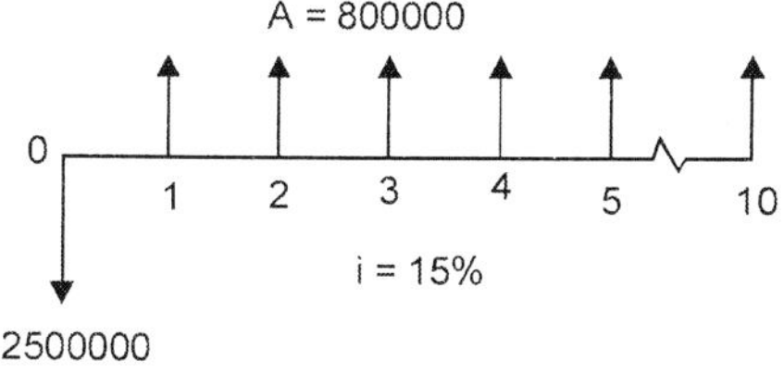

$$AEM = 25,00,000(A/P, i\% \ N) + 800,000$$

$$= 25,00,000\frac{i(1+i)^N}{(1+i)^N - 1} + 8,00,000$$

$$= 25,00,000\frac{0.15(1.15)^{10}}{(1.15)^{10} - 1} + 8,00,000$$

AEM = 1298130.2

Alternative 2

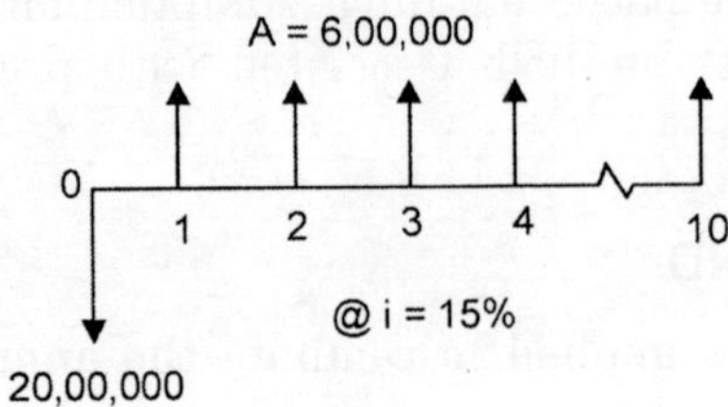

AEM = 20,00,000(A/P, *i*% N) + 6,00,000

= 20,00,000 (A/P, 15%, 10) + 6,00,000

= 20,00,000 × 0.19925 + 6,00,000

AEM = 9,98,500

Alternative 3

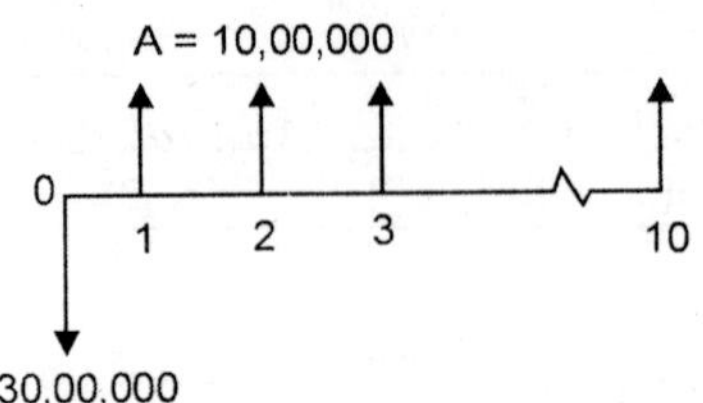

AEM = 30,00,000 (A/P, 15% 10) + 10,00,000

= 30,00,000 (0.19925) + 10,00,000

AEM = 15,97,750

Problem 2. *An investment of Rs. 20,000 in an off gas monitoring system will have a salvage value of Rs. 6000 after an economic life of 5 years. Maintenance and operating costs are 4400/year and the firm cost of capital is 10% what is the average annual cost of this investment.*

Solution:

Give Data: I = Rs. 20,000

Operating and Maintenance Costs = 4,400/year

S = Rs. 6,000

i = 0.1

Economic Life = 5 years

A.A.C = ?

We know that Capital Recovery 'C.R.' = $\frac{I - S}{n}$

(*i*) $\text{C.R.} = \frac{20{,}000 - 6{,}000}{5} = \text{Rs. } 2{,}800$

(*ii*) $i(I_{av}) = \frac{I + S}{2} \times i$

$= \frac{20{,}000 + 6{,}000}{2} \times 0.1 = \text{Rs. } 1300.$

(*iii*) Operating and Maintenance Cost = Rs. 4,400.

Average Annual cost = *i* + *ii* + *iii*

= 2800 + 1300 + 4400

A.A.C. = Rs. 8500.

Problem 3. *The following 2 m/c's are being considered for purchase since they are multipurpose, it is impossible to allocate revenues to them, which of the 2 should be purchased.*

Description	*Machine A*	*Machine B*
Capital cost	*Rs. 13,000*	*Rs. 8,000*
Labour cost	*8,000/year*	*9,000/year*
Maintenance cost	*300/year*	*400/year*
Economic life	*7 years*	*7 years*
Salvage value	*2000*	*2000*
Cost of capital	*10%*	*10%*

Solution. *For m/c A:*

(*i*) $\text{C.R.} = \frac{I - S}{n} = \frac{13{,}000 - 2{,}000}{7} = \frac{11{,}000}{7} = 1571.428.$

(*ii*) $i(I_{av}) = \frac{I + S}{2} \times i = \frac{13{,}000 + 2{,}000}{2} \times 0.1 = 750$

(*iii*) Maintenance and Operating Costs

and Labour Cost = 300 + 8000 = 8300/year and labour cost.

∴ The average annual cost for m/c A

= *i* + *ii* + *iii*

= 1571.428 + 750 + 8300

= Rs. 10,621.4285.

For m/c B:

(*i*) $\text{C.R.} = \frac{I - S}{n} = \frac{8{,}000 - 2000}{7} = 857.1428.$

(*ii*) $i(I_{av}) = \frac{I+S}{2} \times i = \frac{8,000+2,000}{2} \times 0.1 = 500$

(*iii*) Operating, Maintenance and Labour Cost

= 9000 + 400

= 9400/year

∴ The average annual cost for m/c B

= *i* + *ii* + *iii*

= 857.1428 + 500 + 9400

= Rs. 10,757.1428

Conclusion : The average annual cost of m/c A being less than the m/c B, So m/c A should be preferred.

Problem 4. *J.M. Wal Company is a large industrial sewing shop which manufacturers such garments as aluminised asbestus suits for the iron industry heavy duty car top carrier covers, industry approns and so on. Essentially they have several different kinds of machine through many different jobs pass.*

At present they are considering the purchase of a new m/c. The first alternative is lower in price but will incur a higher labour cost while the second is higher in price but requires less labour. In addition, the maintenance cost for their lower priced machine is estimated to be Rs. 1000 for the first year increasing by Rs. 100/year over the life of the machine. The higher priced machine has an estimated first year maintenance cost of Rs. 1500 and an increase of Rs. 200/year. The other economic data are as follows:

Description	*Machine A*	*Machine B*
Capital cost	*15,000*	*25,000*
Labour cost	*14,000*	*7,000*
Salvage value	*2,000*	*1,000*
Economic life	*10 yrs.*	*10 yrs.*

If their cost of capital is 15%, which machine should they purchase.

Solution. Step I : To calculate the average annual maintenance cost

For m/c A:

Average annual cost = [1000 + 1100 + 1200 + 1300 + 1400 + 1500 + 1600 + 1700 + 1800 + 1900]/10.

∴ Average Annual Maintenance Cost = 1450/year.

For m/c B:

Average Annual Maintenance Cost = [1500 + 1700 + 1900 + 2100 + 2300 + 2500 + 2700 + 2900 + 3000 + 3100]/10

Average Annual Maintenance Cost = 2400/year

Step II : To calculate the capital recovery, $i(I_{av})$, and operating and maintenance cost for m/c A and m/c B.

For m/c A,

(*i*) $$\text{Capital Recovery} = \frac{I - S}{n} = \frac{15,000 - 2,000}{10} = \frac{13,000}{10}$$

$\therefore$ C.R. = 1300

(*ii*) $$i(I_{av}) = \frac{I + S}{2} \times i = \frac{15,000 + 2,000}{2} \times 0.15$$

$\therefore$ $i(I_{av})$ = 1275

(*iii*) Operating and Maintenance Costs

= Labour costs + average annual maintenance cost

= 14,000 + 1450 = 15,450

Now Average Annual Cost for *m/c* A

= (*i*) + (*ii*) + (*iii*)

= 13,000 + 1,275 + 15,450

A.A.C. for *m/c* A = 18,025.

For m/c B,

(*i*) $$\text{Capital Recovery (C.R.)} = \frac{I - S}{n} = \frac{25,000 - 1,000}{10}$$

C.R. = 2400.

(*ii*) Interest on Average Investment $i(I_{av})$

$$= \frac{I + S}{2} \times i$$

$$= \frac{25,000 + 1,000}{2} \times 0.15 = 1950.$$

(*ii*) (I_{av}) = 1950.

(*iii*) Operation Labour and Maintenance Costs

= 7000 + 2400 = 9400

Now Average Annual Cost for *m/c* B

= *i* + *ii* + *iii*

= 2400 + 1950 + 9400

A.A.C, for *m/c* B = 13,750.

Conclusion : The average annual cost of *m/c* B being less than that of *m/c* A, so we will prefer m/c B only.

Problem 5. *The Campbell Company is considering the purchase of one of 2 overhead cranes for their fabrication shop. For crane A the capital cost would be Rs. 25,000 while for B it would be Rs. 18,000. Both have an expected life of 10 years. The labour costs for both would be the same but maintenance cost would not be. For A, the maintenance cost would be Rs. 1000/yr while for B it would be Rs. 500 the first year and increases by Rs. 200 each year there after. The salvage value for A at the end its economic life would be Rs. 8000 while for B it would be 3000. If the cost of capital to the firm is 15% which crane should be purchased.*

Solution. Data Given:

Capital Cost of Crane A = Rs. 25,000

Capital cost of crane B = Rs. 18,000

Expected Life = 10 years

To calculate Maintenance Costs for *m/c* or Crane A and B :

For Crane A:

Average Annual Cost for Crane A = [1000 + 1000 + 1000 + 1000 + 1000 + 1000 + 1000 + 1000 + 1000 + 1000]/10

= 1000/year.

For Crane B:

Average Annual Maintenance Cost = [500 + 700 + 900 + 1100 + 1300 + 1500 + 1700 + 1900 + 2100 + 2300]/10

= 1400/year.

Salvage value for crane A = Rs. 8000

Salvage value for crane B = Rs. 3000

Interest = 10%

Crane A:

(*i*) $$\text{C.R.} = \frac{\text{I}-\text{S}}{n} = \frac{25,000-8,000}{10} = \frac{17,000}{10} = 1700$$

(*ii*) $$i(\text{I}_{av}) = \frac{\text{I}+\text{S}}{2}\times i = \frac{25,000+8,000}{2}\times 0.15 = 2475$$

(*iii*) Maintenance Cost = 1000/year

Average Annual Cost = (*i* + *ii* + *iii*) = 1700 + 2475 + 1000

A.A.C. = 5175

Crane B:

(*i*) $$\text{C.R.} = \frac{\text{I}-\text{S}}{n} = \frac{18,000-3,000}{10} = 1500$$

(*ii*) $$i(\text{I}_{av}) = \frac{\text{I}+\text{S}}{2}\times i = \frac{18,000+3,000}{2}\times 0.15 = 1575$$

(*iii*) Maintenance Cost = 1400/year

Average Annual Cost = (*i* + *ii* + *iii*)

= 1500 + 1575 + 1400

A.A.C. = 4475

Conclusion : As the average annual cost of crane B is less than that of A, so we will purchase crane B only.

Problem 6. *A public utility district, (P.U.D.) has found that it can lease a data a processing service at an annual cost of Rs. 9100/year for 5 years. The cost includes all maintenance and operating cost. Alternatively it can purchase Rs. 10,000 worth of equipment* *which will have a salvage value of Rs. 3000 in 5 years and will require Rs. 7000/years in labour cost to operate. In addition maintenance is expected to be Rs. 500 for the first year and increased by Rs. 100/year thereafter. The public utility is limited by the federal power consumption to 8% return on any investment. Using an average investment criteria, determine whether it is more economical for the P.U.D. to lease the service or purchase the equipment.*

Solution. Data:

Lease Data

Cost of data processing servicing machine = Rs. 9100

(*i.e.*,) C = Rs. 9100

Purchase Data

Cost of the data processing servicing machine if purchased

I = Rs.10,000.

Salvage value = Rs.3000

Life = 5 years

Labour costs = Rs.7000.

Maintenance and operating

$$\text{costs} = \frac{500 + 600 + 700 + 800 + 900}{5}$$

$$= 700/\text{year}.$$

i = 8% = 0.08.

Average annual cost of m/c purchased:

(*i*) $$\text{C.R.} = \frac{\text{I} - \text{S}}{n} = \frac{10{,}000 - 3{,}000}{5} = 1400/\text{year}.$$

(*ii*) $$i(\text{I}_{\text{av}}) = \frac{\text{I} + \text{S}}{2} \times i = \frac{10{,}000 + 3{,}000}{2} \times 0.08 = 520/\text{year}.$$

(*iii*) Operating and Maintenance Cost

= Rs. 7700/year

Average Annual Cost = (*i*) + (*ii*) + (*iii*) = 1400 + 520 + 7700

Conclusion : The average annual cost for the data processing servicing unit purchased is greater than that of leased. So we will go in for the leased one only.

Advantages

(*i*) It is relatively simple and quick to calculate.

(*ii*) It is not difficult to visualize an investment costing an equal amount each year.

(*iii*) Alternative with different life times can be effectiveness compared.

(*iv*) Cost comparison can be made without knowledge of income or considering the taxes on incomes.

Disadvantages

(*i*) An approximation is used which ignores the timing of cash flows by computing the interest on the average investment.

(*ii*) Uniform or straight line depreciation of assets is usually followed.

(*iii*) Total costs are not explicitly considered.

4.6 EQUIVALENT ANNUAL COST METHOD

Equivalent annual cost method is a time adjusted (more accurate) method of calculating an equal amount cost over the life of an investment. It is similar to average annual cost method. It is more exact than A.A.C. method. In this method the compounded interest is taken into account. The costs included in this equivalent annual cost method are:

(*i*) Capital Recovery and Return (*i.e.*,) (C.R. and R)

(*ii*) Interest on the Salvage ($i \times s$)

(*iii*) Other Annual Operating and Maintenance Costs

$$\textit{Equivalent Annual Cost} = (\text{C.R. and R} + i(\text{S}) + \text{O and M})$$

$$\text{C.R. and R} = (\text{I} - \text{S}) \times \left[\frac{1}{\left((\text{PV}_a)_i^n\right)}\right]$$

Problem 1. *An investment of Rs. 20,000 in an off gas monitoring system will have a salvage value of Rs. 6,000 after an economic life of 5 years. Maintenance and operating costs are 4400/year and the firms cost of capital is 10%.*

(*a*) *What is the Equivalent Annual Cost of this investment.*

b) *How does this E.A.C. compare with the Average Annual Cost (A.A.C.)*

Ans. Given data :

I = Rs. 20,000

S = Rs. 9,600

Economic Life = 5 years

Maintenance and Operating Costs = 4400/year

Interest = 10% = 0.1

Step I : To find the Capital Recovery and Return.

$$\text{C.R. and Return} = (\text{I} - \text{S}) \times \left[\frac{1}{\left((\text{PV}_a)_i^n\right)}\right]$$

$$= (20{,}000 - 6{,}000) \times \left[\frac{1}{\left[(\text{PV}_a)_{0.1}^5\right]}\right]$$

From the Annuity tables for $i = 0.1$ and $n = 5$

$$\text{PV}_n = 3.791$$

$$\text{C.R. and Return} = 14{,}000 \times \frac{1}{3.791}$$

$$= 3{,}692.95$$

Step II: Interest on Salvage = 6000×0.1

$$= 600.$$

Step III: Operating and Maintenance Costs = 4,400

Now The Equivalent Annual Cost = I + II + III

= 3,692.95 + 600 + 4,400

E.A.C = 8,692.95

Conclusion : When compared with an annual average cost of 8500/year; and is larger because the effect of C.I. on the capital investment is taken into account in the E.A.C. method.

Problem 2 *The following 2 m/c's are being considered for purchase since they are multipurpose, it is impossible to allocate revenues to them, which of the 2 should be purchased.*

Description	*Machine A*	*Machine B*
Capital cost	*Rs. 13,000*	*Rs. 8,000*
Labour cost	*8,000/year*	*9,000/year*
Maintenance cost	*300/year*	*400/year*
Economic life	*7 years*	*7 years*
Salvage value	*2000*	*2000*
Cost of capital	*10%*	*10%*

Solution: For *m/c* A:

(*i*)

$$\text{C.R. and R} = (\text{I} - \text{S})\left[\frac{1}{\left[(\text{PV}_a)i^n\right]}\right]$$

$$= (13{,}000 - 2000)\left[\frac{1}{\left[(\text{PV}_a)^7_{0.1}\right]}\right] \quad \text{from table}$$

$$= (13{,}000 - 2000) \times \frac{1}{4.868}$$

$$\text{C.R. and R} = 2{,}259.46$$

$$(\text{PV}_a)^7_{0.1} = \frac{1}{(1+i)^7}$$

$$= \frac{1}{(1+.1)^1} + \frac{1}{(1+.1)^2} + \frac{1}{(1+.1)^3} + \frac{1}{(1+.1)^4} + \frac{1}{(1+.1)^5} + \frac{1}{(1+.1)^6} + \frac{1}{(1+.1)^7} = 4.868$$

(*ii*) $i \times \text{Salvage} = 2000 \times 0.1 = 200$

(*iii*) Operation, Labour and Maintenance charges

= 8000 + 300 = 8300

Now Equivalent Annual Cost for the Machine A

= 2,259.46 + 200 + 8300 = 10,759.46.

For m/c B:

(*i*) $$\text{C.R. and R} = (8000 - 2000) \times \frac{1}{(PV_a)^7_{0.1}}$$

$$= 6000 \times \frac{1}{4.868} = 1232.53$$

(*ii*) $$i \times \text{Salvage} = 2000 \times 0.1 = 200$$

(*iii*) Operation, Labour and Maintenance Charges

$$= 9000 + 400 = 9400$$

Equivalent Annual Cost for the Machine B

$$= 1232.53 + 200 + 9400$$

$$= 10{,}832.53.$$

Conclusion : The Equivalent Annual Cost of m/c A and m/cB being greater than that of in average annual cost. It is because we consider the compound interest on capital investment in the Equivalent Annual Cost Method. Select m/c A only.

Problem 3. *J.M. Wal Company is a large industrial sewing shop which manufacturers such garments as aluminised asbestus suits for the iron industry heavy duty car top carrier covers, industry approns and so on. Essentially they have several different kinds of machine through many different jobs pass.*

At present they are considering the purchase of a new m/c. The first alternative is lower in price but will incur a higher labour cost while the second is higher in price but requires less labour. In addition, the maintenance cost for their lower priced machine is estimated to be Rs. 1000 for the first year increasing by Rs. 100/year over the life of the machine. The higher priced machine has an estimated first year maintenance cost of Rs. 1500 and an increase of Rs. 200/year. The other economic data are as follows:

Description	***Machine A***	***Machine B***
Capital cost	*15,000*	*25,000*
Labour cost	*14,000*	*7,000*
Salvage value	*2,000*	*1,000*
Economic life	*10 yrs.*	*10 yrs.*

If their cost of capital is 15%, which machine should they purchase.

Solution.

(*i*) For m/c A: $$\text{C.R. and R:} = (I - S) \times \left[\frac{1}{(PV_a)i^n}\right]$$

$$= (15000 - 2000) \times \left[\frac{1}{(PV_a)^{10}_{0.15}}\right]$$

$$= 13{,}000 \times \frac{1}{5.0154}$$

$$(PV_a)_{0.15}^{10} = 0.8695 + 0.7561 + 0.6575 + 0.571 + 0.497 + 0.432 + 0.3759 + 0.326 + 0.284 + 0.2471 = 5.0154.$$

C.R and R = 2,592.01

(ii) $i \times$ salvage $= 2000 \times 0.15 = 300.$

(iii) Operation, Labour and Maintenance Cost

$= 14{,}000 + 1450 = 15{,}450.$

Equivalent Annual Cost for m/c A = 2592.01 + 300 + 15,450 = 18,342.01

For m/c B:

(i) $$\text{C.R and Return} = (I - S) \times \left[\frac{1}{(PV_a)i^n} \right]$$

$$= (25{,}000 - 1000) \times \frac{1}{5.0154} = 4785.26.$$

(ii) $i \times$ salvage $= 1000 \times 0.15 = 150.$

(iii) Operation, Labour and Maintenance Costs

$= 7000 + 2400 = 9400.$

Equivalent Annual Cost for M/C B = 4785.26 + 150 + 9400

= 14,335.26

Result: The E.A.C. being greater than that of A.A.C. of both the m/c, it is because we consider the C.I. on the capital investment in the Equivalent Annual Cost Method.

Alternative

Equivalent Annual Cost method can be done by another *formula also.*

(i.e.) *Equivalent Annual Cost* $= (I - S)\left[\frac{i(1+i)^n}{(1+i)^n - 1} \right] + iS$ + O and M costs

Problem 4.

Machine A	*Machine B*
Capital cost = Rs. 10,000	Rs. 7000
Estimated life = 10 yrs.	10 yrs.
Salvage value = 4000	2000
Annual disbursement = 1000	1500
(Maintenance)	
Cost of Capital = 6%	6%

(a) *Which of the following 2 alternatives is preferred?*

Solution. Considering Machine A:

(i) $$\text{C.R. and R} = (I - S)\left[\frac{i(1+i)^n}{(1+i)^n - 1} \right]$$

$$= (10{,}000 - 4{,}000)\left[\frac{0.06(1.06)^{10}}{(1.06)^{10} - 1}\right]$$

$$= 6{,}000 \times 0.1358 = 815.2077$$

(*ii*) $i \times \text{salvage} = 4{,}000 \times 0.06 = 240.$

(*iii*) Operating and Maintenance Cost = 1000.

Equivalent Annual Cost for m/c A $= i + ii + iii = 2055.2077$

Considering Machine B:

(*i*) $$\text{C.R. and R.} = (\text{I} - \text{S})\left[\frac{i(1+i)^n}{(1+i)^n - 1}\right]$$

$$= (7{,}000 - 2{,}000)\left[\frac{0.06(1.06)^{10}}{(1.06)^{10} - 1}\right]$$

$$= 5000 \times 0.1358 = 679.$$

(*ii*) $i\,\text{S} = 2000 \times 0.06 = 120.$

(*iii*) Operation and Maintenance Cost = Rs. 1500.

Now the Equivalent Annual Cost for m/c B:

$$= i + ii + iii = 679 + 120 + 1500$$

$$\text{E.A.C.} = 2299.$$

Conclusion : It is clear from the calculations that the Equivalent Annual Cost for the m/c A is less than that of B, so we will prefer only m/c A.

4.7 PRESENT VALVE METHOD

Present value method tells the worth of the future income or expense flows in terms of present value whereas the future valves increase due to the compounding effect of interest on progressively large amounts. The reverse of this are the reduction of future values to lesser present values, is what we have referred to discounting. The cash flows are typically the investment value, the maintenance cost and the income flows.

Present Value Cost = (P.V of Investment + P.V. of Other Costs) – P.V of Salvage

$$\text{Present Value Cost} = \left[\text{PV(I)} + \text{PV}\sum(\text{O\&M})\right] - \text{P.V.S.P.(Salvage)}$$

$$\text{Present Value Factor} = \frac{1}{(1+i)^n}$$

Problem 1. *An investment of Rs. 20,000 in an off gas monitoring system will have a salvage value of Rs. 6000 after an economic life of 5 years. Maintenance and operating costs are Rs. 4000 for the 1st year and increase by 200/year thereafter. The firms cost of capital is 10% what is the net present value of this method or investment?*

Solution. I = Rs. 20,000

S = Rs. 6000 after an economic life of 5 years.

Operation and Maintenance Costs = Rs 4000 for the 1st year and increase by 200/year.

Cost of Capital = 10% = 0.1.

(*i*) Present value of investment = Rs. 20,000.

(*ii*) Present value of salvage will be considered at the last of 5th year as the m/c becomes obsolete after 5th year only.

$$= 6000 \times \frac{1}{(1+0.1)^5}$$

$$= 3725.52$$

(*iii*) Present value of Maintenance and Operating Cost: In this case we have to consider the maintenance cost of each year and will multiply it with the corresponding present value factor of that year and then sum all the 5 years together to get the

Present Value of Maintenance Cost:

$$\text{P.V. of Operation and Maintenance Costs} = 4000 \times \frac{1}{(1+.1)^1} + 4200 \times \frac{1}{(1+.1)^2} + 4400 \times \frac{1}{(1+.1)^3} + 4600 \times \frac{1}{(1+.1)^4} + 4800 \times \frac{1}{(1+.1)^5}$$

$$= 16{,}535.50$$

Present value of the cost = [PV(I) + PV (O and M)] – PV Salvage.

= [20,000 + 16,535.50] – 3725.52

= 32809.98

Problem 3. *The Campbell Company is considering the purchase of one of 2 overhead cranes for their fabrication shop. For crane A the capital cost would be Rs. 25,000 while for B it would be Rs. 18,000. Both have an expected life of 10 years. The labour costs for both would be the same but maintenance cost would not be. For A, the maintenance cost would be Rs. 1000/yr while for B it would be Rs. 500 the first year and increases by Rs. 200 each year there after. The salvage value for A at the end its economic life would be Rs. 8000 while for B it would be 3000. If the cost of capital to the firm is 15% which crane should be purchased.*

Solution: *Crane A:*

Expected Life '*n*' = 10 years

Cost of Capital = 15% = 0.15.

(*i*) Present Value of the Capital = Rs. 25,000.

(*ii*) P.V. of Maintenance Cost

$$= 1000\left[\frac{1}{(1+.15)^1} + \frac{1}{(1.15)^2} + \frac{1}{(1.15)^3} + \frac{1}{(1.15)^4} + \frac{1}{(1.15)^5} + \frac{1}{(1.15)^6} + \frac{1}{(1.15)^7} + \frac{1}{(1.15)^8} + \frac{1}{(1.15)^9} + \frac{1}{(1.15)^{10}}\right]$$

$$= 1000 \times 5.0187 = 5018.768.$$

(*iii*) Present Value of the Salvage Value at the end of 10th year

$$= 8000 \times \frac{1}{(1.15)^{10}} = 1977.4776$$

Now the present value of the crane A

$$= (25000 + 5018.768) - 1977.4776$$

$$= 28041.29035$$

Crane B :

(*i*) P.V. of the Capital = Rs. 18,000

(*ii*) P.V. of Maintenance Cost $= 500 \times \frac{1}{(1.15)^1} + 700 \times \frac{1}{(1.15)^2} + 900 \times \frac{1}{(1.15)^3} + 1100 \times \frac{1}{(1.15)^4}$

$$+1300 \times \frac{1}{(1.15)^5} + 1500 \times \frac{1}{(1.15)^6} + 1700 \times \frac{1}{(1.15)^7} + 1900 \times \frac{1}{(1.15)^8}$$

$$+2100 \times \frac{1}{(1.15)^9} + 2300 \times \frac{1}{(1.15)^{10}} = 5905.279$$

Note: All the problems of A.A.C. (method can be solved by E.A.C. method and Present Value Method.)

(*iii*) Present Value of the Salvage $= 3000 \times \frac{1}{(1.15)^{10}} = 741.554$

Net Present Value of the Crane B = *i* + *ii* + *iii*

$$= 18,000 + 5905.279 - 741.554$$

$$= 23,163.725$$

Result : Crane B will be choose as the Net Present Value is less than that of Crane A.

Problem 4. *An instrument transformer manufacturer in long Island is considering purchase of an ultrasonic welding m/c to replace an existing manually operated m/c. The existing m/c costs Rs. 12000, 2 years ago and has been depreciated down to 10,000 book value using a 12 years life and no salvage value. However, the market value of the m/c is only about Rs. 4000 now. The ultrasonic welder would improve product quality enough to boost revenue from an existing Rs. 80,000/year. to 1,00,000/year. It would cost Rs. 44,000 and have a 10 year life. Any salvage value on it would be consumed in the removal expense and advantage of the ultrasonic m/c is that by reducing annual labour cost, it would cut operating expenses from Rs. 8000 to 3000 annually. The manufacturer is in a 50% tax bracket and estimates the firms cost of capital at 12%. Use present value analysis to determine whether they should purchase the ultrasonic m/c.*

Solution.

Existing Machine	*New Ultrasonic Machine*
The cost of old m/c was Rs. 12,000, 2 years ago and now it has depreciated down to 10,000 (*i.e.,*) in 2 years Its depreciation has come down to Rs. 2,000 or 1000/year	The cost of new m/c is Rs. 44,000. In 10 years of life its depreciation/year will be Rs. 4400.

Given data:	
Revenues = Rs. 80,000/year	Revenues = 1,00,000/year.
Operating cost/year = Rs. 8000	Operating cost/year = Rs. 3000
Cost of m/c = Rs. 4000	Cost of m/c = 44,000
Depreciation = 1000/year	Depreciation = 4,400/year
n = 10 years.	n = 10 years.

Considering after 2 years

Determine the after tax profit under each alternative and select the most favourable one. It will be most convenient to do calculations on annual basis and then convert the same to present value.

W.K.T.

Cash Inflow + Salvage = Revenue – O.P. costs – Taxes.

Existing Machine : Revenues = 80,000

Less Depreciation = 1,000

Less Operating Costs = 8,000

Income = Rs. 71,000

Tax at 50% = –35,500

Rs. 35,500

Now Cash Inflow/year = 80,000 – 8000 – 35,500

= Rs. 36,500.

$$\text{P.V. of cash in flow for 10 years} = 36,500\left[\frac{1}{(1+.12)^1}+\frac{1}{(1+.12)^2}+\frac{1}{(1+.12)^3}+\frac{1}{(1+.12)^4}+\frac{1}{(1+.12)^5}+\frac{1}{(1+.12)^6}+\frac{1}{(1+.12)^7}+\frac{1}{(1+.12)^8}+\frac{1}{(1+.12)^9}+\frac{1}{(1+.12)^{10}}\right]$$

= Rs. 2,06,233.1405

Net Present Value of Cash Inflow Gain

= P.V. of Cash Inflow – Cost of m/c

= 2,06,233.14 – 4000

= 2,02,233.14

New machine : Revenues : Rs. 1,00,000/year.

Operating cost/year : Rs. 3,000/year.

Cost of the machine : Rs. 4400

Revenue = 1,00,000

Less Operation Cost = –3,000

Less Depreciation = –4,400

= 92,600

$$\text{Tax at } 50\% \text{ } i.e., = 46{,}300$$

$$\text{Income} = 46{,}300$$

$$\text{Cash Flow/year} = 1{,}00{,}000 - 3{,}000 - 46{,}300$$

$$= \text{Rs. } 50{,}700.$$

$$\text{P.V. of Cash Inflow for 10 years} = 50{,}700\left[\frac{1}{(1+.12)^1} + \frac{1}{(1+.12)^2} + \frac{1}{(1+.12)^3} + \frac{1}{(1+.12)^4} + \frac{1}{(1+.12)^5} + \frac{1}{(1+.12)^6} + \frac{1}{(1+.12)^7} + \frac{1}{(1+.12)^8} + \frac{1}{(1+.12)^9} + \frac{1}{(1+.12)^{10}}\right]$$

$$= 50{,}700 \times 5.65022$$

$$\text{P.V of Cash Inflow for 10 years} = 2{,}86{,}466.154$$

$$\text{Net Present Value Gain} = 2{,}86{,}466.154 - 44000$$

$$= 2{,}42{,}466.154.$$

Net Present Value Gain after taxes from the ultrasonic m/c installation exceeds the existing arrangement by Rs. 40,233 and thus the new m/c should be installed.

Advantages

1. It considers the total return.
2. It includes time adjusted considerations.
3. It can easily handle fluctuation in costs or revenues.
4. Total cost comparison can be made without the knowledge of income or considering the effect of taxes.

Disadvantages

1. It considers the total amount but does not explicitly consider the rate of return.
2. It inherently assumes the capital can be invested at the current cost of capital.

EXERCISE PROBLEMS

1. Find the best alternative using the annual equivalent method of comparison. Assume on interest rate of 15% compounded annually.

Alternative	A	B	C
Initial Cost (Rs.)	5,00,000	8,00,000	6,00,000
Annual receipt (Rs.)	2,00,000	1,50,000	1,20,000
Life (years)	10	10	10
Salvage value	1,00,000	50,000	30,000

2. A small scale industry is in the process of buying a milling machine. The purchase value of the milling machine is Rs. 60,000. It has identified two banks for loan to purchase the milling machine. The banks can give only 80% of the purchase value of the milling machine as loan. In urban bank the loan is to be repaid in 60 equal monthly instalments of Rs. 2500 each. In State Bank the loan is to be repaid in 40 equal monthly instalments of Rs. 4500 each suggest the most economical loan scheme for the company based on the annual equivalent method of comparison. Assume a nominal rate of 24% compounded monthly.

3. There are two alternatives of replacing a machine. The details of the alternatives are as follows:

Alternative 1

Purchase value of the new machine	Rs. 2,00,000
Life of the machine	10 years
Salvage value of the new machine at the end of life	Rs. 20,000
Annual operation and maintenance cost	Rs. 40,000
Buy back price of the existing machine	Rs. 25,000

Alternative 2

Purchase value of the new machine	Rs. 3,00,000
Life of the machine	10 years.
Salvage value of the new machine at the end of its life	Rs. 15,000
Annual Operation and Maintenance Cost	Rs. 35,000
Buy back price of the existing machine	Rs. 5,000.

Suggest the best replacement option for the company using the annual equivalent cost method of comparison by assuming 20% interest rate compounded annually.

4. A company receives two options for purchasing a copier machine for it's office.

Option 1 : Make a down payment of Rs. 30,000 and take delivery of the copier machine. The remaining money is to be paid in 24 equal monthly instalments of Rs. 4,500 each.

Option 2 : Make a full payment of Rs. 100,000 and take delivery of the copier machine. Suggest the best option for the company to buy the copier machine based on the annual equivalent method of comparison by assuming 15% interest rate, compounded annually.

5. Megabit Electronics is considering the purchase of a new programmable circuit tester in order to improve its product quality. The equipment has a first cost of 85,000 and the salvage value is predicted to be Rs. 6000 after a service life of 5 years. Maintenance and operating costs are expected to be Rs. 8000 the first year of operation and to increase by Rs. 1500 per year for each additional year of use. Using an interest rate of 10% determine what annual savings must be obtained through the use of this equipment to make it economically justifiable.

6. A company can purchase a piece of equipment for 20,000 and sell it for 4000 at the end of a 6 year service life or it can lease the unit for the same period by making first of the year payments of Rs. 3000. Compare the equivalent annual costs of the alternative, using an interest rate of 15 percent.

RATE OF RETURN CALCULATIONS

5.1 INTRODUCTION

The rate of return is the last of the discounted cash flow comparison methods. The rate of return of a cash flow pattern is the interest rate at which the present worth of the cash flow pattern reduces to zero. In this method of comparison, the rate of return for each alternative is computed. Then the alternative which has the highest rate of return is selected as the best alternative. The expenditures are always assigned with a negative sign and the revenue/inflows are assigned with a positive sign.

A generalized cash flow diagram to demonstrate the rate of return method of comparison is presented in the Fig. 5.1.

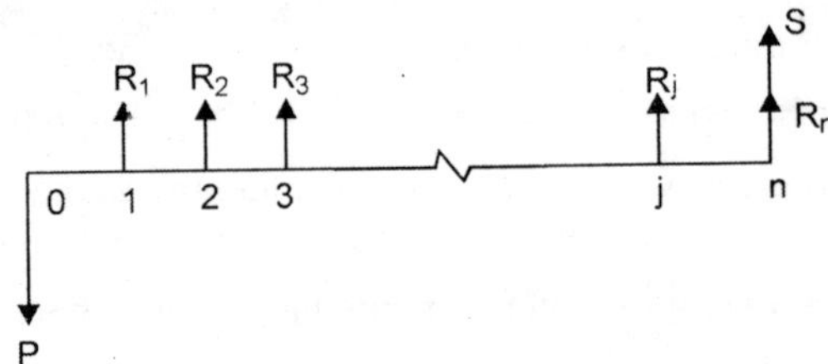

Fig. 5.1 Generalized cash flow diagram.

In the above cash flow diagram, P represents an initial investment, Rj the net revenue at the end of the *j*th year and S the salvage value at the end of the *n*th year.

P ⇒ Initial Investment

Rj ⇒ Net revenue @ the end of *j*th year.

S ⇒ Salvage value @ the end of the *n*th year.

We need to consider the alternative, a minimum acceptable rate of return (MARR) is the lowest at which independent alternative is attractive. It varies among and within organizations. The purpose of establishing a minimum acceptable rate of return is to ration of capital to the most deserving proposals.

Calculation of an internal rate of return (IRR) (it is also called as rate of return adjusted after taxes) will allow us to determine, under possible reinvestment constraints, whether an alternative meets the MARR value.

The internal rate of return (IRR) of an investment is the rate of interest earned on the unrecovered balance of an investment where the terminal balance is zero.

To find P value of the Fig. 5.1. The first step is to find the net present worth of the cash flow diagram using the following expression at the given interest rate i.

$$PW(i) = -P + R_1/(1 + i)^1 + R_2/(1 + i)^2 + \dots$$

$$\dots + Rj/(1 + i)^j + \dots + Rn/(1 + i)^n + \frac{S}{(1 + i)^n}.$$

Now, using above function is to be evaluated for different values of i until the present worth function reduces to zero as shown in Fig. 5.2.

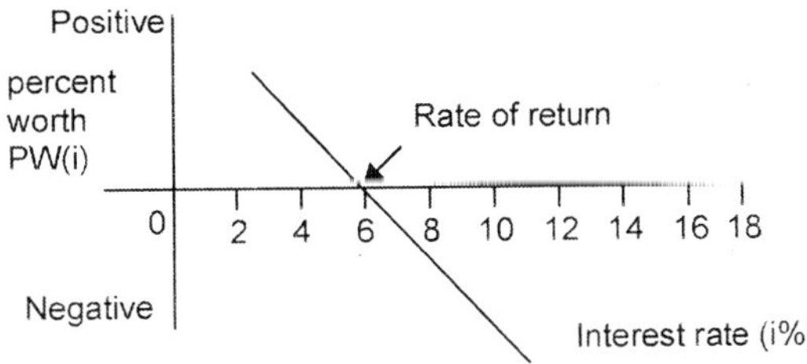

Fig. 5.2 Present worth function graph.

If present worth function value is positive. If so, increase the value of i until $Pw(i)$ becomes negative. The rate of return is determined by interpolation method in the range of values of i for which the sign of the present worth function changes from positive to negative.

5.2 RATE OF RETURN

The rate of return is a percentage that indicates the relative yield on different uses of capital. Interest rates are well understood in the world of commerce.

There are three rates of return appear frequently in Engineering Economy studies.

- The minimum acceptable rate of return (MARR) is the rate set by an organization to designate the lowest level of return that makes an investment acceptable.
- The internal rate of return (IRR) is the rate on the unrecovered balance of the investment in a situation where the terminal balance is zero.
- The external rate of return (ERR) is the rate of return that is possible to obtain for an investment under current economic conditions.

5.3 MINIMUM ACCEPTABLE RATE OF RETURN

The minimum acceptable rate of return is also known as the minimum alternative rate of return.

- It is the device designed to make the best possible use of a limited resource *i.e.,* money.
- Rates vary widely according to the type of organizations and they vary even within the organizations. These variations usually reflect the risk involved.

  For example: The rate of return required for cost reduction proposals may be lower than that required for research and development projects in which there is less certainty about prospective cash flows.

- An MARR is to be used with constant rupees, is an inflation-free interest rate that represents the earning power of capital when inflation effects have been removed. An MARR value that includes the effect of inflation is referred to as the market interest rate. Use of this rate will require that all cash flows be in actual rupees.

The purpose of establishing a minimum attractive rate of return higher than the cost of capital is to ration capital.

5.4 INTERNAL RATE OF RETURN

The IRR is the best-known and most widely used rate of return method and the discounted cash flow method. The internal rate of return can be calculated by equating the annual, present or future worth of cash flow to zero an^d solving for the interest rate (IRR) that allows the equality.

Although both the EAW and FW approaches are legal, the rate of return is often defined in terms of present worth, under the constraints of possible i^* roots where the IRR is

1. The Interest rate at which the present worth of the cash flow of a project is zero or to restate this in another way.
2. The rate which when employed in computing the present worth of all costs and present worth of all returns will make them equal.

Because rate of return computations usually begin with a problem expressed in terms of present worths or annual worth, it is necessary to pay attention to the guideliner for the EAW and PW methods.

5.4.1 Calculation of IRR8

Determining the IRR is a function of the type of investment and the characteristics of the alternatives. The cash flows of several independent alternatives that are being considered as a group may be summed to form the group's composite cash flow. Analysis is carried out, where capital limitations are apparent in a department and several independent alternatives are competing for funding, combinations of alternatives may be formed, where each combination's first costs have to be equal to or less than the capital available.

We can have ranking alternatives according to their IRR values is not consistent with PW, FW or AW rankings. Mutually exclusive alternatives may be analyzed by incremental IRR analysis and the results will be found to be completely consistent with the PW, FW or AW. Incriminated analysis assumes that it starts with a satisfactory low investment alternative. Analysis of a higher-investment alternative is then based on the differences between the cash flows of the second alternatives and the acceptable alternative. These differences in cash flows are incremental cash flows. The cash flow of the second alternative is equal to the cash flow of the first alternative plus the incremental cash flows.

If the incremental cash flow is acceptable when compared to MARR the larger investment has to be a better investment than the first alternative, which was also acceptable. Otherwise, remove the larger investment from consideration. This type of evaluation is continued until all alternatives have been evaluated. One of the mutually exclusive alternatives is then determined to be the best investment.

5.4.2 Single Simple Investment

The rate of return method of a single simple investment is determined by getting the present worth (or EAW) of receipts equal to the present worth of (EAW) of disbursements. This type is the most widely used method for performing engineering economic analysis.

The interest rate is sought that makes the discounted cash flows conform to the equality find i.

$$\text{PW (receipts)} = \text{PW (disbursements)}$$

When the discounted flows are substracted form each other to equal zero.

PW (receipts) – PW disbursements) = Net PW = 0.

For either PW formulation, the manual calculations of i is usually a trial and error procedure.

When a single proposal is for a cost reduction project, the receipts have the form of net savings from method of operation used before the cost reduction investment.

5.4.3 Clues for IRR Calculations

Most analysts will use CHEER or spread Sheet Functions to compute IRR values. We should make how we might possibly simplify the trial and error manual process. There is really no way to avoid the trial and error search procedure for manually determining the IRR in problems with complex cash flows.

For the simple investment the size of a positive net sum, with respect to the amount and length of investment, gives a rough suggestion of the rate of return.

IRR can be calculated easily, when major cash flows at the beginning and end of the study period are same. **Ex. :** When the savage value is close to 100 percent of the first cost the net annuity divided by the first cost gives a close approximation to i : *i.e.* = A/P = i.

When irregular cash flows can be rounded to approximate an ordinary annuity or individual transactions within short time intervals can be lumped together to allow gross preliminary calculations that suggest the vicinity of the IRR.

It is possible to determine the IRR directly for very simple situations.

$$\text{PW} = -\text{CI} + \text{FW (P/F}, i, \text{N)} = 0.$$

When searching for IRR (i), where CI = Capital Investment at time period 0 and FW = Future worth at time period N,

$$\text{CI} = \text{FW(P/F}, i, \text{N)}$$

A graphical approach can greatly simplify the problem. It is used with a variety of IRR values in conjunctions with the text's computer program to determine intermediate values of PW. The computer program may be used directly to determine the solution for problems where only one IRR value exists.

5.5 CONSISTENCY OF IRR WITH OTHER ECONOMIC COMPARISON METHODS

The acceptability of alternative courses of action will be identical whether they are evaluated according to their annual worth, present worth or incremental IRR. Ranking alternatives by individual IRR values will usually not to be consistent with PW rankings. The important points to understand are the meaning of the measures of acceptability and the assumption upon which they are based. They will also illustrate now the rate of return is calculated to compare investments when only the disbursements are known and how to interpret the outcomes.

5.6 IRR MISCONCEPTIONS

The consistency of AW and PW comparisons always agree with IRR evaluations when done correctly and there are some misconceptions that should be clarified.

5.6.1 Ranking Alternatives by Individual IRR Values

The incremental analysis was used that ranking individual alternatives IRR values would be the way to go. Ranking alternatives on their individual IRR values can conflict with PW rankings.

For Example: Suppose two projects with the cash flows are equal initial investment.

If X project as increasing revenue over the years and project Y has decreasing revenues over the years. By calculating present worth of project X is higher than project Y until $i = 10\%$. But, for project Y, if i is > 10% IRR of project Y is higher than that of X.

Therefore, decision should be careful in comparing with PW and IRR, since, they also are associated with i range of values, as in the above case.

End of year cash flow Rs.

Project	0	1	2	3	4
X	–1000	100	350	600	850
Y	–1000	1000	200	200	200

$$Pw_x = -1000 + [100 + 250(A/G, 10,4)](P/A, 10,4)$$
$$= 411.52$$
$$Pw_y = -1000 + [1000 + 200(P/A, 10,3)]\ (P/F, 10,1)$$
$$= 361.24$$

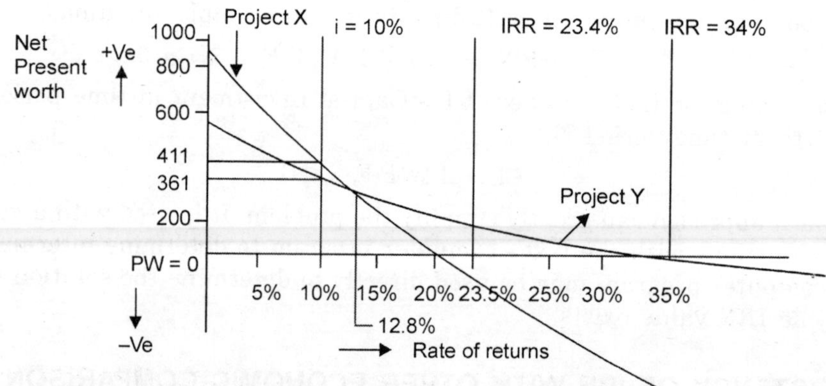

5.6.2 More than One Possible Rate of Return

When the cash flow or cumulative cash flow of a project switches from negative to positive (or the reverse) more than once the project may have more than one roots of the present-worth equation PW (i) = 0. In such cases we have to determine which root is the true IRR value. In single project such situations will occur relatively rarely in practice. When incremental analysis is used to compare mutually exclusive alternatives, especially for projects with unequal lives, this becomes much more common.

5.6.3 Explicit Investment Rate

Approximate answer to the IRR is developed by applying an explicit interest rate to a limited portion of the cash flow that will disturb the total cash flow pattern as little as possible while eliminating one of the sign reversals.

An explicit reinvestment rate is a designated interest percentage appropriate for a specific application.

The explicit reinvestment rate may be the minimum attractive rate of return employed by the organization or rate suggested by the PW profile.

The use of an explicit interest rate may be like an artificial device to alleviate a mathematical difficulty, but the concept is both realistic and reasonable.

One major problem with the use of the explicit interest rate is the rate applied somewhat arbitrarily.

5.6.4 Historical External Rate of Return Method

The occurrence of multiple i^* roots with nonsimple investment return can be avoided by using the historical external rate-of-return (HERR) method where the main assumption is that receipts are actually reinvested at a generally available interest rate. This rate is taken as MARR. In this method it is based on project cash flow balances (receipts).

An unknown rate of return e' is found by equating the future worth of receipts (positive cash flows) compounded at an explicit interest rate (i %) to the future worth of disbursements (negative cash flows) compounded at e'.

FW (receipts compounded at i %) = FW (disbursements compounded at e')

When i% is the MARR and e' exceeds i%, the investment is assumed to be attractive because it promises a yield greater than the lower limit of acceptability.

5.6.5 Project Balance Method

An approach more logical than the traditional HERR methodology is the project balance method (PBM). The PBM has evolved over many years but Park has nicely summarised the steps in the reinvestment problem:

1. Determine i^*. If there are multiple i^* values it is suggested that the one closest to the MARR value be used in the subsequent analysis.
2. Determine the current balance for each period using any of the determined i^* values. These we will call $(i^*)_t$ where t is the end of the period for which CB is being calculated.
3. It all CB $(i^*)_t$ are equal to or less than zero, receipts in all periods t are being used to pay off project investment and so are assumed to be used internally, thus i^* is the true IRR value.

4. If any CB (i^*) is positive and some are negative, then IRR is still not known. The MARR or other acceptable external rate of return will be applied to the positive CB (IRR/ERR). When funds are excess will be applied to the project's investment, i' will be applied to the negative CB(IRR/ERR).
5. Iterate through the period with an assumed IRR & the explicit-external interest rate. Apply the IRR to negative balance and the explicit rate to positive balances, if CB $(\text{IRR/ERR})_N$ is equal to zero. The true IRR values which should be used in project evaluations.

The complete information to determine IRR is summarized in the flow chart shown in Fig. 5.3 and also this gives information regarding the incremental cash flow analysis and for independent alternatives analysis. Mutually exclusive alternative can be accepted based on highest investment whose incremental IRR is greater than MARR.

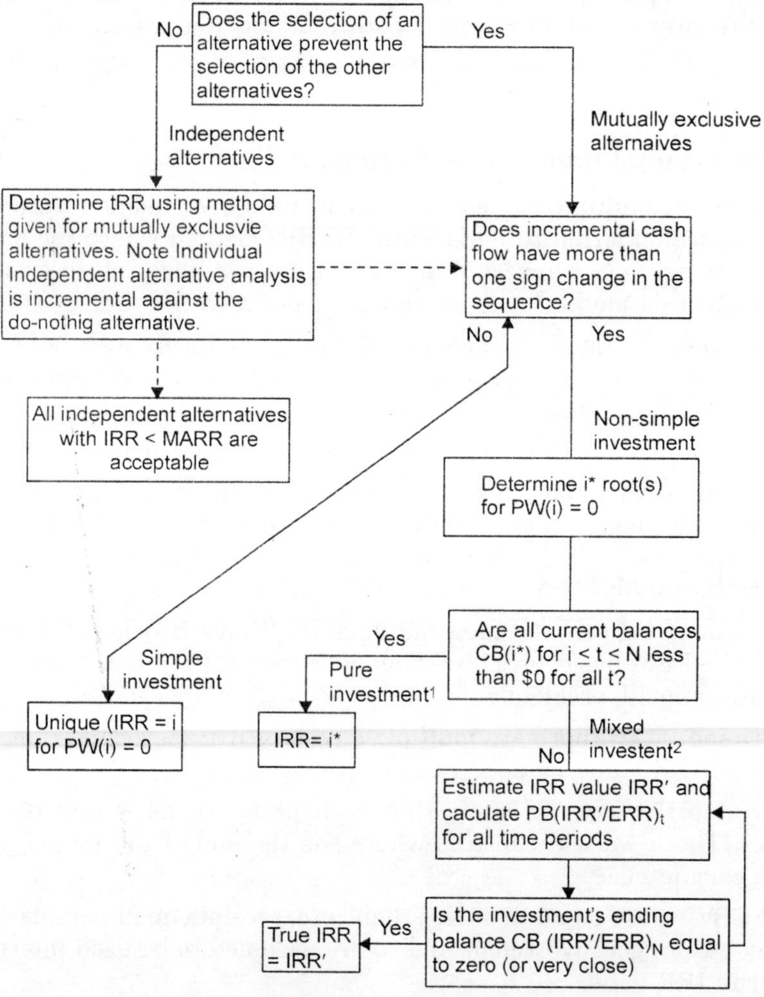

Fig. 53 Flow chart of IRR determination.

5.6.6 Reinvestment Question

Several approaches are available for reinvestment question as it affects the calculation of the IRR. The basic question boils down to whether it is fair to assume that external balances from a project can be invested at a rate equal to the IRR. In general, it is not fair. The PBM, HERR, and Explict Interest Rate ($i\%$) approaches attempt to offset the problems in ways that give, in general, different IRR value. All recommend using two rates of return in the calculations, the IRR that is being determined & an external rate which may or may not be equal to a more logical & current value.

The logical arguments are

(1) Multiple i' roots, the cause of many problems, rarely occur in the most common engineering economy project evaluations.

(2) Where they do occur is generally for multiple project evaluations using the incremental IRR approach.

5.6.7 Alternatives with Unequal Lives

The last potential difficulty with IRR computations relates to projects with unequal lives. It is feasible to use rate of return analyses with different lives. It is necessary to find an overall time lengths that is a multiple of the lives of the alternative or to truncate lives to a common analysis period, the period of need.

5.7 COMMENTS ON THEORY AND PRACTICE BEHIND INTEREST RATES

Interest rates are different. Reasons for the differences among loans include the following:

- Higher interest charges are applied when there is a greater risk that a loan will not be paid.
- Charges for long-term loans are usually greater than those for short-term loans because a lender forgoes the opportunity to take advantage of alternative uses of the money for a longer period.
- Administrative costs of lending are a higher percentage of a smaller loan than a larger loan.
- Local money markets vary. Owing to regional differences, interest rates are often higher in small towns than in large cities because borrowers find it less convenient to shop for better rates.

5.7.1 Cost of Capital Concepts

The cost of capital is seldom. Cost of capital will be determined by the chief financial officer in conjunction with the accounting department. The cost of capital in derived from the composition of the capital pool. The term pool is suggested of capital from many sources, pooled for finding purposes. The proportion of capital from different costs to a firm is represe nted by a weighted cost of capital.

The actual rate of return expected from new investments is normally greater than the cost of capital. Riskier proposals are subject to higher discount rates to compensate for the chance that they will not meet net return expectations.

The cost of capital is troublesome to estimate despite the apparent precision of its formula. There are differences of opinion about both costs and proportions and naturally a number of variations to the general formula have been proposed.

5.7.2 Sources of Funds

Engineers will generally not be responsible for obtaining funds for their enterprise, it will be advantageous for them to have an overview of fund sources. Engineers and managers of operating units have responsibility for proposing and evaluating investment that support the productivity of operations, be they highway construction, steel production, office services, mail delivery or any other productive function. Proposals are usually generated & evaluated on the assumption that a given level of funding is available to carry out the best ones.

The size of a pool of investment funds available at a particular time is a function of many financial decisions made previously. As shown in Fig. 5.4 the pool is fed by gross revenue resulting from sales, income as returns from investments and capital obtained by borrowing or selling equity in the organization, debt and equity sales which are external sources of funds.

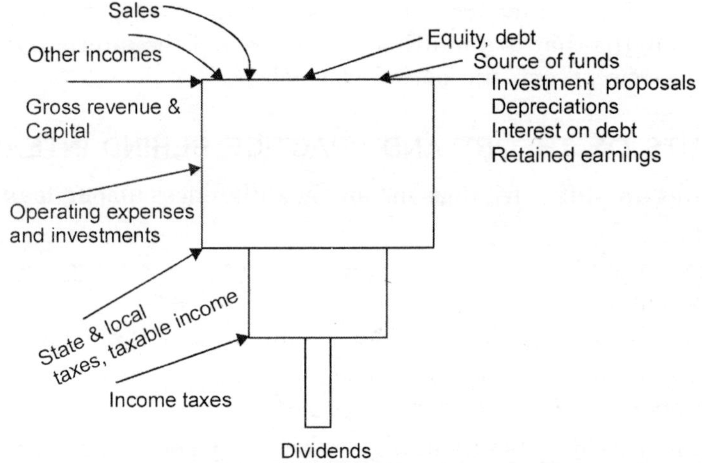

Fig. 5.4. Flow of funds

The management of an industrial enterprise works for the owners. In smaller firms the managers may be the owners. Owners receive gains from successful corporate operations in two forms: dividend payments and increases in the value of their stock. Each of the four main sources of funds shown in Fig. 5.4 has its own effect on returns to equity holders.

Solved Problems

Income producing Proposals

Problem 1. *A piece of land adjacent to a highway road is likely to increase in value. The present value now is Rs. 60,000 and is expected to be worth Rs. 1,30,000 within in 5 years. During that period it can be rented for pasture at Rs. 1300 per year. Annual taxes presently are Rs. 650 and will likely remain constant. What rate of return will be earned on the investment if the estimates are accurate?*

Solution.

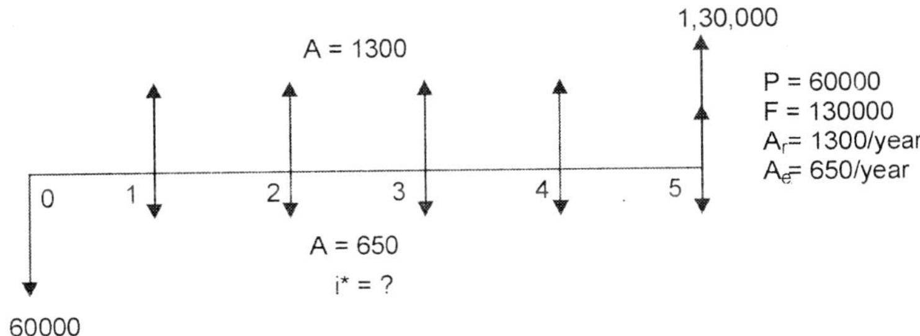

$$\text{PW}_{\text{income}} = \text{PW}_{\text{disbursement}}$$

1,30,000 (P/F, i, 5) + 1300 (P/A, i, 5) = 60,000 + 650 (P/A, i, 5)

1,30,000 (P/F, i% 5) + 1300 (P/A, i, 5) – 650 (P/A, i% 5) – 60,000 = 0

1,30,000 (P/F, i% 5) + 650 (P/A, i% 5) – 60,000 = 0.

i, i^* confirms to the equation above is the rate of return on the remaining balance of the investment.

The value is determined by trial and error needed for a simple investment.

Put, $i = 0$, at $i = 0$

$\therefore$ 1,30,000 – 60,000 + 650 (5) = 73250.

This value indicates that the investment will produce a positive rate of return because the total income is much greater than the outgo.

For instance, the 72 rule suggests that a sum doubles in value every $72/i$ years.

Since 60,000 initial investment results in almost twice as much income at the end of 5 years, i should be near 72/5 = 14.4 percent.

Let i = 15% as first trial.

1,30,000 (P/F, 15% 5) + 650 (P/A, 15% 5) – 60,000 = 0

1,30,000 (0.49718) + 650 (3.35216) – 60,000 = –3,244.10 < 0.

The negative value indicates that the trial i used is too large, now it is known that i lies between 0 and 15 percent.

Let i = 14%

1,30,000 (P/F, 14% 5) + 650 (P/A, 15% 5) – 60,000 = 0

1,30,000 (0.51937) + 650 (3.43308) – 60,000 = 137.00 > 0.

Which shows that IRR lies between 14 and 15%. The approximate value of i is determined by linear interpolation.

i	14%	?	15%
PW	137.00	0	–3244.10

$$i^* = 14\% + \frac{(15-14)\%\ (137.00-0)}{137.00-(-3244.10)}$$

$$= 14\% + \frac{137 \times 1\%}{3381.10}$$

$$\boxed{i^* = 14.04\%}$$

Cost Reduction Proposal

Problem 2. *The subassembles for a product are purchased for Rs. 81 a piece. The annual demand is 450 units and is expected to be continued for 3 years. At this time the new product under development will be ready for manufacturing. The initial investment is Rs. 22,000, the production costs of subassemblies will be Rs. 19,500 for the first-year and Rs. 13,250 in each of the last 2 years. The equipment will have no salvage. Should the company have or buy the subassemblies?*

Solution.

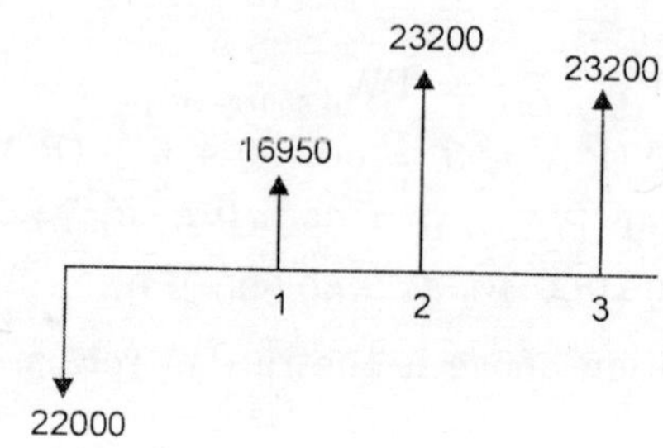

$$\text{Present annual cost} = 450 \times 81 = 36450.$$

$$\text{Net savings (1 year)} = 36450 - 19500 = 16950$$

$$\text{Net savings (2 \& 3 years)} = 36450 - 13250 = 23200.$$

$$\therefore \quad \text{PW} = -22000 + 6350\ (\text{P/F}, i, 1) + 23200\ (\text{P/F}, i, 2) + 23200\ (\text{P/F}, i, 2) = 0$$

$$\text{Trying higher interest rate } i = 10\%$$

$$\text{PW} = -22000 + 6350\ (\text{P/F}, 10, 1) + 23200\ (\text{P/F}, 10, 2) + 23200\ (\text{P/F}, 10, 3)$$

$$= -22000 + 6350\ (0.90909) + 23200\ (0.82645) + 23200\ (0.75131) \quad \text{using table value}$$

$$\text{PW} = 20376 > 0$$

Similarly, using table value we find

$i = 15\%$ PW = 165550.6

$i = 20\%$ PW = 12828.5

$i = 25\%$ PW = 9806.4

$i = 30\%$ PW = 7108.9

$i = 40\%$ PW = 2822.5

$i = 50\%$ PW = –1654.84.

The PW with $i = 50\%$ is negative while PW with $i = 40\%$ is positive, Therefore, we can interpolate to find the rate of return on the 22000 investment

$$\text{IRR} = 40\% + (50\% - 40\%)\frac{2822.5 - 0}{2822.5 - (-1654.89)}$$

$$= 40\% + 6.3\%$$

$$\text{IRR} = 46.3\%$$

Problem 3. *A construction firm can lease a crane required on a project for 3 years for Rs. 1,80,000 payable now, with the maintenance cost included. The alternative is to buy a crane for Rs. 2,40,000 and sell it at the end of 3 years for Rs. 100,000. Annual maintenance costs are expected to be Rs. 5000 the first 2 yrs. and 10,000 the third year. At what interest rate would the two alternatives be equivalent?*

Solution. I: Lease = Rs. 180,000 for 3 years

II:

$$I = 240,000$$
$$N = 3 \text{ yrs.}$$
$$F = 100,000$$
$$A_1 = 5000$$
$$A_2 = 5000$$
$$A_3 = 10,000$$
$$i = ?$$

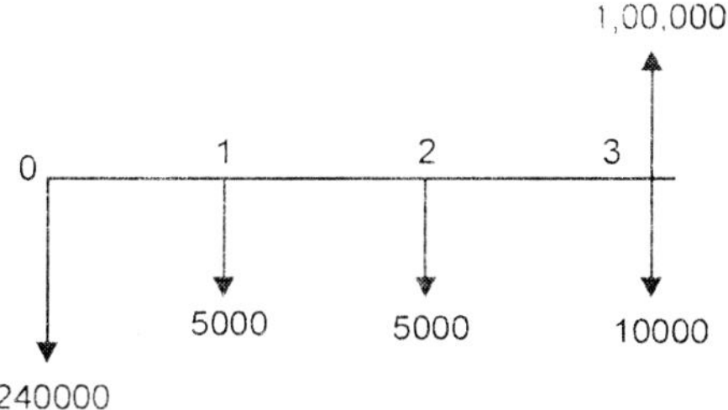

$$PW = -240000 - 5000\,(P/F,\, i\%\; 1) - 5000\,(P/F,\, i\%\; 2) - 10000\,(P/F,\, i\%\; 3) - 100,000\,(P/F,\, i\%\; 3)$$

$$PW = -240000 - 5000\frac{1}{(1+i)^1} - 5000\frac{1}{(1+i)^2} - 10000\frac{1}{(1+i)^3} + 100000\frac{1}{(1+i)^3}$$

@ i = 5%

PW = –168539.4

@ i = 10%

PW = –181945.5

∴ The Interest rates of the two alternatives are equal @ 10%.

4. *Consider the following cash flow of a project.*

Year	*0*	*1*	*2*	*3*	*4*	*5*
Cash flow	*–10,000*	*4,000*	*4,500*	*5,000*	*5,500*	*6,000*

Find the rate of return at the project.

Solution. Initial investment = 10,000

$$G = 500$$
$$A' = 4000$$

$$N = 5 \text{ years.}$$

Cash flow diagram

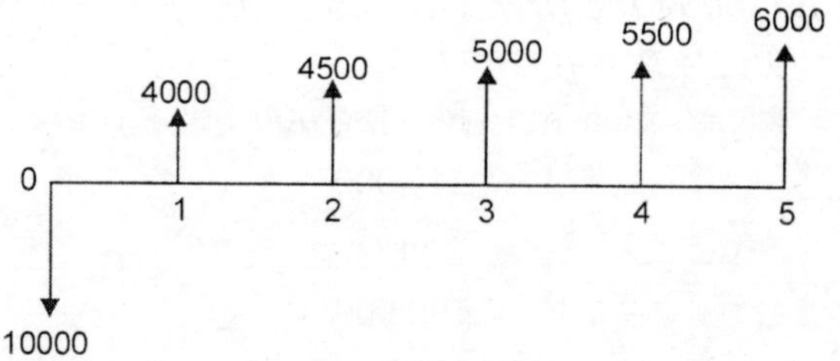

$$PW = -10000 + [4000 + 500(A/G, i\%, 5)](P/A, i\%5)$$

$$= -10000 + \left[4000 + 500\left(\frac{1}{i} - \frac{N}{(1+i)^N - 1}\right)\right]\frac{(1+i)^N - 1}{i(1+i)^N}$$

$$@\ i = 10\%$$

$$PW = -10000 + \left[4000 + 500\left(\frac{1}{0.1} - \frac{5}{(1.1)^5 - 1}\right)\right]\left[\frac{(1.1)^5 - 1}{0.1(1.1)^5}\right]$$

$$= -10000 + 18590.3$$

$$PW = 8590.32$$

$$@\ i = 15\%$$

$$PW = 6296.3$$

$$@\ i = 25\%$$

$$PW = 2859.0$$

$$@\ i = 40\%$$

$$PW = -478.23$$

By Inter polation

$$\therefore \quad i = 25\% + (40\% - 25\%)\left[\frac{2859 - 0}{2859 - (-478.23)}\right]$$

$$i = 25 + 12.85$$

$$\therefore \quad i = 37.85$$

Problem 5. *A company is in the process of selecting the best alternative among the following three mutually exclusive alternatives.*

Alternative	*Initial investment Rs.*	*Annual revenue Rs.*	*Life Years*
A_1	*5,00,000*	*1,00,000*	*10*
A_2	*8,00,000*	*1,40,000*	*10*
A_3	*3,00,000*	*70,000*	*10*

Find the best alternative based on the rate of return method of comparison.

Solution.

Alternative: A_1

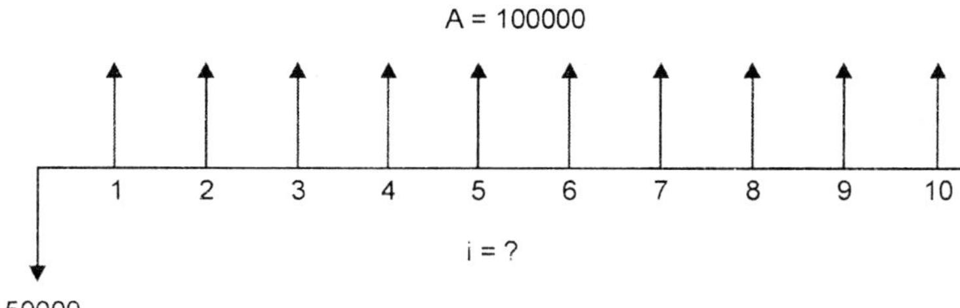

$$PW = -500,000 + 100,000\ (P/A,\ i\%,\ 10)$$

$$= -500,000 + 100,000 \frac{(1+i)^{10} - 1}{i(1+i)^{10}}$$

When $i = 10\%$

$$PW = 114460$$

When $i = 15$

$$PW = 1880$$

When $i = 18\%$

$$PW = -50600$$

$$\therefore \quad i = 15\% + \frac{(18\% - 15\%)\ (1880 - 0)}{1880 - (-50600)}$$

$$= 15 + 0.10$$

$$i = 15.1\%$$

Alternative: A_2

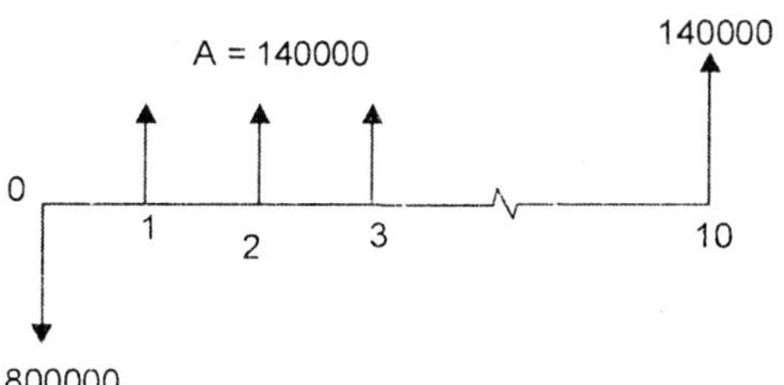

$$PW = -800,000 + 140,000\ (P/A,\ i\%\ 10)$$

$$= -800,000 + 140,000 \left[\frac{(1+i)^{10} - 1}{i(1+i)^{10}} \right]$$

When $i = 10\%$

$$PW = -800,000 + 140,000 \left[\frac{(1+0.1)^{10} - 1}{0.1(1.1)^{10}} \right]$$

$$= -800,000 + 140,000 \times 6.1446$$

$$= 60244$$

When $i = 15\%$

$$PW = -97368$$

$$i = 10\% + \frac{(15\% - 10\%)\,(60244 - 0)}{60244 - (-97368)}$$

$$= 10\% + 1.9$$

$$i = 11.9\%.$$

Alternative: A_3

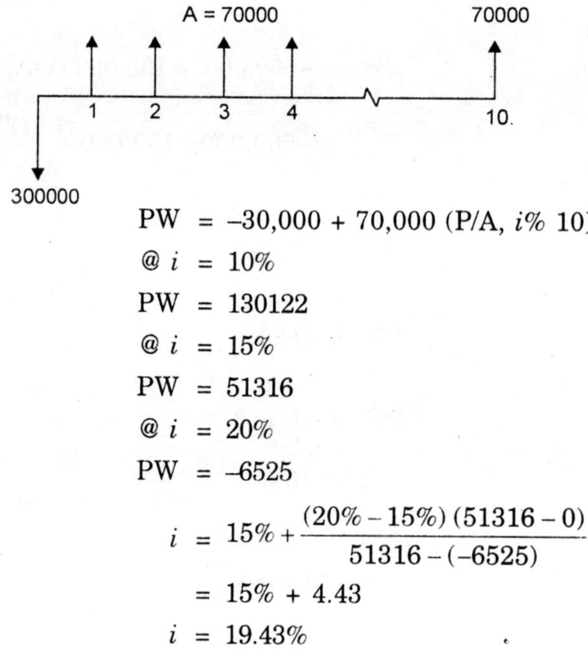

$$PW = -30{,}000 + 70{,}000\ (P/A,\ i\%\ 10)$$

$$@\ i = 10\%$$

$$PW = 130122$$

$$@\ i = 15\%$$

$$PW = 51316$$

$$@\ i = 20\%$$

$$PW = -6525$$

$$i = 15\% + \frac{(20\% - 15\%)\,(51316 - 0)}{51316 - (-6525)}$$

$$= 15\% + 4.43$$

$$i = 19.43\%$$

The rates of return for the three alternatives are now tabulated

Alternative	A_1	A_2	A_3
Rate of return	15.1%	11.9%	19.43%

All 3 should be selected, since it has highest rate of return of all the alternative.

Problem 6. *An Automobile company is planning to buy a robot for its forging unit. It has identified two different companies for the supply of the robot. The details of cost & incremental revenue of using robots are summarized in the following table.*

	Brand	
	Speedex	*Giant*
Initial cost Rs.	*500,000*	*900,000*
Annual incremental revenue Rs.	*80,000*	*250,000*
Life yrs.	*8*	*8*
Life and Salvage value Rs.	*40,000*	*60,000*

The MARR for the company is 12%. Suggest the best brand of robot to the company based on the ROR method.

Solution.

Speed ex

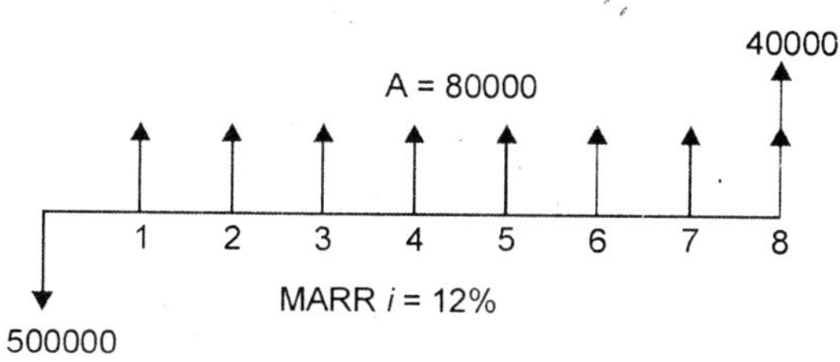

PW = –500,000 + 80,000 (P/A, i% 8) + 40,000 (P/F, i% 8)

@ i = 15%

$$= -500,000 + 80,000\left[\frac{(1+0.15)^8 - 1}{0.15(1+0.15)^8}\right] + 40,000\left(\frac{1}{(1+0.15)^8}\right)$$

= –500,000 + 358984 + 13076

PW = –141016

@ i = 10%

PW = –500,000 + 460720 + 18660 = –20620

@ i = 7%

PW = –500,000 + 477704 + 23280

= –984

@ i = 6%

PW = –500,000 + 496784 + 25096 = 21880

$$i = 6\% + \frac{(7\% - 6\%)\,(21880 - 0)}{21880 - (-984)}$$

i = 6.95%

Giant

60000
A = 250000
1 2 3 4 5 6 7 8
900000

PW = –900,000 + 250,000 (P/A, i% 8) + 60,000 (P/F, i% 8)

@ i = 15%

$$\text{PW} = -900{,}000 + 250{,}000 \times \frac{(.1+0.15)^8 - 1}{0.15(1+0.15)^8} + 60{,}000 \times \frac{1}{(1+0.15)^8}$$

$$= -900{,}000 + 1121830.3 + 19614.1$$

$$= 241444.4$$

$$@\, i = 20$$

$$\text{PW} = -900{,}000 + 959289.9 + 13954.1$$

$$= 73244$$

$$@\, i = 25$$

$$\text{PW} = -900{,}000 + 832227.8 + 10066.3$$

$$= -57705.9$$

$$i = 20 + \frac{(25\% - 20\%)\,(73244 - 0)}{73244 - (-57705.9)}$$

$$i = 20 + 2.79$$

$$i = 22.79\%$$

Alternative 1 cannot be considered for selection, since it is less than MARR.

Alternative 2 is the best, since it is more than MARR

$$i = 22.79\%$$

Problem 7. *Proposal 1 has an initial cost of Rs. 1500 and a positive cash flow that returns Rs. 200 the first year and increases by Rs. 200 each of the following years until the end of the 5 year study period. Proposal 2 also has a 5 year life and an initial cost of Rs. 1500. Its positive cash flow is constant at Rs. 200 for the last 4 years. It also has another receipt in year 1. All receipts occur at the end of the year. What is the rate on proposal 1?*

Solution.

Proposal 1

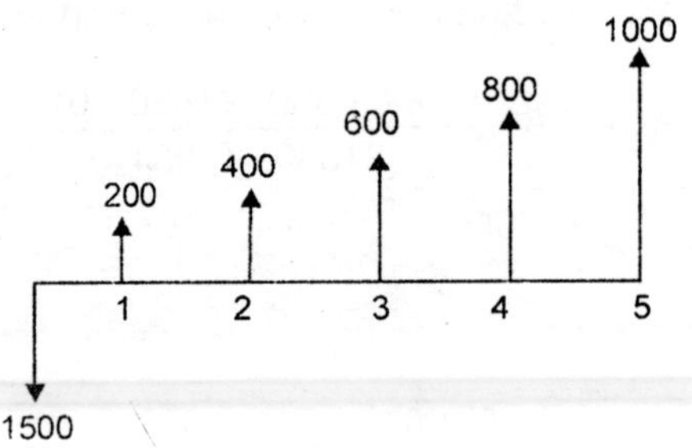

$$\text{PW} = -1500 + [200 + 200\ (\text{A/G}, i\%\ 5)]\ (\text{P/A}, i\%\ 3)$$

$$@\, i = 10\%$$

$$\text{PW} = -1500 + \left[200 + 200\left(\frac{1}{i} - \frac{N}{(1+i)^N - 1}\right)\right]\frac{(1+i)^N - 1}{i(1+i)^N}$$

$$= -1500 + \left[200 + 200\left(\frac{1}{0.1} - \frac{5}{(1+0.1)^5 - 1}\right)\right]\frac{(1+0.1)^5 - 1}{0.1(1+0.1)^5}(5620)3.79$$

$$= 630.1$$

$$@\ i = 15\%$$

$$\text{PW} = -1500 + (544.4)\ 3.36 = 329.1$$

$$@\ i = 20\%$$

$$\text{PW} = -1500 + (528)\ 2.96 = 63.24$$

$$@\ i = 25\%$$

$$\text{PW} = -1500 + 512.6 \times 2.68928 = -121.43$$

$$i = 20\% + \frac{(25\% - 20\%)\ (63.24 - 0)}{63.24 - (-121.43)}$$

$$= 20\% + 1.71$$

$$i = 21.71\% \text{ for proposal 1}$$

Problem 8. *An Investor has an opportunity to purchase a commercial rental property for Rs. 300,000. The current occupants have signed a 10 year lease at a constant annual rent of Rs. 48000 and maintenance costs and taxes on the structure are currently 12000 and are expected to increase at a rate of Rs. 1500 per year over the 10-year period. Assuming that the property can be sold for at least the purchase price when the current lease expires, determine the investor's minimum expected rate of return.*

Solution.

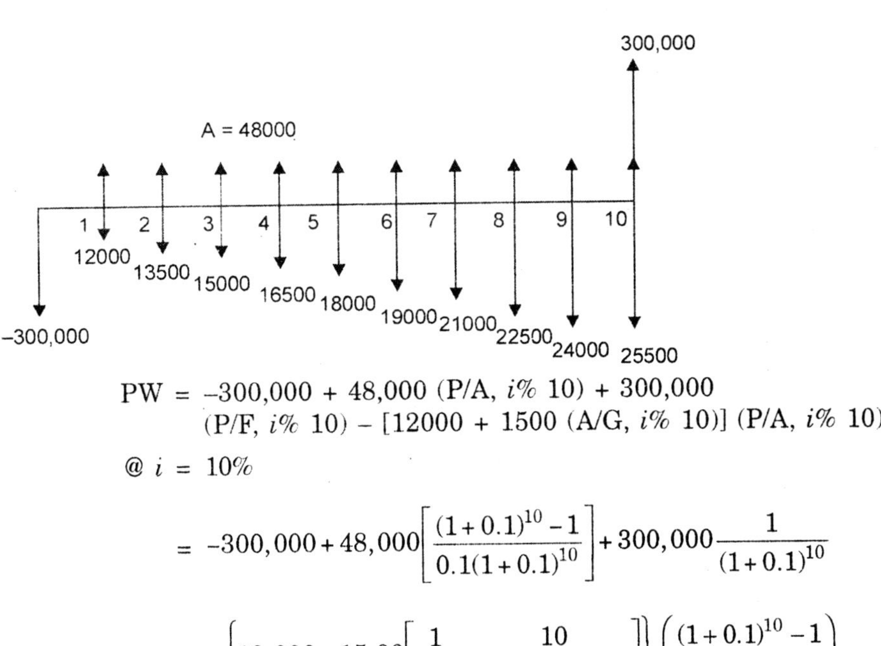

$$\text{PW} = -300{,}000 + 48{,}000\ (\text{P/A}, i\%\ 10) + 300{,}000\ (\text{P/F}, i\%\ 10) - [12000 + 1500\ (\text{A/G}, i\%\ 10)]\ (\text{P/A}, i\%\ 10)$$

$$@\ i = 10\%$$

$$= -300{,}000 + 48{,}000\left[\frac{(1+0.1)^{10} - 1}{0.1(1+0.1)^{10}}\right] + 300{,}000\frac{1}{(1+0.1)^{10}}$$

$$-\left\{12{,}000 + 15{,}00\left[\frac{1}{0.1} - \frac{10}{(1+0.1)^{10} - 1}\right]\right\}\left(\frac{(1+0.1)^{10} - 1}{0.1(1+0.1)^{10}}\right)$$

$$= -300,000 + 294939.2 + 115662.9 - 108071.9$$
$$= 2530.2$$
$$@\ i = 15$$
$$PW = -300,000 + 240900.9 + 74154 - 85694.5$$
$$= -70639.6$$
$$\therefore \quad i = 10\% + \frac{(15\% - 10\%)\,(2530.2 - 0)}{2530.2 - (-70639.6)}$$
$$= 10\% + 0.17$$
$$i = 10.17\%$$

Problem 9. *Two mutually exclusive projects are being considered. Project X requires Rs. 500 now and results in a return amounting to a one-time only profit of Rs. 1000 in 5 years from now. Project Y also requires 500 now but will return Rs. 170 per year for each of the next 5 years. Given a MARR of 14% which project should be adopted?*

Solution.

Project X

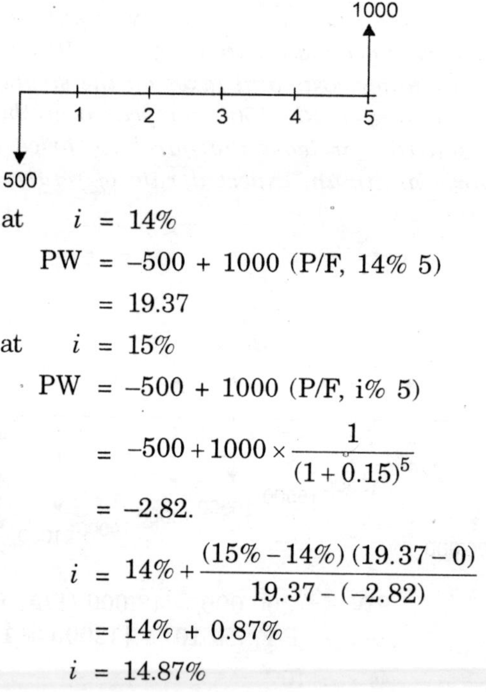

at $i = 14\%$

$$PW = -500 + 1000\ (P/F, 14\%\ 5)$$
$$= 19.37$$

at $i = 15\%$

$$PW = -500 + 1000\ (P/F, i\%\ 5)$$
$$= -500 + 1000 \times \frac{1}{(1 + 0.15)^5}$$
$$= -2.82.$$
$$\therefore \quad i = 14\% + \frac{(15\% - 14\%)\,(19.37 - 0)}{19.37 - (-2.82)}$$
$$= 14\% + 0.87\%$$
$$i = 14.87\%$$

Project Y

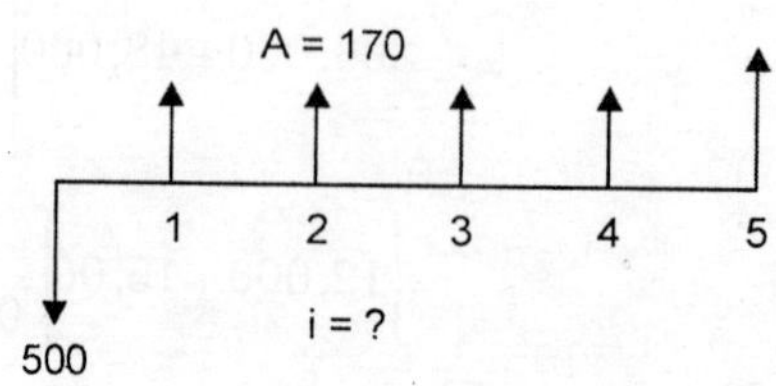

$$\text{PW} = -500 + 170\ (\text{P/A},\ i\%\ 5)$$

$$@\ i = 15$$

$$\text{PW} = -500 + 170\left[\frac{(1+0.15)^5 - 1}{0.15(1+0.15)^5}\right] = 71.56$$

$$@\ i = 20\%$$

$$\text{PW} = -500 + 170\ (\text{P/A},\ 20\%\ 5)$$

$$= -500 + 170\ (2.99061) = 8.4037$$

$$@\ i = 25\%$$

$$\text{PW} = -500 + 170\ (\text{P/A},\ 25\%\ 5)$$

$$= -500 + 170 \times 2.68929 = -42.82$$

$$\therefore \quad i = 20\% + \frac{(25\% - 20\%)\ (8.4037 - 0)}{8.4037 - (-42.82)}$$

$$= 20\% + 0.82$$

$$i = 20.82\%$$

Problem 10. *A company is considering the purchase of a new piece of testing equipment that is expected to produce Rs. 8000 additional profit during the first year of operation. This amount will probably decrease by Rs. 500 per year for each additional year of ownership. The equipment costs Rs. 20,000 and will have an estimated salvage value of Rs. 3000 after 8 years of use. How does the proposal match up against a MARR of 18%?*

Solution.

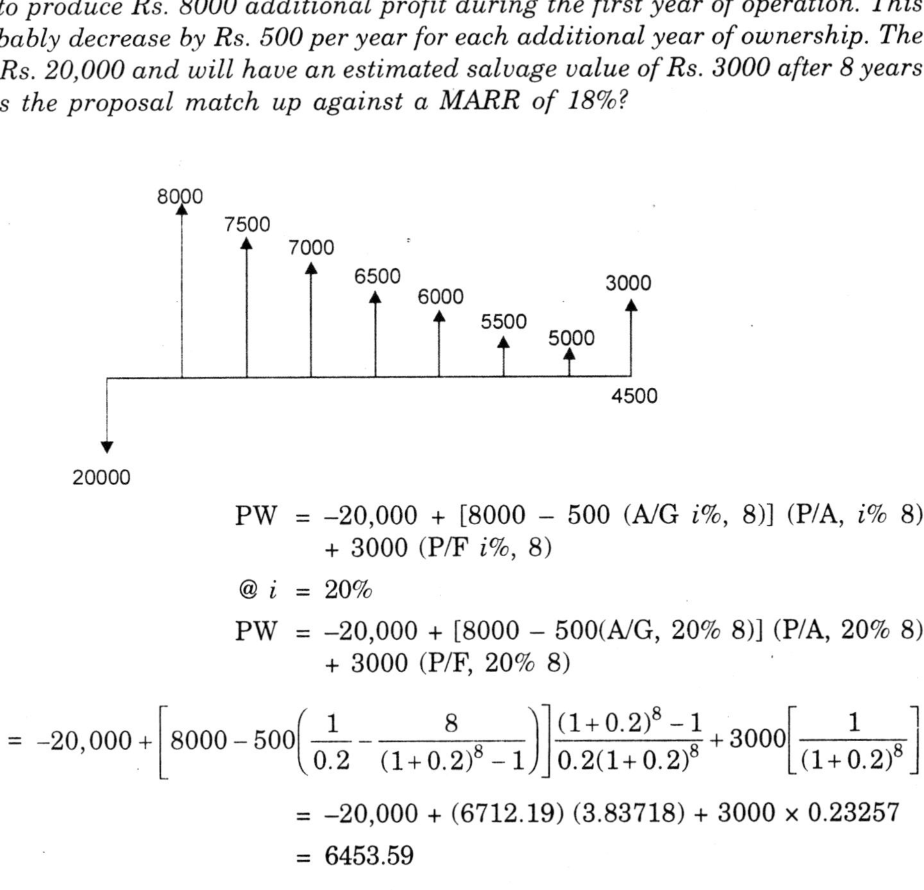

$$\text{PW} = -20{,}000 + [8000 - 500\ (\text{A/G}\ i\%,\ 8)]\ (\text{P/A},\ i\%\ 8) + 3000\ (\text{P/F}\ i\%,\ 8)$$

$$@\ i = 20\%$$

$$\text{PW} = -20{,}000 + [8000 - 500(\text{A/G},\ 20\%\ 8)]\ (\text{P/A},\ 20\%\ 8) + 3000\ (\text{P/F},\ 20\%\ 8)$$

$$= -20{,}000 + \left[8000 - 500\left(\frac{1}{0.2} - \frac{8}{(1+0.2)^8 - 1}\right)\right]\frac{(1+0.2)^8 - 1}{0.2(1+0.2)^8} + 3000\left[\frac{1}{(1+0.2)^8}\right]$$

$$= -20{,}000 + (6712.19)\ (3.83718) + 3000 \times 0.23257$$

$$= 6453.59$$

$$@\, i = 25\%$$

$$\text{PW} = -20{,}000 + (8000 - 500\ (A/G,\ 25\%\ 8)]\ (P/A,\ 25\%\ 8) + 3000\ (P/F,\ 25\%\ 8)$$

$$= -20{,}000 + (6806)\ (3.32891) + 3000\ (0.16777)$$

$$= 3161.1$$

$$@\, i = 30\%$$

$$\text{PW} = -20{,}000 + (6892.2)\ (2.92470) + 3000 \times 0.12259$$

$$= -20{,}000 + 20157.6 + 367.77$$

$$= 525.37$$

$$@\, i = 40\%$$

$$\text{PW} = -20{,}000 + 7040.7 \times 2.33060 + 3000 \times 0.06776$$

$$= -3387.66$$

$$i = 30\% + \frac{(40\% - 30\%)\ (525.37 - 0)}{525.37 - (-3387.66)}$$

$$i = 30\% + 1.34$$

$$i = 31.34\%.$$

5.8 RATE OF RETURN (R.O.R.) CAN ALSO BE CALCULATED IN THE FOLLOWING MANNER

Rate of Return expresses the percentage of profit to investment.

R.O.R. = Profit to Investment.

$$\text{R.O.R.} = \frac{P}{I} \text{ and is expressed as an interest rate (i)}$$

Assumption: Is that, the time factors are not considered or alternatively that the investment and profits all occur within the first time period. Profits consist of the net operating advantage, less the depreciation and taxes if applicable.

The unadjusted R. O. R. has not been so adjusted (uses an approximation much like that used in the A.A.C. method).

Whereas in the unadjusted R.O.R., does take compound interest into account.

(*a*) R.O.R (Unadjusted before taxes).

$$\underset{\text{(before taxes)}}{\text{R.R.U}_{\cdot BT}} = \frac{\text{O.A / yr.} - \text{Dep / yr.}}{I_{\text{average}}}$$

$$\underset{\text{(after taxes)}}{\text{R.R.U}_{\cdot BT}} = \frac{\text{O.A / yr.} - \text{Dep / yr.} - \text{Tax / yr.}}{I_{\text{average}}}$$

Problem 1. *A Nuclear utility in Chio has an opportunity to install an irradiation tube in its reactor during a forthcoming shut down. The tube will cost Rs. 10,000 initially and will be of no value (highly contaminated) after use. During the 4 year life of the project, the tube will be used to collect radioactive isotopes which can be sold to yield an operating advantage of Rs. 3000/yr.*

(a) *Find before taxes rate of return.*

(b) *If the utility claims at 8% interest on its debt equity capital structure. Is this worthwhile project from an economic standpoint?*

Solution.

Given data :

$$\text{Capital Investment} = \text{Rs. } 10{,}000$$

$$n = 4 \text{ year}$$

$$\text{O.A.} = \text{Rs. } 3000/\text{year}$$

$$\text{Depreciation} = \text{Rs. } 2500/\text{year}$$

$$I_{av} = \frac{I+S}{2} = \frac{10{,}000+0}{2} = 5000.$$

(a) $$\text{Rate of Return before taxes} = \frac{3000-2500}{5000} = 0.1/\text{year}$$

(b) *Result:* The project is worthwhile as the required is less than what we are getting.

Problem 2. *A metal brokerage firm has an opportunity to invest 35,00,000 in* UO_2 *and make an immediate sale of a new England Electric Utility 43,50,000. What is the simple rate of return?*

Solution. $$\text{Simple rate of return} = \frac{\text{Profit}}{\text{Investment}}$$

$$\text{Profit} = \text{Utility} - \text{Investment}$$

$$= 43{,}50{,}000 - 35{,}00{,}000$$

$$= 8{,}50{,}000.$$

$$\text{R.O.R simple} = \frac{8{,}50{,}000}{3{,}500{,}000} = 0.24 \; (i.e.) \; 24\%.$$

Problem 3. *A Rs. 20,000 m/c is expected to last 10 yrs and have a salvage of Rs. 2000. It will generate increased income (before depreciation) of Rs. 4000/year. but neccessitate that Rs. 1100 be paid in taxes. What is the unadjusted R.O.R assuming straightline depreciation?*

Solution. Given data:

$$\text{Cost of Machine (I)} = \text{Rs. } 20{,}000$$

$$\text{Life of the machine (n)} = 10 \text{ year}$$

$$\text{Operating Advantage} = \text{Rs. } 4000/\text{year}$$

$$\text{Taxes} = \text{Rs. } 1100$$

$$\text{Salvage value 'S'} = \text{Rs. } 2{,}000.$$

$$\underset{\text{(Unadjusted)}}{\text{R.O.R}_{(AT)}} = \frac{\text{OA/year} - \text{Dep/year} - \text{Taxes}}{I_{av}}$$

$$I_{av} = \frac{20,000 - 2000}{20} = 11,000$$

$$\text{Depreciation} = \frac{I - S}{n} = \frac{20,000 - 2000}{10} = 1800.$$

$$\text{R.O.R.}_{AT} = \frac{4000 - 1800 - 1100}{11,000} = 0.1$$

$$\text{R.O.R.}_{AT} = 10\%$$

Conclusion. R.O.R.$_{AT}$ is 10%.

5.8.1 Advantages

1. It defines a rate of return at which profits are earned & can thus be compared with similar rates from other projects.
2. It is a useful measure to compare against the cost of capital.
3. It can be interpreted either on the basis of the total project (*i.e.*) R.O.R. simple or on annual basis (*i.e.*) Rate of Return unadjusted.

5.8.2 Disadvantages

1. It does not adjust for the compounding value of income.
2. It does not consider the total magnitude of the project.
3. It requires the estimates of return as well as cost.

Rate of Return (Adjusted) or Internal Rate of Return)

Defination : The Internal rate of return (I.R.R.) for an investment is the discount rate that equates the present value of expected cash outflows with the present value of the expected cash inflows.

Mathematically it is represented by rate '*r*'

i.e.

$$r = \left[\sum_{t=0}^{n} \frac{A_t}{(1+r)^t}\right] = 0$$

where A is the cash flow for a period t whether it be a net cash outflow or inflow.

n is the number of years.

$$A_O = \frac{A_1}{(1+r)^1} + \frac{A_2}{(1+r)^2} + \frac{A_3}{(1+r)^3} + \frac{A_4}{(1+r)^4} + \ldots.$$

where r is the rate of return.

Problem 1. *Project Investment = Rs. 18,000*

Cash Inflows/year = 5600 for 5 yrs.

Required Rate of Return = 14%

Calculate I.R.R.

Solution. W.K.T. $A_O = \frac{A_1}{(1+r)^1} + \frac{A_2}{(1+r)^2} + \frac{A_3}{(1+r)^3} + \frac{A_4}{(1+r)^4} + \frac{A_5}{(1+r)^5}$

Discount Rate	*Discount Factor*	*Cash flows in each year*	*Present value of cash inflows*
17%	3.199	5600	17,916.33851
16%	3.2742	5600	18,336.04
16.88%		5600	18,000
		5600	
		5600	

Assuming Discount Rate = 17%

$$18{,}000 = \frac{5600}{(1+.17)^1} + \frac{5600}{(1+.17)^2} + \frac{5600}{(1+.17)^3} + \frac{5600}{(1+.17)^4} + \frac{5600}{(1+.17)^5}$$

$$= 5600\left[\frac{1}{(1+.17)^1} + \frac{1}{(1+.17)^2} + \frac{1}{(1+.17)^3} + \frac{1}{(1+.17)^4} + \frac{1}{(1+.17)^5}\right]$$

$$18{,}000 = 5600 \times 3.1993 = 17{,}916.33851$$

Now the L.H.S $\neq$ R.H.S, so we will interpolate it now to get the exact value of Discount rate of Rs. 18,000. First we will see the value for 16% discount rate, then for that percentage in which the value will lie, will be interpolated.

$$18{,}000 = 5600\left[\frac{1}{(1+.16)^1} + \frac{1}{(1+.16)^2} + \frac{1}{(1+.16)^3} + \frac{1}{(1+.16)^4} + \frac{1}{(1+.16)^5}\right]$$

$$= 5600 \times 3.2742$$

$$= 18{,}336.04$$

Again L.H.S. $\neq$ R.H.S

i.e. The value of 18,000 lies in between 17% & 16%. We will see the difference of the two for 1% of discount factor and then find for the corresponding more money for 16% which will be subtracted in the present value of the cash inflows for 16% but difference will be added as the cash inflows are decreasing from 16% to 17%.

For 16%, Cash inflows = 18,336.04

For 17%, Cash inflows = 17,916.33

For 1% of difference = 419.71

But we need only 336 to subtract into the 16% discount factor to get the proper % of discount factor

For Rs. 419.71, diff. = 1

$$\text{For 1 Re. diff.} = \frac{1}{419.71}$$

$$\text{For Rs. 336, diff.} = \frac{1}{419.71} \times 336 = 0.80.$$

i.e. 0.8% discount factor for Rs. 336.

$$18{,}000 = 5600\left[\frac{1}{(1+.168)^1}+\frac{1}{(1+.168)^2}+\frac{1}{(1+.168)^3}+\frac{1}{(1+.168)^4}+\frac{1}{(1+.168)^5}\right]$$

$$= 5600 \times 3.214$$

$$18{,}000 \simeq 17999.01$$

Acceptance criteria : If I.R.R. exceeds the required rate of return, the project is accepted, if not, it is rejected. The required rate of return is 14% and the I.R.R. is 16.8%, so we will accept the proposal.

Problem 2. *For 4 years, the operating advantage is Rs. 3000/year. Investment is Rs. 10,000, required rate of return = 10%. Determine the I.R.R. and find out whether the proposal is accepted or rejected.*

Solution. Given Data:

n = 4 yrs.

O.A./year = Rs. 3000

Investment = Rs. 10,000

R.O.R. = 10%

I.R.R. = ?

Discount Rate	*Discount Factor*	*Cash flows in each year*	*Present value of cash inflows.*
12%	3.0373	3000	9112.04
10%	3.16986	3000	9509.596
8%	3.3121	3000	9936.38056
7%	3.3872	3000	10,161.633
7.717%	3.3330	3000	9999.2954

W.K.T.

$$A_O = \frac{A_1}{(1+r)^1}+\frac{A_2}{(1+r)^2}+\frac{A_3}{(1+r)^3}+\frac{A_4}{(1+r)^4}$$

$$10{,}000 = 3000\left[\frac{1}{(1+.12)^1}+\frac{1}{(1+.12)^2}+\frac{1}{(1+.12)^3}+\frac{1}{(1+.12)^4}\right]$$

$$10{,}000 = 3000 \times 3.0373$$

$$10{,}000 = 9112.04$$

$$\text{L.H.S.} \neq \text{R.H.S.}$$

Again doing the same with different discount rates till we get the L.H.S = R.H.S. Here for the 7% discount rate the cash flows are more and for 8% discount rate present value of the cash inflows are less, that means the value lies between 7% and 8% of discount rate and this can be found out only by interpolation.

i.e. For present value of cash inflows Rs. 225.25, diff = 1%

For present value of cash inflows of Re. 1, diff $= \dfrac{1}{225.25}$

For present value of cash inflows of Rs. 161, diff. $= \frac{1}{225.25} \times 161 = 0.717\%$

This difference will be added in the 7% discount rate to get the exact value of L.H.S. *i.e.* for Rs. 10,000 of investment, the discount rate is 7.717.

Illustration:

$$10{,}000 = 3000\left[\frac{1}{(1+.0771)^1} + \frac{1}{(1+.0771)^2} + \frac{1}{(1+.0771)^3} + \frac{1}{(1+.0771)^4}\right]$$

$$10{,}000 \simeq 9999.29$$

Result : The proposal is rejected as the rate of return is very less compared to the given one.

Problem 3. *An investment of Rs. 5650 is expected to yield an operating advantage of Rs. 4000 at the end of 1st year, Rs. 2000 at the end of the 2nd year and Rs. 1000 at the end of the 3rd year. What is the time adjusted rate of return?*

Solution.

Discount Rate	*Discount Factor*	*Cash Flow in Each year*	*Present value of the cash inflow*
10%		4000	6040.57
11%		2000	5958.09
13%		1000	5799.16
14%			5722.67
15%			5648.06
14.9739			5650.15

$$A_0 = \frac{A_1}{(1+r)^1} + \frac{A_2}{(1+r)^2} + \frac{A_3}{(1+r)^3}$$

$$5650 = \frac{4000}{(1+.1)^1} + \frac{2000}{(1+.1)^2} + \frac{1000}{(1+.1)^3}$$

$$= 4000 \times 0.909 + 2000 \times 0.8264 + 1000 \times 0.7513$$

$$5650 \neq 6040.57$$

$$\text{L.H.S.} \neq \text{R.H.S.}$$

Similarily, finding the discount factor till we get a 1% diff. b/w 2 value. Here the value lies b/w 14% and 15% discount factor.

To find the exact value of discount factor we will interpolate the value.

Diff. b/w 14% and 15% discount factor = 74.61

For present value of the cash inflow, diff. = 1%.

For Rs. 72.67 of present value of the cash inflow,

$$\text{diff.} = \frac{1}{74.61} \times 72.67 = 0.9739\%.$$

Add this difference of discount factor to the 14% discount factor to get the exact value of present value of the cash inflows.

$$5650 = \frac{4000}{(1+.1497)^1} + \frac{2000}{(1+.1497)^2} + \frac{1000}{(1+.1497)^3}$$

$$5650 = 5650.15$$

$$\text{L.H.S.} \simeq \text{R.H.S.}$$

Conclusion : The rate of return for present value of the cash inflows will be 14.9739%.

Problem 4. *Koonz & Donnel Co. has an opportunity to invest Rs. 7680 in a plant modification which is expected to yield an operating advantage of Rs. 5000 at the end of 1st year and Rs. 1000 at the end of each of the next 5 yrs. What is the Internal rate of return (I.R.R.)?*

Solution.

Discount Factor (d.F.)	*Discount Rate*	*Cash flow in each year*	*Present value of the cash inflow*
10%		5000	7991.624
11%		1000	7834.14
12%		1000	7682.835
13%		1000	7537.3727
12.019489%		1000	7679.9459

$$A_O = \frac{A_1}{(1+r)^1} + \frac{A_2}{(1+r)^2} + \frac{A_3}{(1+r)^3} + \frac{A_4}{(1+r)^4} + \frac{A_5}{(1+r)^5} + \frac{A_6}{(1+r)^6}$$

$$7680 = \frac{5000}{(1+.1)^1} + \frac{1000}{(1+.1)^2} + \frac{1000}{(1+.1)^3} + \frac{1000}{(1+.1)^4} + \frac{1000}{(1+.1)^5} + \frac{1000}{(1+.1)^6}$$

$$7680 = 7991.62$$

$$\text{L.H.S.} \neq \text{R.H.S.}$$

Similarly, doing till we get the nearest value of the cash outflow and then by interpolation finding the exact value of the discount factor.

Now for present value of 145.4623, difference in discount factor = 1%

$$\text{For Present value of Re 1, difference in d.F.} = \frac{1}{145.4623}$$

$$\text{For Present value of Rs. 2.835 difference in discount factor} = \frac{1}{145.4623} \times 2.835$$

$$= 0.019489.$$

Adding this difference of discount factor in the 12% to get the exact value of present value of cash inflows

$$7680 = \left[\frac{5000}{(1+.120194)^1} + \frac{1000}{(1+.120194)^2} + \frac{1000}{(1+.120194)^3} + \frac{1000}{(1+.120194)^4} + \frac{1000}{(1+.120194)^5} + \frac{1000}{(1+.120194)^6}\right]$$

$$7680 = 4463.509 + 796.9165 + 711.4087 + 635.0759 + 566.9334 + 506.1024$$

$$7680 \simeq 7679.9459.$$

Result: The rate of return for the present value of cash inflows will be 12.0194%.

EXERCISE PROBLEMS

1. In 1994 a small apartment building was purchased for 200,000. Receipts from rent have averaged Rs. 30,200 a year, taxes, maintenance and repair costs have totalled Rs. 8620 annually. The owner intends to hold the property until she retires in 2004. If at that time the property sales for Rs. 200,000, what rate of return will be obtained on the investment?

2. A bank introduces two different investment schemes whose details are as follows : Find the best investment alternative from the investor's point of view.

	Alpha Bank	*Beta Bank*
Deposit amount (Rs.)	1,00,000	2,00,000
Period of deposit (years)	5 years	3 years
Maturity amount (Rs.)	3,00,000	4,50,000

3. Company is planning for its expansion programme which will take place after five years. The expansion requires an equal sum of Rs. 5,00,000 for consecutive three years. Gamma Bank has recently introduced a scheme in this line. If the company invests Rs. 7,00,000 now with this bank, it will make equal repayments of Rs. 5,00,000 for three consecutive years starting from the end of the fifth year from now. The minimum attractive rate of return for the company is 12% Suggest whether the company should invest with the Gamma Bank for its expansion programme.

4. Consider the following table which summarises data of two alternatives.

	First Cost	*Annual return*	*Life*
Alternative 1	Rs. 5,00,000	Rs. 1,50,000	10 yrs.
Alternative 2	Rs. 8,00,000	Rs. 2,50,000	10 yrs.

Find the best alternative based on the rate of returned method comparison.

5. A Company is planning to expand its present activity. It has two alternatives for the expansion programme and the corresponding cash flows are given in the following table. Each alternative has a life of five years and a negligible salvage value. The minimum attractive rate of return for the company is 15%. Suggest the best alternative for the company.

	Initial Investment (Rs.)	*Yearly Revenue (Rs.)*
Alternative 1	4,50,000	1,50,000
Alternative 2	7,50,000	2,50,000

6. A person invests a sum of Rs. 200,000 in a business and receives equal net revenue of Rs. 50,000 for the next 10 years. At the end of the 10th year, the salvage value of the business is Rs. 25000. Find the rate of return of the business.

7. West Texas Oil has paid 300,000 for a producing oil well field. Engineers estimate that net receipts will be 120,000 for the first year of operation with a reduction of 15% per year in the following years it plans to sell the well after 5 years for 80,000. How does this seem financially if their MARR is 20 percent?

8. A bioengineering research laboratory has a patent that it is considering leasing for 10 years at Rs. 40,000 for the first year with increments of Rs. 4000 per year for the following 9 years. The accounting office says the company has Rs. 250,000, in development investment costs that are considered to be a first cost. Is the lease reasonable if the laboratory's MARR is 20 percent ?

9. Consider the three mutually exclusive alternative below and use the internal rate of return method to make a selection. The feasible alternative chosen must provide service for a 10 year period. The MARR is 12%. State all assumptions you make.

	A	*B*	*C*
Investment	Rs. 2000	Rs. 8000	Rs. 20,000
Revenue less costs	Rs. 600/year	Rs. 2220/year	Rs. 3600/year
Salvage value	0	0	0
Project Life (years)	5	5	10

10. In the Rawhide Co. Inc, decisions regarding approval of proposals for plant investment are based upon a stipulated minimum attractive rate of return of 20% before income taxes. The following five packaging devices were compared assuming a 10 year life and zero salvage value for each. Which one should be selected? Make any additional calculations you think are needed to make a comparison using the IRR method.

PACKING EQUIPMENT

	A	B	C	D	E
Investment (first-cost)	Rs. 38000	50,000	55,000	60,000	70,000
Net annual return	11,000	14,100	16,300	16,800	19,200
Rate of return	26.1%	25.2%	26.9%	25.0%	24.3%

(Incremental analysis problem).

DEPRECIATION AND INCOME TAX CONSIDERATIONS

6.1 INTRODUCTION

Depreciation is the decrease in value of practical properties with the passage of time. Depreciation is an accounting concept that establishes an annual deduction against income tax. From a business viewpoint, a physical asset has value because one expects to receive future monetary benefits through its possession and use. The benefits are in the form of future cash flows resulting from (1) the use of the asset to produce valuable goods or service (2) the ultimate sale of the asset.

Depreciation charges are not actual cash flows and the true decrease in market value of an asset may not correspond to the allowable deductions. Major changes in depreciation accounting occurred with the passage of the Economic Recovery Act of 1981, which introduced the accelerated cost recovery system (ACRS) and the Tax Reform Act of 1986, with the help of MACRS (Modified accelerated cost recovery system) specifies percentages for the recovery of investments in various property classes that determine the amount which can be deducted each year for tax purposes.

Before 1981, depreciation methods like straightline method and declining balance method were based on the tax life of individual assets rather than classes of property, forming the basis for the MACRS depreciation rates and along with units of production method.

Taxes are a major factor in any profit seeking venture. Property, sales, excise and income taxes affect net returns for both individuals and corporations. Types and amounts of taxes vary as governments pursue their fiscal policies.

An effective income tax rate that represents total corporate tax liability can be developed for after-tax economic evaluations. Special tax provisions and charges for depreciation and interest are applied to the before-tax cash flow to determine taxable income.

An after-tax analysis defines the actual cash flows expected from a proposal. It may reveal the tax advantages of one proposal over another—advantages that are not part of a before-tax comparison. However, comparisons that include tax effects are more complicated because tax laws are very complex. Precise tax considerations require expert assistance but attention to basic tax provisions provides an adequate evaluation for most situations in the province of engineering economics.

The purpose of depreciation is to consider for two reasons:

(1) To provide for the recovery of capital that has been invested in physical property.

(2) To enable the cost of depreciation to be charged to the cost of producing products or service that results from the use of the property. Depreciation cost is deductible in computing profits on which income taxes are paid.

6.2 DEPRECIATION

6.2.1 Definition

Depreciation is the dimension in the value of fixed asset due to use/lapse of time.

Buildings equipment and machinery wear and tear with the amount of use and passage of time. Their worth does not remain as much after a few years as it is today. Depreciation may be defined as the reduction in value.

6.2.2 Causes of Depreciation

(1) Usage

(2) Physical causes

(3) Abnormal occurrences

(4) Technological development and changes

(5) Sudden failure and

(6) Depletion

1. Usage

With some type of fixed assets like car and other vehicles there are customs which have been established and the rate of wear and tear normally expected every year. There is definitely a correlation between the price of second hand compared to a new one.

2. Physical Causes

(*a*) Normal physical wear and tear due to friction, pull, impact, fatigue, twisting etc.

(*b*) Lack of maintenance on timely repair of fixed assets.

(*c*) Action of chemical elements on the component parts.

(*d*) Passage of time.

3. Abnormal Occurences

(*a*) Accidents

(*b*) Defects in m/c

(*c*) Excessive wear and tear

(*d*) Contingent occurrence, e.g. appearance of hair-line cracks in pressure vessels.

4. Technological Development and Changes

(*a*) New equipment which supersedes the existing ones start coming into the market. e.g., Calculators have superseded the slide rules to a major extent.

(*b*) Change in manufacturing methods which necessitates use of other type of equipment.

(*c*) Improved and Automated m/c tools which render the use of existing ones uneconomical.

(*d*) Inadequacy of the existing equipment to perform the necessary functions such as increased o/p, more precision and better quality.

5. Sudden Failure

This refers to sudden or catastrophic loss in value due to technological characteristics inherent in the asset. This happens due to accident or misuse. This category of asset includes items used in large numbers with a relatively small cost per item.

6. Depletion

Consumption of an exhaustible natural resource to produce products or service is termed depletion.

Ex. : Removal of oil, timber, rock or minerals.

$$\text{Depletion rate (Rs./Unit)} = \frac{\text{Adjusted basis of resource}}{\text{Remaining units of resource}}$$

6.3 METHODS OF PROVIDING FOR DEPRECIATION

1. Straight Line Method.
2. Reducing Balance or Dimnishing Balance Method.
3. Production based Methods.
   (*a*) Per Unit (*b*) Per Hour.
4. Repair Provision Method.
5. Annuity Method.
6. Sinking Fund Method.
7. Endowment Policy Method.
8. Re-evaluation Method.
9. Sum of Digits Method.

6.4 METHODS OF CALCULATING DEPRECIATION

Each method should adopt the following any one or more of the following:

(*i*) Depreciation is a function of time.

(*ii*) Depreciation is a function of use.

(*iii*) Depreciation is a function of time and use.

(*iv*) Depreciation is a function of time and maintenance.

(*v*) Depreciation is a function of time and interest.

Mehod I: Straight Line Method or Fixed Instalment Method or Proportional Method

This is the method of providing for depreciation by means of periodic charges over the assumed life of the asset.

'C' is the cost of asset or equipment.

'R' is the residual value of the asset (scrap value).

'N' is the estimated life of machine.

$$\text{Depreciation/year} = \frac{\text{I} - \text{S}}{\text{N}} \text{ or } \frac{\text{C} - \text{R}}{\text{N}}$$

PROBLEMS

Problem 1. *A melting unit for a steel foundry was purchased for Rs. 40,000. Rs. 10,000 more were spent in its errection and commisioning. The estimated residual value after 10 yrs. was Rs. 12,000.*

(*a*) *Calculate the annual rate of depreciation.*

(*b*) *Calculate the book value of the m/c at the end of each year using the straight line depreciation and plot the graph of no. of yrs. v/s the depreciation fund.*

(*c*) *Calculate the depreciation fund collected at the end of the 8th year.*

Solution. Given data:

Cost of Capital = Rs. 40,000

Money spent on errection and maintenance

= Rs. 10,000.

Total cost of capital = Rs. 50,000

Salvage value = Rs. 12,000

Estimated Life = 10 yrs.

Note : Book value after the estimated life should be equal to the salvage value.

(*a*)

$$\text{Depreciation/yr.} = \frac{\text{C} - \text{R}}{\text{N}} = \frac{50{,}000 - 12{,}000}{10} = \frac{38{,}000}{10}$$

$$= 3800.$$

(*b*) To calculate the book value after each year.

No. of years.	*Cost and Balance*	*Depreciation /year*	*Book value after Depreciation in each year*
1	50,000	3800	46,200
2	46,200	3800	42,400
3	42,400	3800	38,600
4	38,600	3800	34,800

(Contd....)

5	34,800	3800	31,000
6	31,000	3800	27,200
7	27,200	3800	23,400
8	23,400	3800	19,600
9	19,600	3800	15,800
10	15,800	3800	12,000

(*c*) Depreciation fund collected at the end of 8th year

$$= 3800 \times 8 = 30{,}400.$$

Problem 2. *An investment of Rs. 5000 in a new equipment is expected to have a salvage value of Rs. 1000 after a 4 yrs. life.*

(*a*) *Find a straight line depreciation expense.*

(*b*) *Plot the graph, the depreciation fund v/s the no. of yrs.*

Solution. (*a*) Depreciation/year $= \dfrac{5000 - 1000}{4} = 1000.$

(*b*)

Years	*Cost of capital or balance*	*Depreciation*	*Book value at the end of each year.*
1	5000	1000	4000
2	4000	1000	3000
3	3000	1000	2000
4	2000	1000	1000

Check : Book value at the end of 4th year is equal to the scrap value of the m/c.

Advantages

1. This method is easy to understand and simple to operate.
2. It is frequently used in practice.
3. Uniform annual charge offers better comparative costs.
4. This method requires little work for calculating depreciation amounts.

Disadvantages

1. The fixed assets do not wear out exactly at the same rate during their life.
2. A straight line method in many cases becomes unrealistic.

Method II : Reducing Balance Method or Diminishing Balance Method or Percentage on Book Value Method

Under this method, a fixed percentage of the book value of the asset is deducted each period and the book value decreases at a decreasing rate. This is the method of providing depreciation by means of periodic charges calculated as a constant of the balance asset after deducting the amounts previously provided.

Depreciation Percentage 'P' $= 1-\left(\dfrac{R}{C}\right)^{1/N}$

Note : If R is not given or no salvage, consider it as 1 (not zero).

Problem 1. *A machine costs Rs. 10,000 has a scrap value of Rs. 400 at the end of 4 yrs. of its serviceable life. Using the Reducing Balance :*

(*a*) *Determine the depreciation and book value in each year.*

(*b*) *Plot the graph, no. of yrs. v/s Depreciation Fund.*

(*c*) *What is the amount of depreciation collected at the end of 2nd year or after 2 yrs.*

Solution. (*a*) Percentage Depreciation $= 1-\left(\dfrac{400}{10,000}\right)^{1/4}$

$= 0.552$ or 55.2% of the amount/year.

(*b*)

No. of years	*Cost of Capital at beginning of each year*	*Depreciation*	*Book value at the end of each year*
1	10,000	10,000 × .552 = 5520	4480
2	4,480	4,480 × .552 = 2472.96	2007.04
3	2,007.04	1107.886	899.153
4	899.153	496.332	402.82

(*c*) Amount of depreciation fund collected at the end of 2nd year

= 5520 + 2472.96 = 8262.96 Rs.

Problem 2. *A Car was purchased for Rs. 32,000 and salvage value was estimated Rs. 8000 after 7 yrs. Using the reducing balance method*

(*a*) *Calculate the percentage of depreciation.*

(*b*) *Calculate the book value and depreciation in each year.*

(*c*) *Plot a graph, no. of yrs. v/s depreciation fund.*

Solution. (*a*) Percentage Depreciation $= 1-\left(\dfrac{R}{C}\right)^{1/N}$

$= 1-\left(\dfrac{8000}{32,000}\right)^{1/7} = 0.1796$, *i.e.*, 17.96%/year.

(*b*)

No. of years	*Cost of capital at the beginning of each year*	*Depreciation*	*Book value at the end of each year*
1	32,000	5747.2	26,252.8
2	26,252.8	4715.0028	21,537.7971
3	21,537.7971	3868.1883	17,669.6087
4	17,669.6087	3173.4617	14,496.14697
5	14,496.14697	2603.5079	11,892.638
6	11,892.638	2135.91796	9756.7200
7	9,756.72	1752.306	8004.41308.

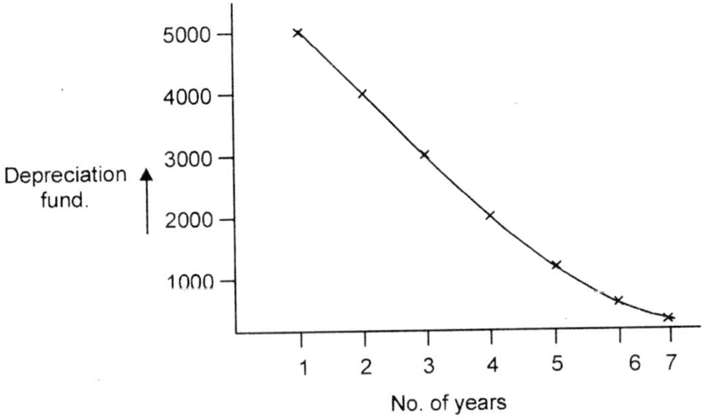

Problem 3. *2 Machine's are purchased each for Rs. 12,000. The estimated useful life of m/c is 5 yrs. The estimated scrap value is Rs. 2000. For machine A, the straight line method and for B the reducing balance method are used to calculate the depreciation every year. Compare the depreciations charged in each year of both i.e. A and B and plot the graph, the no. of yrs. v/s the depreciation fund for both.*

Solution. (*a*) Depreciation for machine A $= \frac{C-R}{N} = \frac{12,000-2000}{5} = 2,000$

Depreciation/year for m/c A = Rs. 2,000.

$$\text{Percentage Depreciation for m/c B} = 1-\left(\frac{R}{C}\right)^{1/N}$$

$$= 1-\left(\frac{2000}{12000}\right)^{1/5}$$

$$= 0.3011 \text{ or } 30.11\%.$$

(b)

No. of years	*Cost of Investment at the beginning of the year*		*Depreciation/year*		*Book value at the end of each year*	
	Machine A	*Machine B*	*St. line Machine A*	*Reducing balance Machine B*	*Machine A*	*Machine B*
1	12,000	12,000	2,000	3613.2	10,000	8386.8
2	10,000	8,386.8	2000	2525.26	8,000	5861.53
3	8,000	5,861.534	2000	1764.9	6,000	4096.625
4	6000	4,096.625	2000	1233.49	4000	2863.13
5	4000	2863.13	2000	862.08	2000	2001.04

Advantages

(1) It is simple to understand and calculate.

(2) Mathematical relationship can be employed to arrive at the appropriate percentage.

(3) The method is more logical as the largest annual amount of depreciation is charged in the first year where repair and maintenance charges are almost negligible.

Disadvantages

(1) It is not simple to fix percentage P accurately.

(2) A standard percentage P for all conditions may produce misleading results.

Method III : Production Unit Method

This is the method of providing for depreciation by means of fixed assets rate/unit of production calculated by dividing the value of the asset by the estimated no. of units to be produced in its life.

$$\text{Depreciation rate/unit} = \frac{\text{Value of the Assets}}{\text{No. of units produced}}$$

Problem 1. *A machine costing Rs. 2,00,000 has a residual value of Rs. 1,00,000 after 10 yrs. of service. The estimated rate of production is 8 units/hr. using the production unit method. Calculate the rate of depreciation. Assume 50 week year and 46 hrs. week.*

$$\text{Depreciation rate/unit} = \frac{2,00,000 - 1,00,000}{8 \times 46 \times 50 \times 10}$$

$$= 0.5437 = 0.54. \text{ rupees/unit}$$

Method IV : Production Hour Method

Solution. $\text{Depreciation rate/hour} = \dfrac{\text{Value of the Asset}}{\text{No. of production hrs.}}$

This is the method of providing for depreciation by means of fixed rate/hour of production calculated by dividing the value of asset by the estimated no. of hrs. of its life.

Problem 1. *A machine costing Rs. 15,000 has a scrap value of Rs. 5000 at the end of 10 yrs. of its serviceable life. If the m/c runs 2100 hrs./year calculate the depreciation rate/hr. of the m/c and the total annual depreciation.*

Solution. Depreciation rate/hour $= \dfrac{15,000 - 5,000}{2100 \times 10} = 0.4761$ or 47.61%.

Depreciation rate/year $= 0.4761 \times 2100 =$ Rs. 1000.

Problem 2. *The estimated life of the lathe is 10 yrs. and works 16 hrs./day. The initial cost of the lathe is Rs. 8000 and scrap value after 10 yrs. is Rs. 2500. If the m/c works for 5840 hrs. in a year find out the rate of depreciation charged annually under m/c hour basis method.*

Solution. Rate of depreciation/hour $= \dfrac{8000 - 2500}{5840 \times 10} = 0.09417.$

Rate of depreciation/year $= 0.09417 \times 5840$

$= 549.9528$

Method V : Sum of Digits Method

This method provided depreciation by means of differing time periodic rates calculated as follows :

If 'n' is the estimated life of the m/c the rate is calculated for each period as a fraction in which the denominator is always the sum of the series 1, 2, 3,......n and the numerator for the first period is n, for the second period is $n - 1$ and so on.

Problem 1. *The cost of the vehicle is Rs. 19,000, the scrap value after a period of 5 yrs. is estimated at Rs. 4000. Using the sum of digits method, calculate the depreciation at the end of each year. Plot a graph, the no. of yrs. v/s the depreciation fund.*

Solution. No. of yrs. 'n' = 1, 2, 3, 4, 5.

Denominator = 1 + 2 + 3 + 4 + 5 (sum of series) = (15)

The depreciation for the first year $= \dfrac{5}{15} \times (19,000 - 4,000)$

$= \dfrac{5}{15} \times 15,000 =$ Rs. 5000.

Depreciation for the 2nd year $= \dfrac{4}{15} \times 15,000 =$ Rs. 4,000.

Depreciation for the 3rd year $= \dfrac{3}{15} \times 15,000 =$ Rs. 3,000.

Depreciation for the 4th year $= \dfrac{2}{15} \times 15,000 =$ Rs. 2,000.

Depreciation for the 5th year $= \dfrac{1}{15} \times 15,000 =$ Rs. 1,000.

Problem 2. *An Investment of Rs. 5900 in new equipment is expected to have a salvage value of Rs. 1000 after 4 yrs. life. Using the sum of digits method, find out the depreciation for each year and what is the depreciation amount to be collected after 2 years.*

Solution. No. of yrs. = 1, 2, 3, 4.

Sum of digits = 1 + 2 + 3 + 4 = 10.

(*a*) Depreciation in the first year $= \frac{4}{10} \times (5900 - 1000)$

$= \frac{4}{10} \times 4900$ = Rs. 1960.

Depreciation in the 2nd year $= \frac{3}{10} \times 4900$ = Rs. 1470.

Depreciation in the 3rd year $= \frac{2}{10} \times 4900$ = Rs. 980.

Depreciation in the 4th year $= \frac{1}{10} \times 4900$ = Rs. 490.

(*b*) Depreciation amount after 2 yrs. = 1st year + 2nd year

= 1960 + 1470

= Rs. 3430.

Method VI : Sinking Fund Method

This method is based on the assumption of setting up a sinking fund in which money accumulates to replace the existing asset at the proper time. *At the end of the useful life of the asset, the total amount in depreciation + compound interest should become equal to original cost of the fixed asset.*

The mathematical relationship used to calculate the rate of depreciation:

$$\textbf{Rate of depreciation} = \frac{i(C - R)}{(1+i)^n - 1}$$

where 'i' is the rate of interest

'n' is the expected life

'C' is the investment

'R' is the salvage value of the equipment

Problem 1. *A machine is purchased for Rs. 10,000, the estimated life of the m/c is 4 yrs. and the scrap value is Rs. 400. The rate of interest on the depreciation fund is 4%. Calculate the book value of the machine at the end of each year using sinking fund method.*

Solution. Depreciation/year $= \frac{0.04(10{,}000 - 400)}{(1 + 0.04)^4 - 1}$ = 2,260.7044.

Book value of each year = 10,000 – corresponding years balance.

1 *No. of years*	2 *Balance*	3 *Interest at 4%*	4 *Annual Provision 4 = 2 + 3*	5 *Annual Investment*	6 *Balance 6 = 4 + 5*	7 *Book Value 7 = 10,000 – 6*
1	–	–	2260.70	2260.70	2260.70	7739.3
2	2260.70	90.4	2351.1	2260.70	4611.8	5388.2
3	4611.8	184.47	4796.27	2260.70	7056.97	2943.03
4	7056.97	282.27	7339.24	2260.70	9599.94	**400.06**

Problem 2. *A m/c is purchased for Rs. 40,000. The estimated life of the m/c is 5 years. The scrap value is Rs. 15,000. The rate of interest on the depreciation is 10%. Calculate the book value of the m/c at the end of each year using Sinking Fund Method and plot a graph of number of years v/s depreciation fund.*

Solution. Depreciation/year $= \dfrac{1(40,000-15,000)}{(1+.1)^5-1} = 4,094.937$

No. of yrs.	*Balance*	*Interest at 10%*	*Annual Provision*	*Annual Investment*	*Balance*	*Book Value*
1.	–	–	4094.937	4094.937	4094.937	35905.06
2.	4094.937	4094.937	4504.4307	4094.937	8599.3677	31400.6323
3.	8599.3677	859.9367	9459.3044	4094.937	13554.2414	26445.156
4.	13554.2414	1355.4241	14909.6655	4094.937	19004.6025	20995.1975
5.	19004.6025	1900.4602	20905.0628	4094.937	24999.999	**15000**

Requirements of a depreciation method

The requirements for a depreciation method are

1. Provide for the recovery of invested capital as rapidly as is consistent with economic facts involved.
2. Not be too complex.
3. Ensure that the book value will be reasonably close to the market value at any time.
4. Be accepted by the Internal Revenue Service (IRS), if the method is also to be used for determining federal income tax.

Method VII : Endowment Policy Method or Insurance Method

This is the method of providing for depreciation by means of fixed periodic charges equivalent to the premium on an endowment policy for the amount required to provide at the end life of the asset.

$$\textbf{Annual Depreciation} = \frac{(C-R)xi}{(1+i)^{n+1}-1-i}$$

Problem 1. *The cost of the m/c is Rs. 10,000 and life of the m/c is 4 yrs. Salvage value after 4 yrs. is zero. Calculate the rate of depreciation at an interest of 4%.*

Solution. Depreciation/year $= \dfrac{(10{,}000 - 0) \times 0.04}{(1+0.04)^{4+1} - 1 - .04}$ = Rs. 2264.327.

In this, the premium should be paid periodically and a charge equal to the premium payable must be made to cost. This is similar in effect to the Sinking Fund Method. The only difference is that while under the sinking fund method the interest is received and reinvested in each period, under the endowment policy method any earning of interest is not considered.

6.5 TAX CONCEPTS

In Engineering Economy Studies, we need to distinguish between income taxes and several other types of taxes.

1. *Income taxes* are assessed as a function of gross revenues minus certain allowable deductions and exemptions.
2. *Property taxes* are assessed as a function of the 'value' of real estate, business and personal property. Hence, they are independent of the income or profit of an individual or a business.
3. Sales taxes are assessed on the basis of purchases of goods and or services and are independent of gross income or profits. They are normally levied by state, municipal or country governments.
4. Excise taxes are federal taxes assessed as a function of the sale of certain goods or services often considered and are independent of the income or profit of an individual or a business.

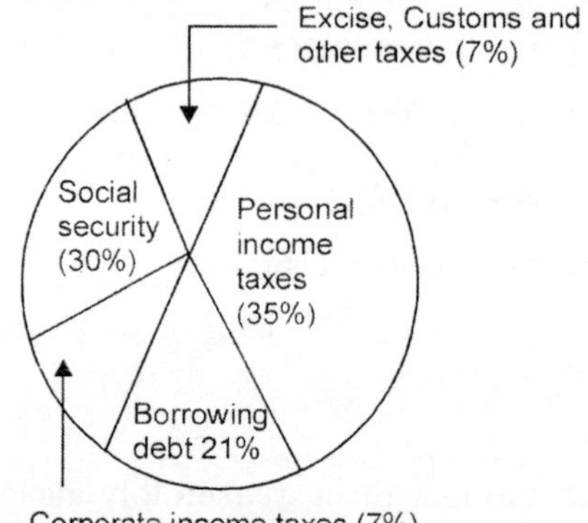

Fig. 6.1 Sources of federal incomes.

6.6 CHANGING TAXES

The federal government controls the monetary and fiscal policy of the nation to influence the level of economic activity. Monetary Policy influences the availability and cost of credit and fiscal policy deals with government receipts and expenditures.

Taxation is the key instrument in fiscal policy.

The principal methods for altering government receipts income are

(1) Changing the tax rate.

(2) Changing the depreciation requirements.

(3) Allowing tax credits.

Tax rates imposed on incomes may be raised to dampen the level of economic activity when rapid expansion threatens inflationary consequences.

The reduction in disposable incomes reduces the purchasing power of individuals and thereby decreases demand for goods and services.

There are charges in tax laws almost every year. The changes take place due to current fiscal needs.

Several tax-reducing measures legislated in 1981 file very generous tax credits and safe harbour leasing rules that allowed ailing corporations to sell their unused tax breaks, or eliminated by the Tax Equity and Fiscal Responsibility Act of 1982 made by the Deficit Reduction Act of 1984 which reformed fringe benefit and real estate taxation, among other amendments.

Engineering Economic studies ideally would be based on the tax rates in effect during the lives of the assets being evaluated.

6.7 CORPORATE INCOME TAXES

A corporation for income tax purposes includes associations joint stock companies, insurance companies and trusts, partnerships will operate as associations or corporations. Organizations of doctors, lawyers and engineers are recognized as corporations. There organizations have the following characteristics.

* Associates organized to carry on business.
* Gains from the business that are divided.
* Continuity of life and centralized management.
* Limited liability and free transferability of interests.

Organizations possessing a majority of these characteristics must file corporate tax returns. Income taxes are due from corporations and businesses, whenever revenue exceeds allowable tax deductions. Revenue includes sales to customers of goods and services, dividends received on stocks. Interest from loans and securities, rents, royalties and other gains from ownership of capital or property. Deductions are expenses incurred in the production of revenue: wages, salaries, rents, repairs taxes, interest, employee benefits, advertising etc.

The special provisions are losses from fire and theft, contributions, depreciation and depletion, bond interest, research and development and outlays satisfy legislated objectives such as pollution control.

The difference between revenue and deductions is taxable income.

Taxable income = Gross income – Expenses
– Interest on debt – Depreciation
– Other allowable deduction.

Other Corporate Income Taxes are

(1) Federal income tax
(2) Effective income tax
(3) Investment credit
(4) Expense deduction
(5) Capital gain and losses
(6) Carry backs and carry overs.

REPLACEMENT ANALYSIS

7.1 INTRODUCTION

All industries and military equipment gets worn with time and usage and function with decreasing efficiency. For example a machine requires higher operating cost, a transport vehicle such as a car or aeroplane requires more and more maintenance cost, a railway timetable becomes more and more out of date with the passage of time. The ever-increasing repair and maintenance cost necessitates the replacement of the equipment. However, there is no sharp, clearly defined time which indicates the need for this replacement. The replacement policy, in this case, consists of calculating the increased operating cost, maintenance cost, forced idle time cost together with cost of replacing new equipment.

The objective is to minimize the sum of the cost of the item, cost of replacing the item and the cost associated with failure of item.

The equipment needs replacement not because it no longer performs to the designed standards, but because more modern equipment performs higher standards. For example an equipment may have an economic life of 20 years yet it may become obsolete after 10 years because of better technical development.

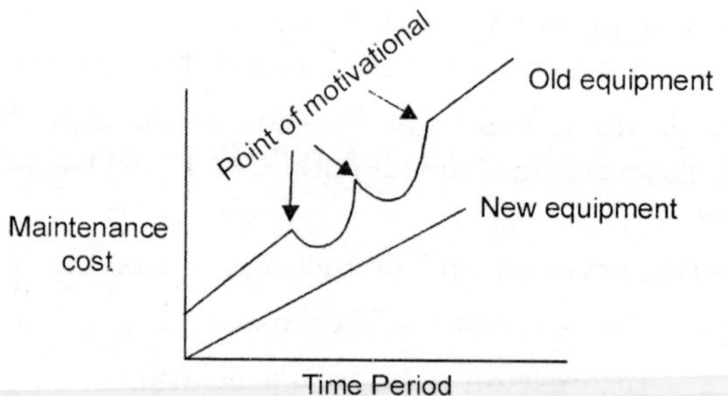

Fig. 7.1 Replacement of machine.

Capital equipment that deteriorates with time: It is concerned with equipment and machinery that deteriorates with time. Many people feel that an equipment must not be replaced until it is physically worn out. But it is not correct, preferably equipment must constantly be renewed and updated otherwise there is increase in the risk of failure or it becomes absolete.

7.2 REASONS FOR REPLACEMENT

(1) Deterioration

(2) Obsolescence

(3) Inadequacy

Deterioration is the decline in the performance of the equipment as compared to the new equipment identical to the present one.

Deterioration may occur due to wear and tear due to the

(*a*) Increases the maintenance cost.

(*b*) Reduces the product quality.

(*c*) Decreases the rate of production.

(*d*) Increases labour cost.

(*e*) Reduces efficiency of equipment.

7.3 REPLACEMENT MODELS

I. Replacement of items whose maintenance cost increases with time and the value of money remains same during the period.

II. Replacement of items whose maintenance cost increases with time and the value of money changes with time.

III. *Group Replacement Policy*

Considering the Case I

Value of money remains same during the period

Notations : 'C' is the capital or purchase cost of the machinery

'S' is the scrap value or resale value of the machinery.

Total cost incurred on the item during period 'Y' = capital cost of the machine + total maintenance cost during period 'Y' – the scrap value.

$$T_C = C + M(Y) - S.$$

Average cost/unit of time incurred during the period 'Y' on the item

$$G(Y) = \frac{C + M(Y) - S}{Y}$$

where Y is the no. of years, and M(Y) is the cumulative maintenance cost in that year.

PROBLEMS

1. *The cost of the machine is Rs. 6100 and its scrap value is Rs. 100. The maintenance cost found from experience are as follows :*

Year	*1*	*2*	*3*	*4*	*5*	*6*	*7*	*8*
Maintenance cost in years	*100*	*250*	*400*	*600*	*900*	*1200*	*1600*	*2000*

Solution. (*a*) Where should the machine be replaced?

$$C = 6100, S = 100.$$

No. of years	*Maintenance cost*	*Cumulative maintenance cost M(y)*	*Total cost C – S + M(Y)*	*Average cost/unit* $\frac{C - S + M(Y)}{Y}$
1	100	100	6100	6100
2	250	350	6350	3175
3	400	750	6750	2250
4	600	1350	7350	1837.5
5	900	2250	8250	1650
6	**1200**	**3450**	**9450**	**1575**
7	1600	5050	11050	1578.57
8	2000	7050	13050	1631.25

Conclusion: We should replace the machinery by the end of the 6th year or at the beginning of the 7th year as the maintenance cost is more than the average total cost of the machine in the 7th year.

Running cost = Operating cost = Maintenance cost.

2. *A fleet owner finds from his past experience records that cost of the machine is Rs. 6,000 and the running costs are given below, at what stage the replacement is due?*

Year	*1*	*2*	*3*	*4*	*5*	*6*	*7*	*8*
Maintenance Cost 'Y'm	*1000*	*1200*	*1400*	*1800*	***2300***	*2800*	*3400*	*4000*
Scrap or Resale Value 'S'	*3000*	*1500*	*750*	*375*	*200*	*200*	*200*	*200*

C = 6000.

Solution.

Year (Y)	*Cumulative maintenance cost M(Y)*	*Cost-scrap value = C – S*	*Total cost (C – S) + M(Y)*	*Average cost/unit G(Y) = C – S + M(Y)/Y*
1	1000	3000	4000	4000
2	2200	4500	6700	3350
3	3600	5250	8850	2950
4	5400	5625	11025	2756.25
5	7700	5800	13500	**2700.00**
6	10500	5800	16300	2716.66
7	13900	5800	19700	2814.28
8	17900	5800	23700	2962.5

Conclusion : As the table indicates the Average cost/unit is greater than the maintenance cost till the end of the 5th year but increases suddenly in the 6th year, so till the 5th year ending we will use the same fleet but will have to replace it by a new one at the beginning of the 6th year.

Case II

Replacement of items whose maintenance cost increases with time and the value of the money also changes with time.

The maintenance cost varies with time and we want to find the optimum value of time at which the item should be replaced. The value of money decreases with a constant rate —which is known as depreciation ratio or discount factor.

Given by: $$v = \frac{1}{(1+i)^{n-1}} \text{ for the value of Re. 1.}$$

where i is the interest or discount factor,

n is the year and

1 the value of Re. on which discount is considered in that year.

In this we made an assumption that the maintenance cost spent on the equipment or an item is at the beginning of each year.

1. *A company buys a machine for Rs. 6000 and gives us a 20% declining method of depreciation. Maintenance costs are expected to be Rs. 300 in each of the first 2 years and then to go up annually as follows:*

Rs. 700, Rs. 1000, Rs. 1500, Rs. 2000, Rs. 2500.

When should the machine be replaced?

Solution:

Sample calculation : For the first year.

$$\text{Discount factor first year} = v^{n-1} = v^{1-1} = 1.$$

$$\text{Discount factor for 2nd year} = v^{2-1} = v^1.$$

$$v = \frac{1}{(1+0.2)^1} = 0.833 \text{ for Re. 1.}$$

Present value of the operating cost = Discount × Operating cost

for Rs. 300 for first year. = 1 × 300 = 300.

0.833 × 300 = 249.9 (second year).

Cost of the machine 'C' = Rs. 6000

1	2	3	4	5	6	7	8
Year 'Y'	*Main-tenance cost*	*Discount Factor* v^{n-1}	*Present value of the main-tenance cost (P.V.M.C.)*	*Cumu-lative of P.V.M.C. M(Y)*	*6 = 5 + C 6 = M(Y) + cost of machine*	*Cumu-lative discount factor*	*Weighted average cost = 6/7*
1	300	1.0	300	300	6300	1.00	6300
2	300	0.833	249.9	549.9	6549.9	1.833	3573.34
3	700	0.6944	486.08	1035.95	7035.95	2.5274	2,783.86
4	1000	0.5784	578.7	1614.68	7614.68	3.1061	2451.52
5	1500	0.482	723	2337.68	8337.68	3.5881	2323.25
6	2000	0.4018	803.6	3141.28	9141.28	3.9899	**2291.10**
7	2500	0.3348	837	3978.28	9978.28	4.3247	2307.2
8	–	–	–	–	–	–	–

Conclusion: It is clear from the table that the maintenance cost is less than the weighted average cost till the end of the 6th year but gets increased in the beginning of the 7th year, so it is advisable to replace the machine with a new one at the beginning of the 7th year in order to overcome it.

2. *A machine costs Rs. 10,000, operating costs are Rs. 500/year for the first 5 years, in the 6th year and the subsequent years operating cost increases by Rs. 100 each year. Assuming money is worth 10% year, find the optimum length of time to hold the machine before replacement.*

C = Cost of the machine = 10,000.

Solution.

1 Year 'Y'	2 main-tenance cost	3 Discount factor v^{n-1}	4 Present value of the Main-tenance cost	5 Cumu-lative Present value of machine M(Y)	6 6 = 5 + C	7 Cumu-lative discount factor	8 8 = 6/7
1	500	1	500	500	10,500	1	10,500
2	500	0.909	454.5	954.5	10,954.5	1.909	5738.33
3	500	0.826	413	1367.5	11,367.5	2.735	4156.30
4	500	0.7513	375.65	1743.15	11,743.15	3.486	3368.66
5	500	0.6830	341.5	2084.65	12,084.65	4.1693	2898.43
6	600	0.6209	372.54	2457.19	12,457.19	4.790	2600.66
7	700	0.5644	395.08	2852.27	12,852.27	5.3546	2400.02
8	800	0.5131	410.48	3262.75	13,262.75	5.8677	2260.29
9	900	0.4665	419.85	3682.6	13,682.6	6.334	2160.18
10	1000	0.4240	424	4106.6	14,106.6	6.758	2087.39
11	1100	0.3855	424.05	4530.65	14,530.65	7.1435	2034.10
12	1200	0.3504	420.48	4951.13	14,951.13	7.4939	1995.10
13	1300	0.3186	414.18	5365.31	15,365.31	7.8125	1966.75
14	1400	0.2896	405.44	5770.75	15,770.75	8.1021	1946.50
15	1500	0.2633	394.95	6165.7	16,165.7	8.3654	1932.44
16	1600	0.2393	382.88	6548.58	16,548.58	8.6047	1923.20
17	1700	0.2176	369.92	6917.5	16,917.5	8.8223	1917.5
18	1800	0.1978	356.12	7273.62	17,273.62	9.0201	1917.01
19	**1900**	**0.1798**	**341.731**	**7615.351**	**17,615.35**	**9.1999**	**1914.73**
20	2000	0.1635	327.015	7942.366	17,942.36	9.3634	1916.22
21	2100	0.1486	312.151	8254.51	18,254.51	9.512	1919.10
22	2200	0.1351	297.28	8551.797	18,551.797	9.6471	1923.04
23	2300	0.1228	282.54	8834.33	18,834.33	9.7699	1927.79
24	2400	0.1116	268.02	9102.35	19,102.35	9.8815	1933.14
25	2500	0.10152	253.81	9356.16	19,356.16	9.9830	1938.91
26	2600	0.0922	239.96	9596.12	19,596.12	10.075	1945.02
27	2700	0.0839	226.54	9822.66	19,822.66	10.159	1951.24
28	2800	0.0762	213.577	10,036.244	20,036.24	10.235	1957.61
29	2900	0.0693	201.095	10,237.33	20,237.33	10.340	1963.90

$$v = \frac{1}{(1+i)^n} = \frac{1}{(1+.1)^2} \text{ (for the 3rd year).}$$

Result: Till the 19th year, average cost is greater than the maintenance cost but increases on 20th year *i.e.*, maintenance cost is greater than the average weighted cost, so we will replace the machinery at the beginning of the 20th year.

3. *A manufacturer is offered 2 machine's A and B. A is priced at Rs. 5000 and running costs are estimated at Rs. 800 at each of the first 5 years increasing by Rs. 200/year in the 6th and subsequent years, machine B which has the same capacity as A costs Rs. 2500 but will have running cost of Rs. 1200/year for 1st 6 years increasing by Rs. 200/year there after. If money is worth of 10%/year which machine should be purchased? Assume that the scrap value is zero at the end).*

Solution: Machine A : C = 5000; i = 0.1

1 Year 'Y'	2 Operating cost	3 Discount v^{n-1} factor	4 Present value of the operating cost	5 Cumulative P.V.M.C. M(Y)	6 6 = C + M(Y)	7 Cumulative discount factor	8 8 = 6/7
1	800	1.0	800	800	5800	1.0	5800
2	800	0.909	727.2	1527.2	6527.2	1.909	3419.07
3	800	0.826	660.8	2188	7188	2.735	2628.15
4	800	0.751	600.8	2788.8	7788.8	3.486	2234.30
5	800	0.683	546.1	3335.2	8335.2	4.169	1999.30
6	1000	0.620	620	3955.2	8955.2	4.789	1869.95
7	1200	0.564	676.8	4632	9632	5.353	1799.36
8	1400	0.5131	718.2	5350	10350.2	5.866	1764.44
9	**1600**	**0.466**	**745.2**	**6095.8**	**11095.8**	**6.322**	**1752.3**
10	1800	0.424	763.2	6859	11859	6.756	1755.32

Machine B C = 2500; i = 0.1

Year	Operating cost	Discount factor	P.V.M.C.	Cumulative P.V.M.C M(Y)	6 = C + M(Y)	Cumulative discount factor	8 = 6/7
1	1200	1.0	1200	1200	3700	1.0	3700
2	1200	0.909	1090.8	2290.8	4790.8	1.909	2509.58
3	1200	0.826	991.2	3282	5782	2.735	2114.07
4	1200	0.757	908.4	4190.4	6690.4	3.486	1919.2
5	1200	0.683	819.6	5009.8	7509.8	4.109	1827.6
6	1200	0.620	744	5753.8	8253.8	4.789	1722.8
7	1400	0.564	789.4	6543.2	9043.2	5.353	1689.3
8	**1600**	**0.5131**	**820.96**	**7364.16**	**9864.16**	**5.866**	**1680.89**
9	1800	0.466	838.8	8202.96	10702.46	6.332	1689.44
10	2000	0.424	848	9050.96	11550.96	6.756	1708

Conclusion: The machine A should be replaced by the end of 9th year and machine B should be replaced by the end of 8th year and the average cost of the machine B is less than A, so purchase machine B.

4. *A company has a 2 year old machine purchased at a cost of Rs. 10,000. Now a new machine is available at Rs. 12,000. The operation and maintenance cost of the existing machine is Rs. 600 for the 1st year, Rs. 800 for the 2nd year and then it increases by Rs. 300/yr. The operation and maintenance cost of the new machine is Rs. 400 for the first year, then it increases by Rs. 150/year. The capacities of the machine's are same and they have no resale value. The company expects a minimum of 12% return on its capital investments. Determine whether the existing machine should be replaced by the new machine.*

Solution.

Machine A (old machine) C = Rs. 10,000; $i = 0.12$

1 Year 'Y'	2 Operating cost	3 Discount factor	4 P.V. of operating cost	5 Cumulative P.V.M.C.	6 6 = C + M(Y)	7 Cumulative discount factor	8 8 = 6/7
1	600	1	600	600	10,600	1	10,600
2	800	0.892	713.6	1313.6	11,313.6	1.892	5979.7
3	1100	0.7971	876.81	2190.41	12,190.41	2.6891	4533.26
4	1400	0.7117	996.38	3186.79	13,186.79	3.4008	3877.55
5	1700	0.6355	1080.35	4267.14	14,267.14	4.0363	3534.7
6	2000	0.5674	1134.8	5401.94	15,401.94	4.6037	3345.55
7	2300	0.5066	1165.18	6567.12	16,567.12	5.1103	3241.9
8	2600	0.4523	1175.98	7743.1	17,743.1	5.5626	3189.71
9	**2900**	0.4038	1171.02	8914.12	18,914.12	5.9664	**3170.10**
10	3200	0.3606	1153.92	10068.04	20,068.04	6.327	3171.80
11	3500	0.3219	1126.65	11194.64	21,194.64	6.6489	3187.69
12	3800	0.2874	1092.12	12286.81	22,286.81	6.9363	3213.06

By the end of 9th year, we will have to replace old machine

New machine B C = Rs. 12,000; $i = 0.12$

1 Year	2 Operating cost	3 Discount factor	4 Present value of operating cost	5 Cumulative P.V.M.C.	6 6 = C + M(Y)	7 Cumulative discount factor	8 8 = 6/7
1	400	1	400	400	12400	1	12400
2	550	0.892	490.6	890.6	12890.6	1.892	6831.2
3	700	0.7971	557.97	1448.57	13448.57	2.6891	5001.14
4	850	0.7117	604.94	2053.51	14053.51	3.4008	4132.41

(Contd...)

5	1000	0.6355	635.5	2689.01	14689.01	4.0363	3639.49
6	1150	0.5674	652.51	3341.52	15341.52	4.6037	3332.93
7	1300	0.5066	658.58	4000.1	16000.1	5.1103	3131.13
8	1450	0.4523	655.83	4655.93	16655.93	5.5626	2994.59
9	1600	0.4038	646.08	5302.01	17302.01	5.9664	2900.10
10	1750	0.3606	631.05	5933.06	17933.06	6.327	2834.37
11	1900	0.3219	611.61	6544.67	18544.67	6.6489	2789.13
12	2050	0.2874	589.17	7133.84	19133.84	6.9363	2758.50
13	2200	0.2566	564.52	7698.36	19698.36	7.1929	2738.58
14	2350	0.2291	538.38	8236.74	20236.74	7.422	2726.58
15	2500	0.2046	511.549	8748.28	20748.2	7.626	2720.71
16	**2650**	0.1826	484.145	9232.42	21232.42	7.8086	**2719.10**
17	2800	0.1631	456.68	9689.1	21689.1	7.9717	2720.76

Conclusion: The new machine should be replaced by the end of the 16th year. and the old machine should be replaced by the end of the 9th year. as their respective maintenance cost increases w.r.t. the weighted average. So we will go for the new machine only as it has got more life than the old machine and the weighted average cost is less.

5. *A truck is priced at Rs. 60,000 and running costs are estimated at Rs. 6000 for each of the 1st 4 years, increasing by Rs. 2000/year in the 5th and subsequent years. If the money is worth 10%/year, when should the truck be replaced? Assume the truck has no scrap value.*

Cost of the truck 'C' = 60,000 Rs. Discount factor = 0.1

1 Year	*2* Operating cost	*3* Discount factor	*4* Present value of operating cost	*5* Cumulative P.V.M.C. M(Y)	*6* 6 = C + M(Y)	*7* Cumulative discount factor	*8* 8 = 6/7
1	6000	1	6000	6000	66000	1.0	66000
2	6000	0.909	5454	11454	71,454	1.909	37430.06
3	6000	0.826	4956	16410	76,410	2.735	27937.84
4	6000	0.751	4506	20916	80916	3.486	23211.70
5	8000	0.683	5464	26380	86380	4.169	20725.56
6	10000	0.620	6200	32580	92580	4.789	19331.8
7	12000	0.564	6768	39348	99348	5.353	18559.31
8	14000	0.5131	7183.4	46531.4	106531.4	5.866	18160.82
9	16000	0.466	74.56	53987.4	1,13,987.4	6.332	18000.80
10	**18000**	**0.424**	**7632**	**61619.4**	**1,21,619.4**	**6.756**	**18000.68**
11	20000	0.3855	7710	69329.4	1,29,329.4	7.1415	18109.55
12	22000	0.3504	7708.8	77038.2	1,37,038.2	7.4919	18470.77
13	24000	0.3186	7646.4	84684.6	1,44,684.6	7.8105	18524.37
14	26000	0.2896	7531.27	92215.87	1,52,215.87	8.1001	18791.85

Result: By the end of the 9th year we will have to replace the truck by a new one as the maintenance cost is greater than the average weighted cost after 10th year onwards.

6. *A person is considering to purchase a machine for his factory. The related data about the alternative machine's are as follows :*

	Machine A	*Machine B*	*Machine C*
Present Investment	*10,000*	*12,000*	*15,000*
Total Annual Maintenance Cost or Running Cost	*2,000*	*15,00*	*1,200*
Life in years	*10*	*10*	*10*
Salvage value in Rs.	*500*	*1000*	*1200*

As an advisor of the company. You have been asked to select the best machine. Consider 12% normal rate of return/year.

Solution:

Step I : To calculate the discount factor for each year.

Year	*Discount factor*
1	0.8928
2	0.7971
3	0.7117
4	0.6355
5	0.5674
6	0.5066
7	0.4523
8	0.4038
9	0.3606
10	0.3219
ΣV	5.6497

(Discount factor for 10 years)

Total maintenance for machine A for the 10 years with discount factor

$$= 2000 \times 5.6497 = 11299.4$$

Similarly total maintenance cost for machine B

$$= 1500 \times 5.6497 = 8474.55$$

Similarly total maintenance cost for machine C

$$= 1200 \times 5.6497 = 6779.64$$

The salvage value of the machine will be considered in the 10th year, so we will consider only the discount factor of 10th year, (*i.e.*,) 0.3219.

Salvage value of machine A = 500 × 0.3219 = 160.95

Salvage value of machine B = 1000 × 0.3219 = 321.9

Salvage value of machine C = 1200 × 0.3219 = 386.28

Total Present value of machine A = Investment + Maintenance cost – Salvage value

= 10,000 + 11299.4 – 160.95

= 21,138.45

Total Present value of machine B = 12,000 + 8475 – 321.9

= 20,153.1

Total Present value of machine C = 15,000 + 6779.64 – 386.28

= 21,393.36

Result: So we will consider the machine B, as the total cost of machine B is less than the costs of A and C.

7.4 GROUP-REPLACEMENT POLICY

Replacement of Items that Fail Completely

We always come across situations where the possibility of failure of any item in any system increases with time. The nature of the system may be such that if the item fails then it may result in complete breakdown of the system. The breakdown implies loss of production, idle inventory, immediate replacement of the item may not be available and many other losses.

Individual Replacement Policy: Whenever an item fails it should be replaced immediately.

Group Replacement Policy: It consists of 2 steps : In the first step it consists of individual replacement at the time of failure of any unit in the system and in the second step there is a group replacement of existing live units at some suitable time. *In the group replacement* we decide all the items of the system should be replaced after a certain period irrespective of the fact that items have failed or not with a provision that if any item fails before this time, it can be replaced individually. It requires to :

(1) The rate of individual replacement during the period.

(2) The total cost incurred for an individual and group replacement during the chosen period.

(3) The period for which the total cost incurred is minimum will be the optimum period for replacement.

Problem 1. *The following mortality rates have been observed for a certain type of light bulbs:*

Week	*1*	*2*	*3*	*4*	*5*
% of bulb failing by the end of the week	*10*	*25*	*50*	*80*	*100*
Convert % into probability	*0.1*	*0.25*	*0.50*	*0.80*	*1.0*

Solution. There are 1000 bulbs in use and its cost is Re. 1 to replace an individual bulb which has burnt out. If all bulbs were to be replaced simultaneously, it would cost 25 paisa/bulb. It is proposed to replace all bulbs at fixed intervals whether or not they have burnt out and counting replacing burnt-out bulbs as they fail. At what intervals should all the bulbs be replaced?

Note : If the probability is given in increasing order, *i.e.*, it is in cumulative form, Subtract one from other to get the individual probability failures/week.

First of all we have to calculate the probability of failure of bulbs/week individually *i.e.*,

Week	*1*	2	3	*4*	5
Probability of Failure of Bulbs/Week	0.1	0.15	0.25	0.30	0.20

Assumption : The bulbs that fail during the week are just replaced before the end of the week.

Let n_i denote the no. of replacements made at the end of the ith week.

Initially 1000 bulbs are available

i.e., $n_0 = 1000.$

n_0 is the no. of bulbs available initially.

n_1 is the no. of bulbs replaced at the end of the first week.

n_1 = No. of bulbs during the 1st week × The probability of failure of the bulbs that fail in the 1st week

$= 1000 \times 0.1 = 100$

$\mathbf{n_1 = 100}$

$n_2 = n_0P_2 + n_1P_1$

$= 1000 \times 0.15 + 100 \times 0.1 = 160$

$\mathbf{n_2 = 160}$

$n_3 = n_0P_3 + n_1P_2 + n_2P_1$

$= 1000 \times 0.25 + 100 \times 0.15 + 160 \times 0.1 = 281$

$\mathbf{n_3 = 281}$

$n_4 = n_0P_4 + n_1P_3 + n_2P_2 + n_3P_1$

$= 1000 \times 0.3 + 100 \times 0.25 + 160 \times 0.15 + 281 \times 0.1$

$= 377$

$\mathbf{n_4 = 377}$

$n_5 = n_0P_5 + n_1P_4 + n_2P_3 + n_3P_2 + n_4P_1$

$= 1000 \times 0.2 + 100 \times 0.3 + 160 \times 0.25 + 281 \times 0.15 + 377 \times 0.1$

$\mathbf{n_5 = 349.85}$

(No. of probability considered zero)

$n_6 = n_0P_6 + n_1P_5 + n_2P_4 + n_3P_3 + n_4P_2 + n_5P_1$

$$= 0 + 100 \times 0.2 + 160 \times 0.3 + 281 \times 0.25 + 377 \times 0.15 + 349.85 \times 0.1$$

$$n_6 = \mathbf{229.78}$$

Conclusion : It will be noticed that the no. of bulbs failing in each week increases until the 4th week and then suddenly decreases.

To calculate expected life of bulbs :

i.e.,
$$\sum_{i=1}^{n} P_i \times i = P_1 \times X_1 + P_2 \times X_2 + P_3 \times X_3 + P_4 \times X_4 + P_5 \times X_5$$

$$= 0.1 \times 1 + 0.15 \times 2 + 0.25 \times 3 + 0.3 \times 4 + 0.2 \times 5$$

$$= 3.35$$

∴ Expected life of each bulb = **3.35** week.

To calculate average no. of failures/week.

$$\text{Average no. of failures/week} = \frac{\text{Total no. of bulbs}}{\text{Expected life}}$$

$$= \frac{1000}{3.35} = \mathbf{299}$$

Now the cost of individual replacement is Re. 1.

∴ Total cost of replacement week = 299 × 1

= **Rs. 299**

To calculate group replacement:

No. of bulbs replace at the end of each week	*Total cost of group replaced*	*Average cost/week*
1.	1000 × 0.25 + 100 × 1 = **350** (Group replacement of 1000 items or bulbs is 0.25 Rs. and Individual Replacement is Re. 1)	$\frac{350}{1}$ = **Rs. 350**
2.	1000 × 0.25 + 100 × 1 + 160 × 1 = **510**	$\frac{510}{2}$ = **Rs. 255**
3.	1000 × 0.25 + 100 × 1 + 160 × 1 + 28 × 1 = **791**	$\frac{791}{3}$ = **Rs. 264**
4.	No need	–
5.	No need	–

Conclusion: As the average cost of group replacement is decreasing and it suddenly increases in the 3rd week so we will stop doing it further and will compare it *i.e.*, 2nd week average cost with the individual cost of replacement, whichever is minimum that will be our final selection.

In this case the group replacement cost is less than the individual replacement, so we will go in for group replacement of bulbs.

By the end of each 2 weeks we will replace the complete group by 1000 bulbs.

Problem 2. *An Automatic safety electric switched to press has the following probability of failures.*

No. of years	*1*	*2*	*3*	*4*	*5*	*6*	*7*
Probability of failures	*0.05*	*0.10*	*0.15*	*0.20*	*0.35*	*0.10*	*0.05*

(Here the probability of failure is given individually, not cumulatively.)

The average cost to replace a single switch is Rs. 15 but this comes to Rs. 3 when replacement is carried out on planned group basis during the annual sales down. Find the minimum cost planned replaced policy. (No. of total switches not given).

Assuming that there are 100 automatic switches in use

$$n_0 = 100$$

$$n_1 = n_0 \times P_1 = 100 \times 0.05 = \mathbf{5}$$

$$n_2 = n_0P_2 + n_1P_1 = 100 \times 0.1 + 5 \times 0.05 = \mathbf{10.25}$$

$$n_3 = n_0P_3 + n_1P_2 + n_2P_1$$
$$= 100 \times 0.15 + 5 \times 0.1 + 10.25 \times 0.05$$

$$\boldsymbol{n_3} = \mathbf{16.0125}$$

$$n_4 = n_0P_4 + n_1P_3 + n_2P_2 + n_3P_1$$
$$= 100 \times 0.2 + 5 \times 0.15 + 10.25 \times 0.1 + 16.01 \times 0.05$$

$$\boldsymbol{n_4} = \mathbf{22.5755}$$

$$n_5 = n_0P_5 + n_1P_4 + n_2P_3 + n_3P_2 + n_4P_1$$
$$= 100 \times 0.35 + 5 \times 0.2 + 10.25 \times 0.15 + 16 \times 0.01$$
$$+ 22.57 \times 0.05$$

$$\boldsymbol{n_5} = \mathbf{40.267}$$

$$n_6 = n_0P_6 + n_1P_5 + n_2P_4 + n_3P_3 + n_4P_2 + n_5P_1$$
$$= 100 \times .10 + 5 \times .35 + 10.25 \times 0.2 + 16.01 \times .15$$
$$+ 22.57 \times 0.1 + 40.267 \times 0.05$$

$$\boldsymbol{n_6} = \mathbf{20.47}$$

Conclusion: It will be noticed that the no. of switches failing in each year increases until the 5th year and then suddenly increases.

Step II : Expected life of each switch : $\sum_{i=1}^{n} P_i \times i$

$$= P_1 \times X_1 + P_2 \times X_2 + P_3 \times X_3 + P_4 \times X_4 + P_5 \times X_5$$
$$+ P_6 \times X_6 + P_7 \times X_7$$
$$= 1 \times 0.05 + 2 \times 0.1 + 3 \times 0.15 + 4 \times 0.2 + 5 \times 0.35$$
$$+ 6 \times 0.1 + 7 \times 0.05$$

= **4.2** Expected life of each switch.

Average no. of failures/year $= \frac{100}{4.2} = 23.80$

= **24** say.

Cost of Individual Replacement = Rs. 15

Total cost of Replacement/year = 24 × 15

= **Rs. 360**

To calculate the Group Replacement :

No. of switches replaced at the end of each year	*Total cost of group replaced*	*Average cost/year.*
1	100 × 3 + 5 × 15 = 375	375
2	375 × 1 + 10.25 × 15 – 528.75	264.37
3	528.75 + 16.0 × 15 = 768.9	**256.3**
4	768.9 + 22.5755 × 15 = 1107.53	276.88
5	1107.53 + 40.267 × 15 = 1711.535	342.307

It is clear from the average cost/year that costs are decreasing till the end of the 3rd year but suddenly increases in the 4th year and the cost of group replacement by the end of 3rd year is less than the individual replacement. So we will make a policy of replacing the complete 100 switches by the end of the 3rd year in order to maximize the profits.

Problem 3. *A Computer contains 10,000 resistors. When resistor fails, it is replaced. The cost of replacing a resistor individually is Re. 1 only. If all the resistors are replaced at the same time the cost/resistor will be reduced to 0.35 Re. The probability of surviving at the end of the month t are given below. What is the optimum replacement plan?*

Month t	*0*	*1*	*2*	*3*	*4*	*5*
Probability of Surviving	1.0	0.97	0.90	0.70	0.30	0.15

Note: In this problem the probability of survival is given so we have to find the probability of failure first, then the probability of failure for individual months.

Probability of Failure = [1 – Probability of survival for each month]. We get the cumulative probability of failure. Then we have to subtract the cumulative probability of failure of previous month to the succeeding month to get the probability of failure in that month only.

Month	*0*	*1*	*2*	*3*	*4*	*5*	*6*
Probability of failure	0	0.03	0.1	0.3	0.7	0.85	1
Probability of failure/ Month	0	0.03	0.07	0.2	0.4	0.15	0.15
	P_0	P_1	P_2	P_3	P_4	P_5	P_6

No. of resistors available n_0 = 10,000.

No. of resistors replaced at the end of 1st month

$$n_1 = n_0 \times P_1 = 10{,}000 \times 0.03 = 300$$
$$\boldsymbol{n_1 = 300}$$
$$n_2 = n_0\ P_2 + n_1\ P_1$$
$$= 10{,}000 \times 0.07 + 300 \times 0.03$$
$$\boldsymbol{n_2 = 709}$$
$$n_3 = n_0P_3 + n_1P_2 + n_2P_1$$
$$= 10{,}000 \times 0.2 + 300 \times 0.07 + 709 \times 0.03$$
$$\boldsymbol{n_3 = 2042.27}$$
$$n_4 = n_0P_4 + n_1P_3 + n_2P_2 + n_3P_1$$
$$= 10{,}000 \times 0.4 + 300 \times 0.2 + 709 \times 0.07 + 204 = 27 \times 0.03$$
$$\boldsymbol{n_4 = 4170.89}$$
$$n_5 = n_0P_5 + n_1P_4 + n_2P_3 + n_3P_2 + n_4P_1$$
$$= 10{,}000 \times 0.15 + 300 \times 0.4 + 709 \times 0.2 + 204.27 \times 0.07 + 4170.89 \times 0.03$$
$$\boldsymbol{n_5 = 2029.88}$$

Result : It is observed that the no. of resistors failing in each month increases until the 4th month but suddenly decreases in the 5th month.

$$\text{Expected life of each resistor} = \sum_{i=1}^{6} \text{P}i \times i$$
$$= P_1 \times 1 + P_2 \times 2 + P_3 \times 3 + P_4 \times 4 + P_5 \times 5 + P_6 \times 6$$
$$= 0.03 \times 1 + 0.07 \times 2 + 0.2 \times 3 + 0.4 \times 4 + 0.15 \times 5 + 0.15 \times 6$$
$$= \mathbf{4.02}$$

$$\text{Average no. of failures/month} = \frac{10{,}000}{4.02}$$

Average no. of failures/month = **2,487.56**

Cost of Individual Replacement = Re. 1

Total cost of Individual Replacement = 2,487.56 × 1

= **2,487.56/month**

To Calculate the Group Replacement :

No. of resistors replaced at the end of each mounth	*Total cost of group replaced*	*Average cost/month*
1	10,000 × 0.35 + 300 × 1 = 3,800	3800
2	10,000 × 0.35 + 300 × 1 + 709 × 1 = 4,509	2254.5
3	10,000 × .35 + 300 × 1 + 709 × 1 + 2042.27 × 1 = 6,551.27	2183.75
4	6551.27 + 4170.89 × 1 = 10,722.16	2680.54

Conclusion: It is clear from the table that the average cost is decreasing till the 3rd month but suddenly increases in 4th month. So we will consider minimum cost *i.e.*, 2183.75 only as it is also less than the individual replacement which is 2487.56. So Group Replacement is preferred.

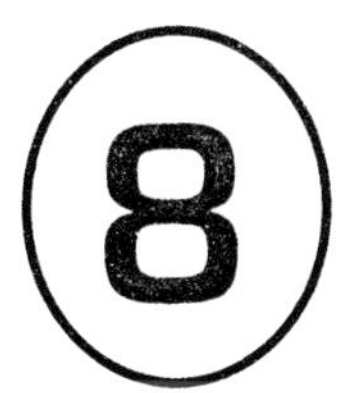

ESTIMATION

8.1 DEFINITION

The process of compiling the statement of the quantity of materials involved, the amount of time involved in the production and the procedures to be followed in putting an order through together with the cost of the articles made or to be made where experience supplies no complete figures.

8.2 OBJECTIVES OF ESTIMATION

1. To help the owner of a factory to decide about the manufacturing and selling price.
2. To help in filling up the tenders so as to estimate the profit.
3. To decide the amount of overheads.
4. To determine the usage rate of the workers.
5. To take a decision whether to manufacture or to buy a particular product from the markets.

8.3 THE FUNCTIONS OF ESTIMATING DEPARTMENT

1. To determine the material cost taking into consideration different allowances given for different manufacturing operations.
2. To find the cost of the parts to be purchased from outside the venders.
3. To determine cost of equipment, tools, machinery, jigs and fixtures etc. (direct expenses).
4. To determine the labour cost considering the labour time with the help of the wage rate, labour cost in values the direct and indirect labour cost based on "WORK-STUDY".
5. To calculate the overhead charges associated with the product.
   The different overhead charges are administrative, office, selling and distribution.
6. To calculate the selling price of the product including the products.
7. To find the most economical procedure of making the product.
8. To conduct time study.
9. To keep control over selling expenses with the help of sales manager.

8.4 QUALITIES OF A GOOD ESTIMATIOR

1. An estimator should be able to read and understand the drawing and also the blue prints which are the part of the many process. Allowances of different material used?
2. He must have a good knowledge about the different machine, the various process output times etc., which are connected to the manufacturing of a product.
3. He should have a sound knowledge about proper tools, jigs and fixtures etc.
4. He should be a qualified technical person.
5. He should possess the knowledge about the wage rate of worker.
6. He should have the knowledge of work study.
7. He should possess a good knowledge of material.
8. He should also know the cost accounting procedures.

8.5 CONSTITUENTS OF ESTIMATION

1. Design time and cost
2. Drafting time and cost
3. Motion and time study
4. Planning and production central time
5. Procurement, design or manufacturing of special tools, jigs and fixture etc.
6. Experimental work
7. Material cost
8. Labour cost
8. Overheads.

1. Design cost

Considerable time is consumed in designing a particular product. In estimating, the usages and other expenditures required to be paid in designing a product are to be considered, for this reason, cost can be estimated on basis of similar products already designed in past or on basis of good judgement of designing. In this the remuneration paid to design office staff and other expenditure incurred during designing a product in a particular period are also added.

2. Drafting cost

After design work is over, the estimated time consumed by draftsman in drawings of individual components is called "drafting time". For calculating drafting cost, remuneration of draftsman is taken as basis.

3. Time and motion studies planning and production control cost

Sufficient time is consumed for such activities therefore estimated time and costs incurred on it are decided by past experience or judgement.

4. Cost of design and arrangement of special time

The estimator must take into account the cost of special items such as patterns, care boxes, flasks, dies, jigs and fixtures and tools etc., whenever they are used.

5. Cost of experimental work

The best and cheapest method of production is determined with the help of researches and experiments. They are performed on old and present methods and sometimes inventions are required to be done. The cost incurred on such estimates is given due considerations.

6. Materials cost

The material cost is found with the help of samples on drawings which show only finished dimensions, while estimation takes into account additional figs. to be provided including policy turning, stamping, moulding, wastage or spoilage in cutting and finishing etc, Therefore estimation must have practical knowledge of various allowances.

An estimation first prepares rough drawing with all allowances and calculates volume and cut is obtained by multiplying density. The material cost is obtained by knowing market rates of material., the scrap value is deducted from material cost.

7. Labour cost

To estimate labour cost, an estimation must have a large experience and through knowledge of all operations carried out during production and he should take advice of production department while deciding about exact time of each operations different allowances like personal, fatiguc, tool changing and funding and checking measurement allowance should also be taken into consideration.

8. Time allowance

The classification of time allowances are as under:

1. Set up time
2. Operation time (*a*) handling time (*b*) machine time
3. Tear down time
4. Miscellaneous allowances

1. **Set up time :** It is the time required for setting and fixing the job and different tools on the m/c and time required to study drawings, blue prints, to set m/c, to inspect job, setting of gauge etc.
2. **Opeation time :** It is the time taken by m/c's for actual o/p's on the job, also known as 'cutting or floor' time', which includes.

   (*a*) **Handling time :** It means time required in physical movements while performing m/c'ing o/p's.

   (*b*) **m/c'ing time :** The time taken by m/c's from start when tool touches job to the end when tool leaves the job.
3. **Tear down time :** It is the time counted from when last element of operation has been completed.

4. **Miscellaneous allowances :** an estimator should consider different allowances for time because it can't be expected that a worker can work for all the 8 hrs. without rest. Therefore, allowances as given which generally consume 10 to 15% of the total time taken. These allowances are :

(*a*) Personal allowances

(*b*) Fatigue allowances

(*c*) Tool changing and grinding allowances

(*d*) Measurement checking allowances

(*e*) Other allowances

(*a*) **Personal allowances :** is time consumed to other his personal needs such as going to lavatour, to take water, to smoke etc. Reduction in efficiency due to falls in this category. This takes 5% of total working time.

(*b*) **Fatigue allowance :** A worker cannot work for full time with same speed and sometimes he feels some tiredness, due to excessive work, poor lighting, machine noice etc., leading to fatigue and allowance for fatigue is nearly 5% of total time.

(*c*) **Tool changing and grinding allowance :** is time required to remove tool from its holder and to fix another tool and sharpen the tool. To sharpen the tool, worker has to remove the tool, rear up to the grinder to grind tool, come back to m/c and set the tool again on the m/c and the allowance is nearly 5-10% of total time.

(*d*) **Measurement and checking allowance :** is time taken for measuring and checking different dimensions of the product and this takes 2 to 3% of total time.

(*e*) **Other allowances :** is time consumed in as like periodic cleaning, oiling, getting stocks, delivering jobs, disposing of scraps and surplus stocks etc., and this takes 15-20% of the operation time.

9. Overheads

These include the expenditure which cannot be definitely charged to a particular product during production. These include expenses such as:

(*a*) **Indirect material cost :** In this, the cost of greases, coolants, oils, cotton waste, contingencies, light, water, etc. are considered.

(*b*) **Indirect labour cost :** In this, the pay of supervisors, draftsmen, designers, research workers, helpers, chowkidars and persons working for material handling etc., are included.

(*c*) **Administrative overheads :** These include salaries of high officials, salaries of persons working in general office, telephone, telex to insurance premium etc.

(*d*) **Selling overheads :** These include salaries of salesmen, commission to salesmen and advertising expenditure etc.

(*e*) **Repairs and maintenance expenses**

(*f*) **Insurance premium on building and plants.**

(*g*) **Depreciation of building, furniture and equipment.**

The sum of all these expenses is found out which will be total estimated cost of product.

8.6 COSTING

It is the determination of an actual cost of an article after adding different expenses incurred in various departments. It is a system which systematically records all the expenses and determines the cost of manufacturing product.

The basic difference between estimation and costing is while estimation estimates the expenses because the actual production of a product cost whereas costing is done after manufacturing the actual product.

The technique and process of ascertaining the lost is known as costing. Every bit of expenditure has to be considered while calculating the actual cost of production.

The elements of cost are as follows :

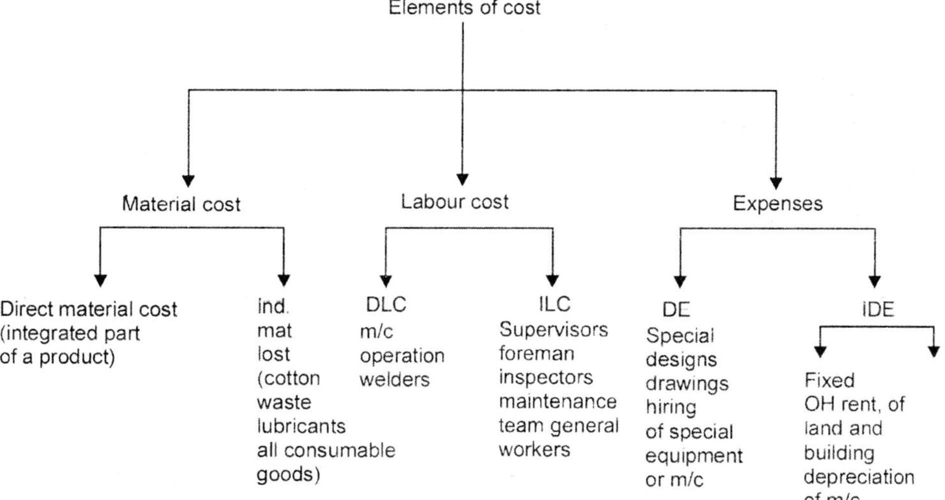

Prime cost = DMC + MLC + DE

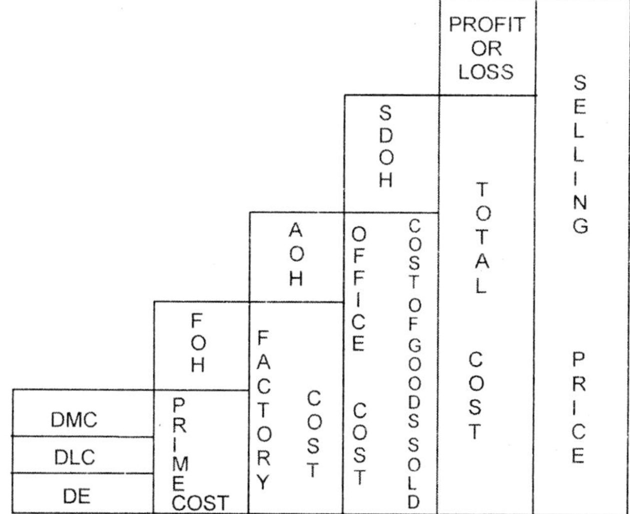

Factory cost (production cost) = P.C. + FOH (IMC + ILC + IE) or production OH.

Office cost (cost of goods sold) = FC + AOH

= (administration OH)

Total cost = Office cost + SDOH + R and D expenses

= (selling and distribution OH)

Selling price = Total cost ± profit.

MENSURATION

Plane figures	*Perimeter*	*Area*
1. **Rectangle** 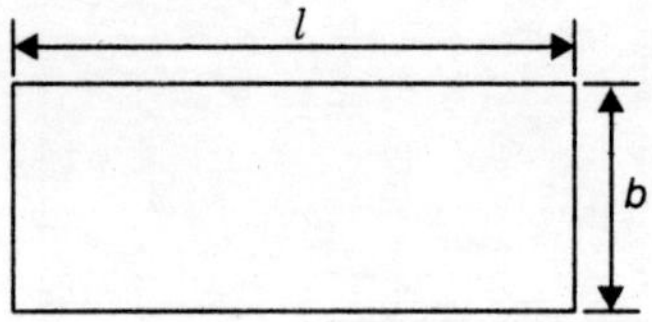 l = length of rectangle b = breadth of rectangle	$P = 2(l + b)$	$A = l \times b$
2. **Square** 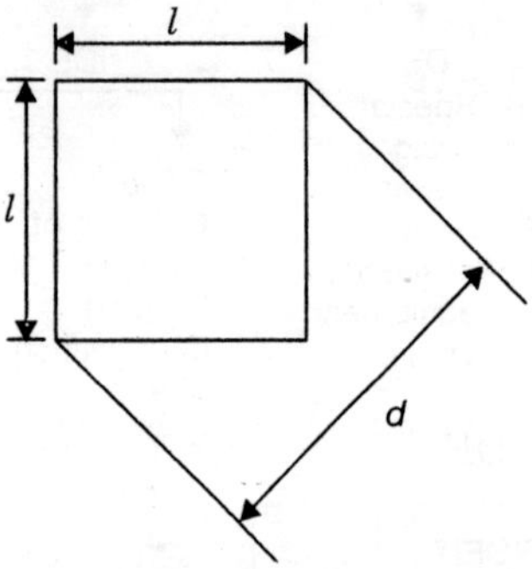 l = length of each side of square d = length of diagonal	$P = 4l$	$a = l^2$
3. **Parallelogram** 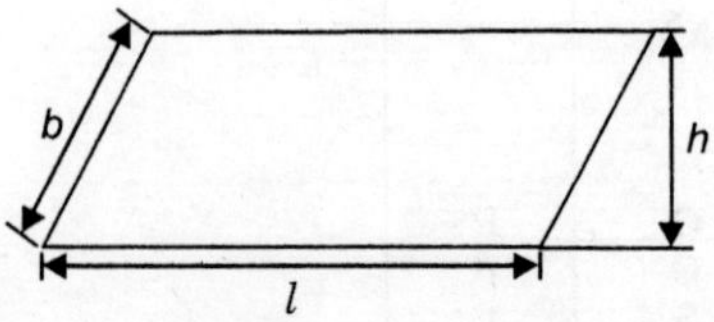 l = length of 1 side of parallelogram	$P = 2(l + b)$	$a = l \times h$

(Contd.....)

b = length of another side of parallelgram. h = height of parallelogram		
4. **Triangle** 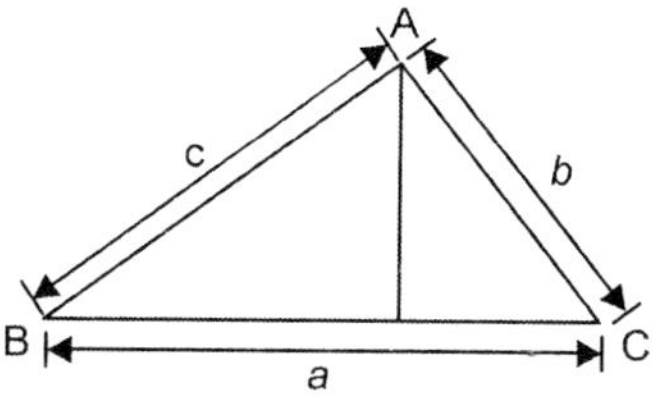 A, B, C, c, b, a abc = length of sides h – ht. of Λ from baoc BC	$P = a + b + c$ if $\left(s = \dfrac{a+b+c}{2}\right)$	$a = 1/2\, a \times h$ $= 1/2 \times \text{base} \perp°$ area = $\sqrt{s(s-a)(s-b)(s-c)}$
5. **Hexagon** 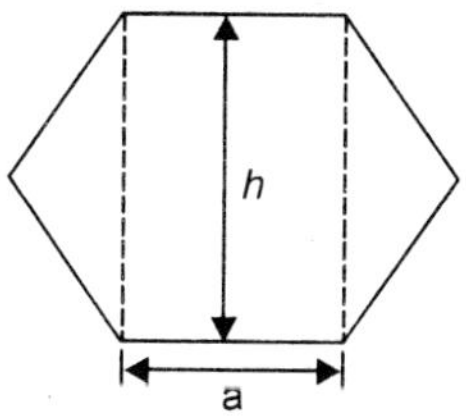 h, a a = length of each side h = height	$P = 6a$ height $h = \sqrt{3a}$	area = $\dfrac{3\sqrt{3}}{2}a^2$
6. **Any regular polyon** n sides of length a units.	$P = na$	area = ½ × perimeter × inner radius.
7. **Trapezium** $a\ b\ c\ d$ = length of 4 sides. 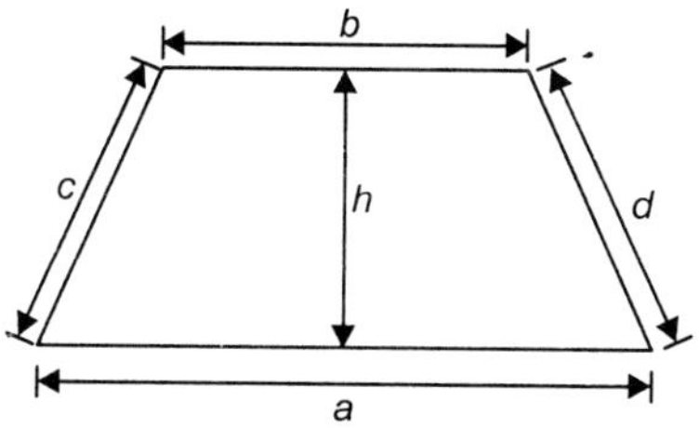 b, c, h, d, a h = parallel dist between parallel sides.	$P = a + b + c + d$	area = $\left(\dfrac{a+b}{2}\right) \times h$

8. **Circle** 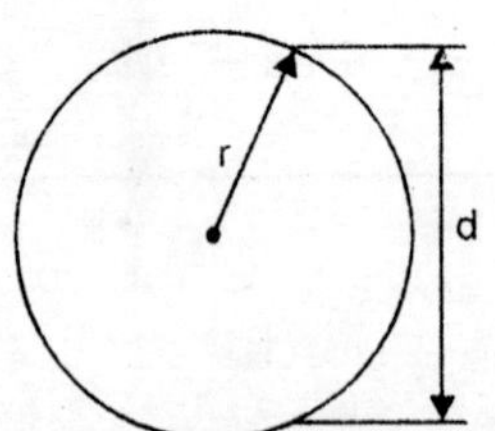 r = radius of circle d = dia of circle	$P = \pi d$ $P = 2\pi r$	$a = \pi / 4d^2$ $a = \pi r^2$
9. **Sector** r = radius of circle θ = L[le] in radious l = length of arc $l = r\theta$. 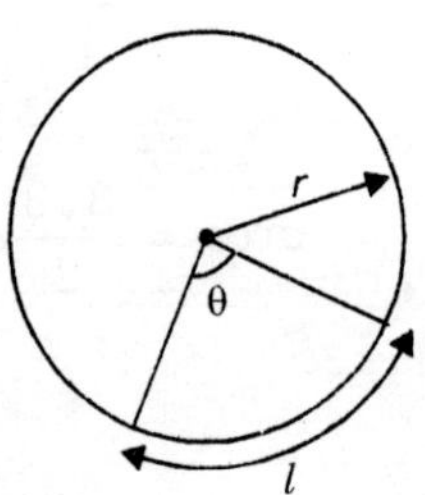		area = $\frac{\text{angle of sectar}}{2\pi}$ × area of circle $a = \frac{\theta}{2\pi} \times nr^2$ $a = \frac{lr}{2}$
10. **Fillet** r = radius of fillet 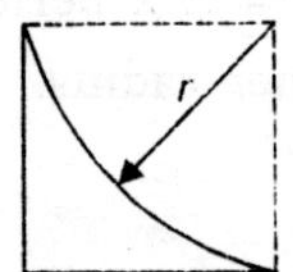		$a = r^2 \frac{-\pi}{4} r^2$ $= r^2\left[1 - \frac{\pi}{4}\right]$ $a = 0.215r^2$ $a = \frac{r^2}{4}$ approx.
11. **Ellipse** a = semi-major axis b = semi-minor axis 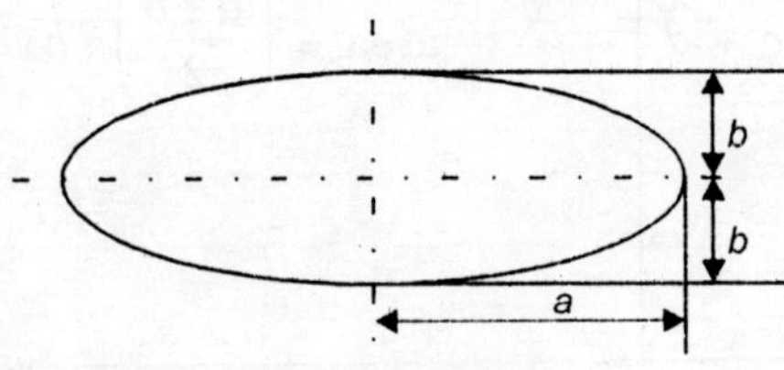	$P = \pi(a + b)$	$a = \pi ab$

12. **Segment of circle**		
		area $= \frac{2}{3} hL$ $a = \frac{4}{3} h \sqrt{\frac{l^2}{4} + \frac{2}{5} h^2}$ h = BD l = AC

AREAS OF IRREGULAR FIGURES

1. By Bumprant's Rule

Let h = distance between each ordinate

then area $= h/3[(y_1 + y_a) + 2\,(y_3 + y_5 + y_7) + 4(y_2 + y_4 + y_6 + y_8)]$

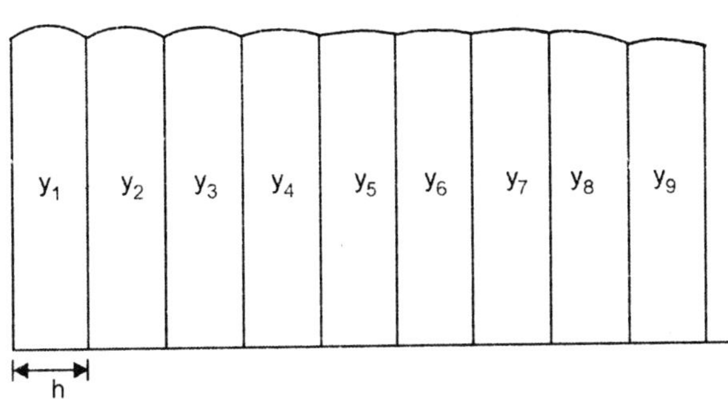

2. By Trapezoidal Rule

If $y_1, y_2 \ldots y_6$ are lengths of ordinates and h = distance between each ordinate.

then

$$a = h\left[\frac{1}{2}(y_1 + y_6) + y_2 + y_3 + y_4 + y_5\right]$$

3. Method of Counting Squares

Area of figure = No. of squares × area of 1 square.

4. Area of Trapezium

= Distance between successive ordinates × sum of half of first and last ordinates together with all remaining ordinates.

5. Mid-ordinate Method

$$\text{Area} = \frac{(y_1 + y_2 + y_3 + y_4 + y_5 + y_6) \times \text{CD}}{6}$$

Average length of line × length of base line.

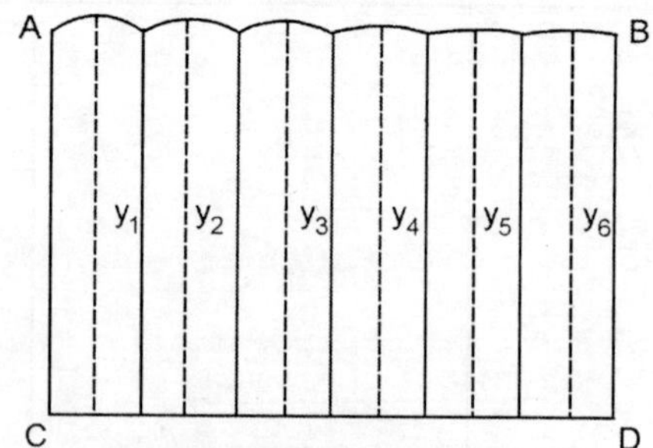

VOLUME AND SURFACE AREA OF SOLIDS

Plane figure	Volume	Surface area
1. **Rectangular solid** Let a, b, c be length, breadth and height of solid.	$V = abc$ (length of diagonal $= \sqrt{a^2 + b^2 + c^2}$)	$s = 2(ab + bc + ca)$
2. **Prism** Let h = height of prism a = breadth of 1 side.	V = area of base × height	s = No. of surface × ah + area of ends.
3. **Cylinder** Let d = dia of cylinder l = length of cylinder	$V = \frac{\pi}{4} d^2 l$	$s = \pi d l + \frac{\pi}{2} d^2$

4. **Pyramids and cones** 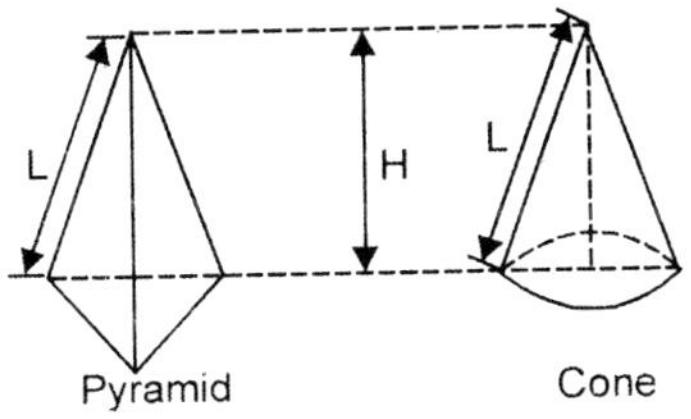 L H L Pyramid Cone	V = area of base $\times \frac{\text{perpendicular } ht \text{ H}}{3}$	s = area of base + perimeter of base × 1/2 length b of slant side
5. **Wledge** 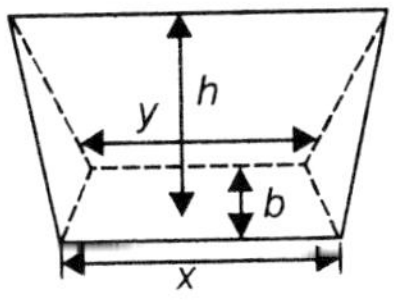 y h b x	V = h/3 × area of base $V = h/3 \times \left(\frac{x+y}{2}\right) b$	
6. **Triangular prism** 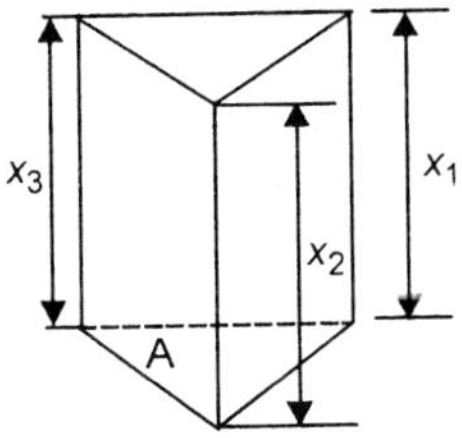 x_3 x_1 x_2 A	V = area of c.s. × mean length of edges $= A \times \left(\frac{x_1 + x_2 + x_3}{3}\right)$ where A = area of c.s. of triangle	
7. **Frustrum of pyramid and cone** a_1 and a_2 are areas of 2 ends. but for frustrum of cone $a_1 = \pi R_1^2$ $a_2 = \pi R_2^2$ 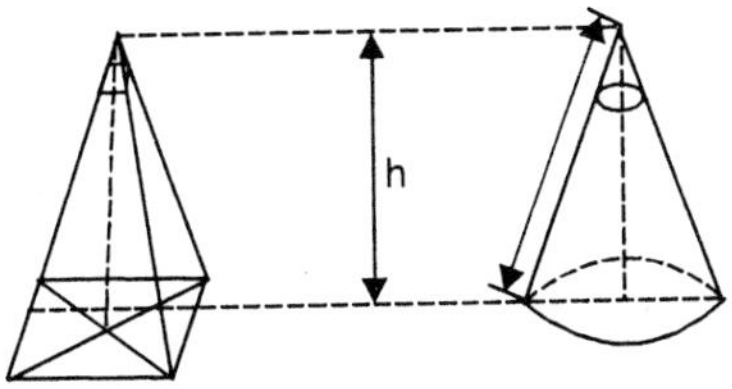 h	$V = h/3 \times \left(a_1 + a_2 + \sqrt{a_1 a_2}\right)$ Cone $V = \frac{\pi h}{3}(R_1^2 + R_2^2 + R_1 R_2)$	

8. **Sphere** 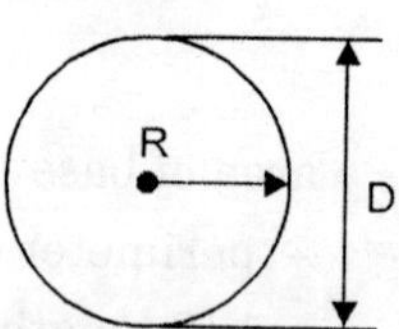	$V = \frac{\pi}{6}D^3$ $= \frac{4}{3}\pi r^3$	$s = \pi D^2$ $= 4\pi r^2$
9. **Hollow sphere** 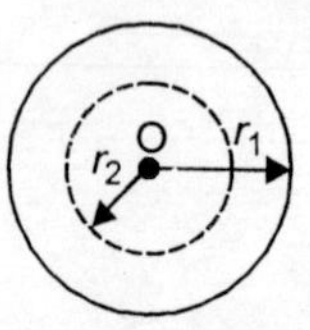	$V = \frac{4}{3}\pi(r_1^3 - r_2^3)$	
10. **Segment of sphere** 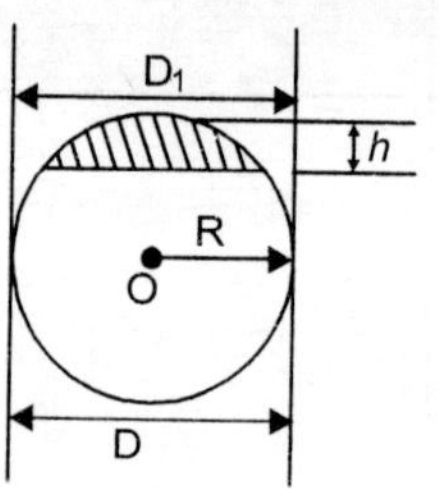	$V = \frac{\pi h^3}{3}(3R - h)$ $= \frac{\pi}{6}h^2(3D - 2h)$ or $V = \frac{\pi}{6}h\left[\frac{3}{4}(D_1^2 + h^2)\right]$	
11. **Circular ring** $a = \pi r^2$, 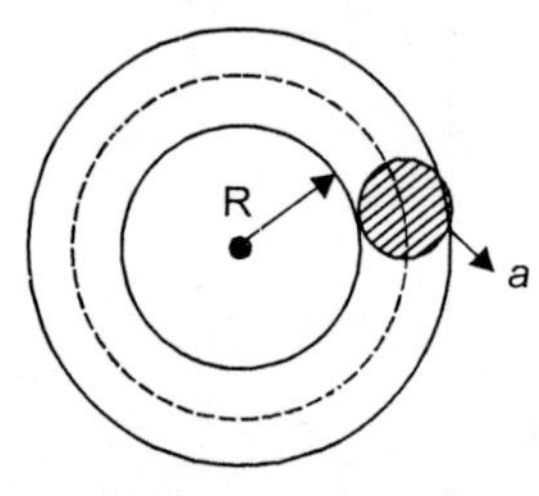 r = radius of circular cross section R = mean radius of ring	V = area of c.s. × mean length $= 2\pi R \times \pi r^2$ $= 2\pi^2 R r^2$	
12. **Spherical sectar** 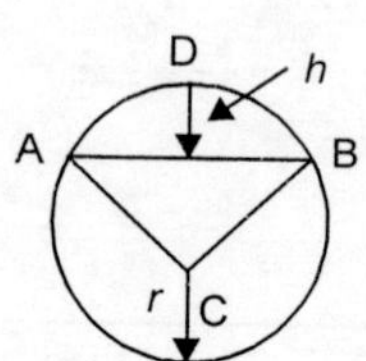	$V = ABCD = \frac{2}{3}\pi r h$	

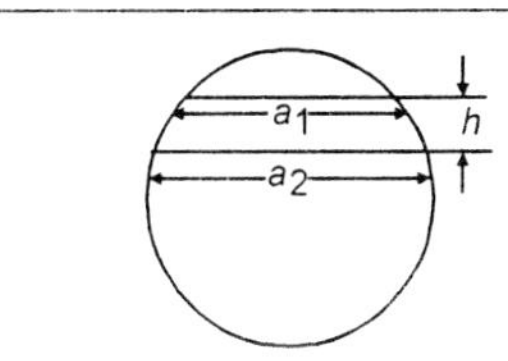	$V = \frac{\pi}{6} h \left[\frac{3}{4} \left(a_1^2 + a_2^2 \right) + h^2 \right]$	

POSITION OF CENTROID

1. **Triangle**	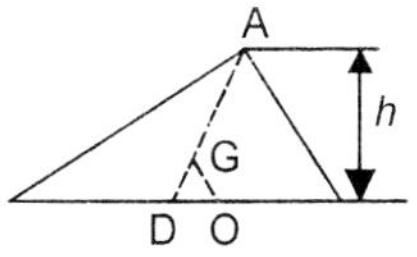	$OG = \frac{1}{3} h$ $GD = \frac{1}{2} AD$ where AD = median
2. **Square**	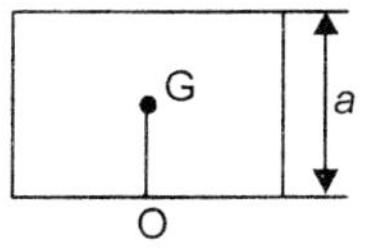	$OG = \frac{a}{2}$ $= \frac{1}{2}$ side of square
3. **Trapezoid**	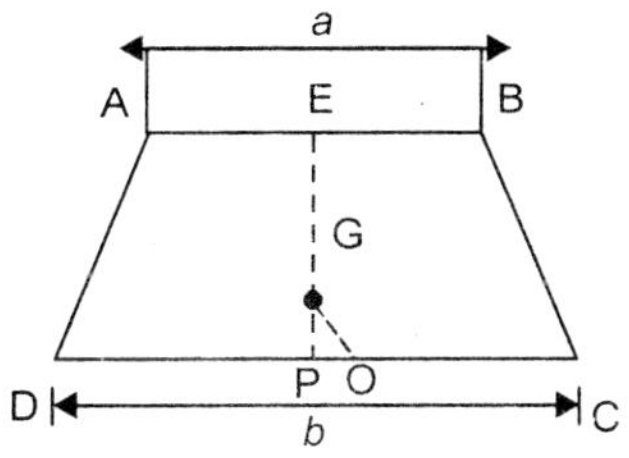	$OC = \frac{h}{8} \times \frac{(2a+b)}{(a+b)}$
4. **Quardrant of circular arc**	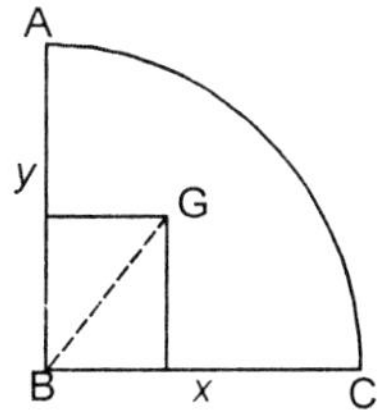	$xG = Gy$ $= 0.637$ R. and BG = 0.9 R where r = radius of circular arc.
5. **Semi circular arc**	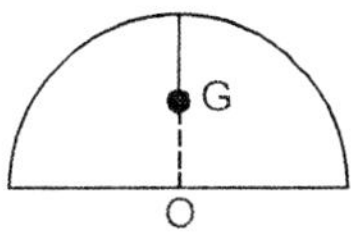	OG = 0.637 R. where R = rad. of circular arc.
6. **Quadrant of Circle**	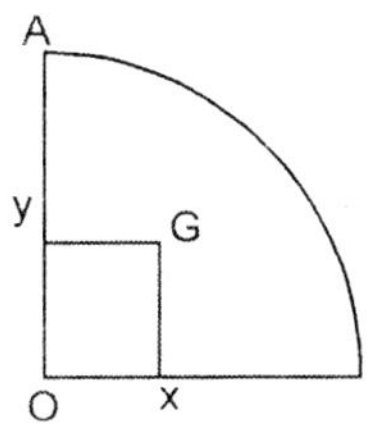	XG = YG = 0.424 R. where R = rad Oa

7. **Semi Circle**	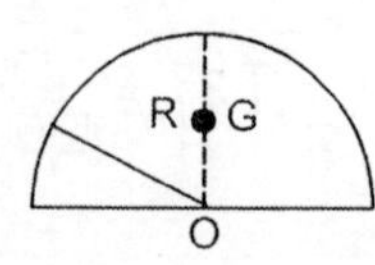	$OG = \frac{4}{3\pi}R$ $= 0.424\ R$ R = radius of circular area

8.7 DIFFERENTIATE BETWEEN ESTIMATING AND COSTING

Estimating	Costing
1. It is the process of determining the probable cost of an article.	1. It is a process of determining the actual lost of an article.
2. It is done before the product is manufactured.	2. It is done after the product is manufactured.
3. Based on assumption.	3. Based on facts.
4. The estimator should be engineer, well qualified.	4. Anybody having accounting knowledge.
5. It requires a lot of experience.	5. It does n't require much experience as that of estimating.
6. It may be overestimation or underestimation as estimation may go wrong as it is based on some assumptions.	6. The cost will never go wrong as it is based on facts.
7. Profits or losses may be forecasted	7. Profits or losses are actually felt.

8.8 ESTIMATION OF MATERIAL COST

Generalized procedure to calculate cost of the material

Step 1 : Observe the component drawing, break up the drawing into simple parts as per your convenience so as to calculate areas and volumes easily.

Give the notations for each of the part x, y, z and so on.

Step 2 : Using formulae calculate areas and volumes of each part.

Step 3 : Add the volumes of all the parts.

$$\text{Total volume} = x + y + z \text{ and so on.}$$

Step 4 : Multiply component vol × density

$$\therefore \text{weight} = \text{density} \times \text{volume.}$$

Problem 1. *A CI stepped cone pulley is shown in the following figure.*

Material cost = 20/kg

Calculate the cut and material cost.

Density of CI = 7.009 gms/cc

All dimensions are in mm.

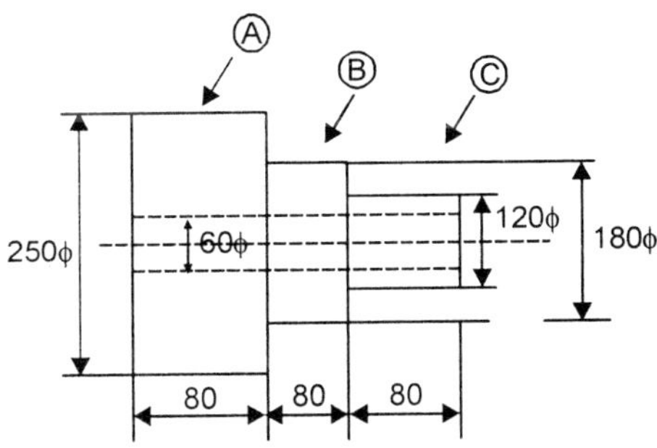

Solution. Volume of part A $= \frac{\pi}{4} d^2 l$

$= \frac{\pi}{4} \times 250^2 \times 80$

$V_A = 3.927 \times 10^6 \text{ mm}^3$

$= \mathbf{3.927 \text{ cm}^3.}$

Volume of part B $= \frac{\pi}{4} d^2 l$

$= \frac{\pi}{4} \times 180^2 \times 80$

$= 2.036 \times 10^6 \text{ mm}^3$

$V_B = \mathbf{2036 \text{ cm}^3.}$

Volume of part C $= \frac{\pi}{4} d^2 l$

$= \frac{\pi}{4} \times 120^2 \times 80 = 9.0478 \times 10^5 \text{ mm}^3$

$= \frac{9.0478 \times 10^5}{1000} \text{cc.}$

$V_C = \mathbf{904.7 \text{ cm}^3.}$

Volume of cone D $= \frac{\pi}{4} d^2 l$

$= \frac{\pi}{4} \times 60^2 \times 240 = 6.786 \times 10^5 \text{ mm}^3$

$V_D = \mathbf{678.6 \text{ cm}^3}$

Total volume of stepped cone pulley

$= V_A + V_B + V_C - V_D = 3927 + 2036 + 904.7 - 678.6.$

$= 6189.1 \text{ cm}^3$ or cc.

Unit of component = density × volume

$= 7.209 \frac{\text{gms}}{\text{cc}} \times 6189.1\text{cc}$

Unit = 44617.2 gms.

$= \frac{44617.2}{1000}$ kg

= **44.617 kg.**

Cost of pulley = 44.617 kg × 20 Rs./kg.

Cost = Rs. 892.34

Problem 2. *A CI conepulley is shown in fig. Taking density of CI as 7.0208 gm/cc. Calculate unit of component. Also what is cost of material if cost per kg is Rs. 15.*

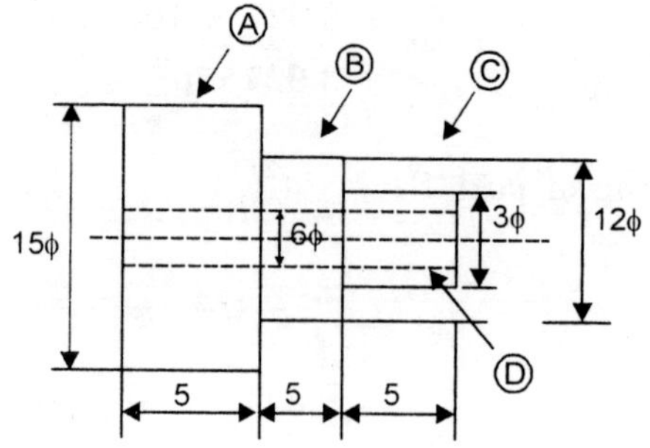

Solution. Volume of part A $= \frac{\pi}{4} d^2 l$

$= \frac{\pi}{4} \times 15^2 \times 5$

$= 883.57 \text{ mm}^3.$

$V_A = 0.8836 \text{ cm}^3.$

Volume of part B $= \frac{\pi}{4} d^2 l$

$= \frac{\pi}{4} \times 12^2 \times 5$

$V_B = 0.565 \text{ cm}^2.$

Volume of part C $= \frac{\pi}{4} d^2 l$

$= \frac{\pi}{4} \times 9^2 \times 5$

$$= 318 \text{ mm}^3.$$

$$\boxed{V_C = 0.318 \text{ cm}^3.}$$

$$\text{Volume of part D} = \frac{\pi}{4} d^2 l$$

$$= \frac{\pi}{4} \times 6^2 \times 15$$

$$= 424 \text{ mm}^3.$$

$$\boxed{V_D = 0.424 \text{ cm}^3.}$$

Total volume of stepped conepulley

$$= V_A + V_B + V_C - V_D$$

$$= 0.8836 + 0.565 + 0.318 - 0.424$$

$$\boxed{V = +1.3426 \text{ cm}^3.}$$

$$\text{Unit of component} = \text{density} \times \text{volume}$$

$$= 7.209 \text{ gm/cc} \times 1.3426 \text{ cc.}$$

$$\text{Unit} = 9.688 \text{ gm.}$$

$$\text{Unit} = 0.0096 \text{ kg.}$$

$$\text{Cost of pulley} = 0.0096 \text{ kg} \times 15 \text{ Rs/kg}$$

$$= \text{Re. } 0.145$$

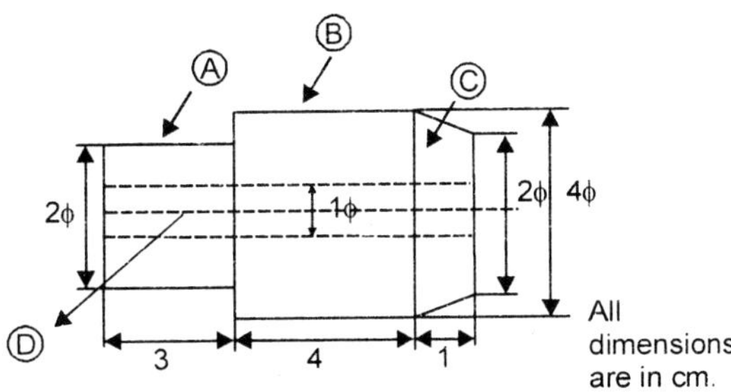

Problem 3. *The spindle has dimensions as shown in the figure above. The spindle is turned from MS rod of 45 mm dia facing and parting off allowances are 1mm and 5mm respectively. Assuming 15 mm length of rod is required for grip in the churk. Calculate the unit of 10 MS spindle taking density of MS = 7.8 gms/cc. Also calculate unt of the scrap.*

Solution. Volume of part A $= \frac{\pi}{4} d^2 l$

$$= \frac{\pi}{4} \times 2^2 \times 3$$

$$V_A = 9.425 \text{ cm}^3.$$

$$\text{Volume of part B} = \frac{\pi}{4} d^2 l$$

$$= \frac{\pi}{4} \times 4^2 \times 4$$

$$V_B = 50.26 \text{ cm}^3.$$

$$\text{Volume of part D} = \frac{\pi}{4} d^2 l$$

$$= \frac{\pi}{4} \times 1^2 \times 8$$

$$V_D = 6.283 \text{cm}^3.$$

Part C is like a frustrum of cone

$$V_C = \left[\frac{\pi}{4} d_1^2 + \frac{\pi}{4} d_2^2 + \sqrt{\frac{\pi}{4} d_1^2 \times \frac{\pi}{4} d_2^2}\right]^{\frac{h}{3}}$$

$$= \left[\frac{\pi}{4} 4^2 + \frac{\pi}{4} 2^2 + \sqrt{\frac{\pi}{4} 4^2 \times \frac{\pi}{4} 2^2}\right]^{\frac{1}{3}}$$

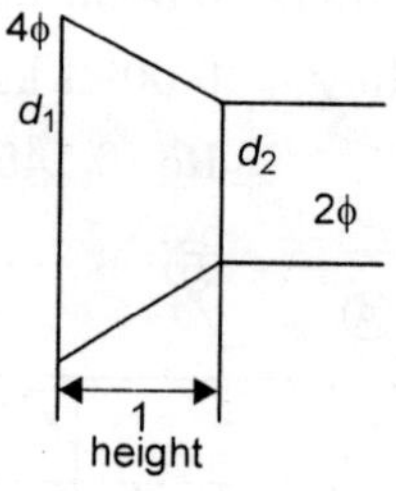

4ϕ
d_1
d_2
2ϕ
1
height

$$V_C = 7.33 \text{ cm}^3.$$

Total Volume of Finished Spindle.

$$V_A + V_B + V_C - V_D = 9.425 + 50.26 + 7.33 - 6.283$$

$$= \mathbf{60.732 \text{ cc.}}$$

Weight of 1 spindle = volume × density

= 60.732 cc × 7.8 gm/cc.

= 473.71 gms.

Unit of 10 spindles = 473.71 × 10

= **4737 gms.**

$$= \frac{4737}{1000} \text{kg.}$$

Unit of 10 spindles = **4.74 kg.**

Calculation of Total material reqd. including scrap.

Length of 1 spindle = 80 mm

Total length for 10 spindles = 80 × 10

= **800 mm.**

Allowance reqd. for facing = 1 mm.

Reqd. for 2 sides.

Total allowance facing = 10 × 2 = 20 mm.

parting off allowance = **5 mm**

for 10 spindles = 5 × 10 = 50 mm.

Length reqd. for holding the job = 15 mm.

Total length of MS rod = 800 + 20 + 50 + 15

= 885 mm.

Diameter of MS rod = 4.5 cm.

$$\text{Volume of rod reqd.} = \frac{\pi}{4}d^2l$$

$$= \frac{\pi}{4} \times (4.5)^2 \times 8.85$$

$$= \mathbf{1407.5}$$

Unit of rod = volume × density

= 1407.5 × 7.8

= 10978.7 gms.

Unit of scrap = 10978.7 – 4734. g

= 6241.7 g

= **6.2417 kg.**

Problem 4. *The lathe centre has dimensions shown in 1. The material density = 7.78 g/cc. The material costs Rs. 30/kg. Estimates the unit and cost of material.*

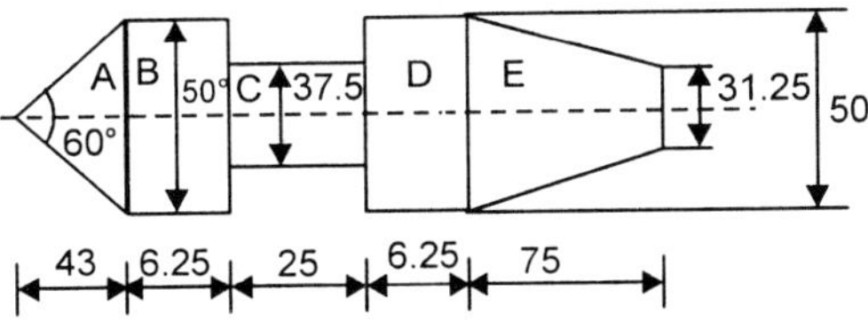

Solution.

$$V_A = \frac{1}{3} \times \pi r^2 \times h \qquad a = 1/2 \times \text{base} \times \text{height}$$

$$= \frac{1}{3} \times \pi \times (25)^2 \times 43 \qquad = 1/2 \times 50$$

$$= 28.14 \text{ cm}^3.$$

$$V_B = \frac{\pi}{4}d^2l$$

$$= \frac{\pi}{4} \times 50^2 \times 6.25$$

$= 12271.8 \text{ mm}^3.$

$= \mathbf{12.27\ cm^3.}$

$$V_C = \frac{\pi}{4} d^2 l$$

$$= \frac{\pi}{4} \times 37.5^2 \times 25 = \mathbf{27.611\ cm^3.}$$

$$V_D = \frac{\pi}{4} d^2 l$$

$$= \mathbf{12.27 cm^3.}$$

$$a_1 = \pi R_1^{\ 2}$$

$$a_2 = \pi R_2^{\ 2}$$

$$= \pi \times (15.625)^2 = 766.99 \text{ mA}^2$$

$$a_2 = \pi \times (25)^2$$

$$a_2 = 1963.4 \text{ mm}^2$$

$$V_E = \frac{h}{3}\left[a_1 + a_2 + \sqrt{a_1 a_2}\right]$$

$$= \frac{75}{3}\left[766.99 + 1363.4 + \sqrt{766.99 \times 1963.4}\right]$$

$$V_E = 98938.6 \text{ mm}^3$$

$$V_E = \frac{98938.6}{1000} \text{cm}^3.$$

$$\boxed{V_E = 98.94 \text{ cm}^3.}$$

Total volume of component $= V_A + V_B + V_C + V_D + V_E$

$= 28.14 + 12.27 + 27.61 + 12.27 + 88.84$

$$\boxed{V = 179.23 \text{ cm}^3.}$$

Unit of material $= V \times \text{density}$

$= 179.23 \times 7.78$

$= 1334.3$ gm.

$= 1.334$ kg.

Cost of material $= 1.394 \text{ kg} \times 30$

$=$ Rs. 41.83

Problem 5. *The following figure shows the dimensions of lathe centre. Calculate weight and cost of material. If material has density of 7.8 gms/cc. and cost of material is 25/kg.*

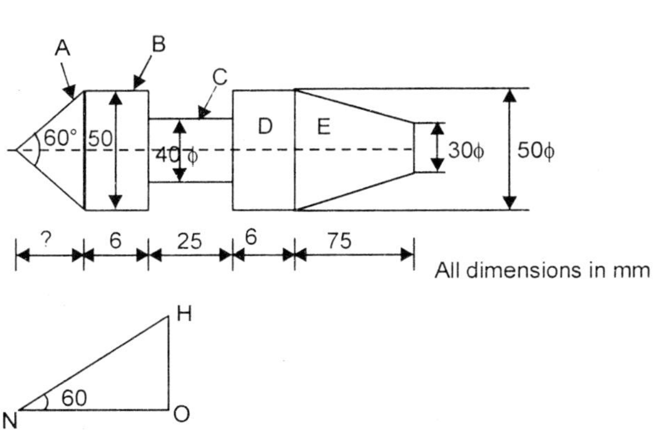

Solution.

$$\text{Tan } 60 = \frac{OM}{ON}$$

$$ON = \frac{OM}{\text{Tan}60}$$

$$ON = \frac{25}{\text{Tan}60} = 14.43 \text{ cm.}$$

$$V_A = \frac{1}{3}\pi r^2 h$$

$$= \frac{1}{3} \times \pi \times (25)^2 \times ON$$

$$= \frac{1}{3} \times \pi \times (25)^2 \times 14.43 = 9444.4$$

$$\boxed{V_A = 9.44 \text{ cm}^3.}$$

$$V_B = \frac{\pi}{4} d^2 l$$

$$= \frac{\pi}{4} \times 50^2 \times 6 = 11780.97$$

$$\boxed{V_B = 11.780 \text{ cm}^3.}$$

$$V_C = \frac{\pi}{4} d^2 l$$

$$= \frac{\pi}{4} \times 40^2 \times 25$$

$$\boxed{V_C = 31.416 \text{ cm}^3.}$$

$$V_D = \frac{\pi}{4} d^2 l$$

$$= \frac{\pi}{4} \times 50^2 \times 6$$

$$V_D = 11.78 \text{ cc.}$$

$$V_E = \frac{h}{3}\left[a_1 + a_2 + \sqrt{a_1 a_2}\right]$$

$$= \frac{15}{3}\left[1963.5 + 706.86 + \sqrt{1963.48 \times 706.86}\right]$$

$$a_1 = \pi \times 25^2 = 1963.49 \text{ mm}^3.$$

$$a_2 = \pi \times 15^2 = 706.86 \text{ mm}^3.$$

$$V_E = 96.208 \text{ cm}^3.$$

$$\text{Total volume} = V_A + V_B + V_C + V_D + V_E$$

$$= 9.44 + 11.78 + 31.41 + 11.78 + 96.21$$

$$\boxed{V = 160.62 \text{ cm}^3.}$$

$$\text{Unit of material} = \text{density} \times \text{volume}$$

$$= 4.8 \text{ gm/cc} \times 160.62 \text{ cc.}$$

$$= 1252.8 \text{ gm}$$

$$= \mathbf{1.253 \text{ kg.}}$$

$$\text{Cost of Material} = \text{Unit} \times \text{Cost/kg}$$

$$= 1.253 \text{ kg} \times 25 \text{ kg.}$$

$$= \text{Rs. } 31.32$$

Problem 6. *Calculate unit of lathe centre if material density is 8 g/cc. Calculate the cost of material if its rate Rs. 20/kg.*

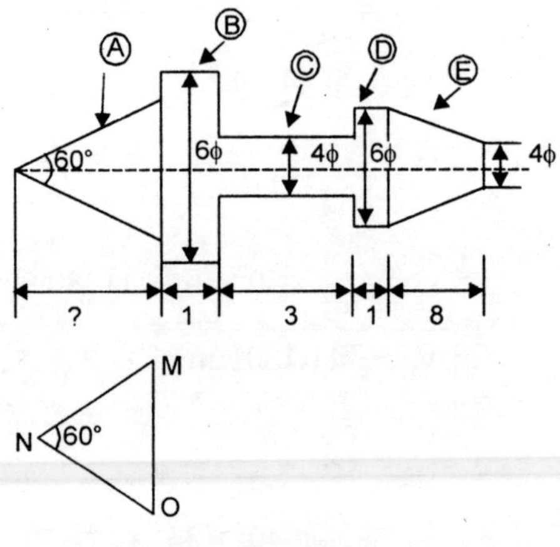

Solution.

$$V_A = \frac{1}{3}\pi r^2 h$$

$$= \frac{1}{3}\pi \times 3^2 \times 7 = \frac{1}{3} \times \pi \times 3^2 \times 5.19 \text{ cm}$$

$$\boxed{V_A = 48.97 \text{ cm}^3.}$$

$$V_B = \frac{\pi}{4} d^2 l$$

$$= \frac{\pi \times 6^2}{4} \times 1$$

$$\boxed{V_B = 28.27 \text{ cm}^3.}$$

$$V_C = \frac{\pi}{4} d^2 l = \frac{\pi}{4} \times 4^2 \times 3$$

$$V_C = 37.69 \text{ cm}^3.$$

$$V_D = \frac{\pi}{4} d^2 l = \frac{\pi}{4} \times 6^2 \times 1$$

$$\boxed{V_D = 28.27 \text{ cm}^3.}$$

$$a_1 = \pi R_1^2$$

$$= \pi \times 3^2$$

$$= 28.27 \text{ cm}^3$$

$$a_2 = \pi R_2^2$$

$$= \pi \times 2^2$$

$$= 12.5 \text{ cm}^3$$

$$V_E = \frac{h}{3}\left[a_1 a_2 + \sqrt{a_1 a_2}\right]$$

$$V_E = \frac{8}{3}\left[28.27 + 12.57 + \sqrt{28.27 \times 12.57}\right]$$

$$\boxed{V_E = 159.17 \text{ cm}^3.}$$

Total volume

$$V = V_A + V_B + V_C + V_D + V_E$$

$$= 48.97 + 28.27 + 37.69 + 28.27 + 159.17$$

$$\boxed{= 302.2}$$

Unit of material = density × volume

= 4.8 g/cc × 302.2 cc.

Unit = **2357.16 gm**

$$\boxed{\text{Unit} = 2.357 \text{ kg.}}$$

Cost = cut × cost/kg.

= 2.357 kg × 20/kg

$$\boxed{\text{Cost} = \text{Rs. } 47.14}$$

Problem 7. *As crank shaft is similar in dimensions on both sides of xy. ∴ volume on one side can he calculated and total volume will be twice of that of one side.*

Solution. Volume of A $= \frac{\pi}{4} d^2 l = \frac{\pi}{4} \times 27^2 \times 27$ = **15.46 cc.**

Volume of B $= \frac{\pi}{4} d^2 l$

$$= \frac{\pi}{4} \times 24^2 \times 30 = \mathbf{13.57\ cc}$$

$$\text{Volume of C} = \frac{\pi}{4} d^2 l$$

$$= \frac{\pi}{4} \times 60^2 \times 22.5 = \mathbf{63.62\ cc.}$$

$$\text{Volume D} = \frac{\pi}{4} d^2 l$$

$$= \frac{\pi}{4} \times 30^2 \times 22.5 = \mathbf{15.90\ cc.}$$

$$\text{Volume E} = \frac{\pi}{4} d^2 l$$

$$= \frac{\pi}{4} \times 60^2 \times 22.5 = \mathbf{63.62\ cc.}$$

$$\text{Volume of F} = \frac{\pi}{4} d^2 l$$

$$= \frac{\pi}{4} \times 24^2 \times 30 = \mathbf{13.57\ cc.}$$

Calculate the unit of the forged crank shaft as shown in figure. It is made up of MS. which has density of 0.008 kg/cm^3. $\Rightarrow$ 8kg / cc.

Total volume of 1/2 crank shaft = VA + VB + VC + VD + VE + VF

= 15.46 + 13.57 + 63.62 + 15.9 + 63.62 + 13.57

Total Volume = 183.74 × 2

V = 371.48 cc.

$\therefore$ Unit of crank shaft = volume × density

= 371.48cc × 0.008 kg/cc.

Unit = 2.971 kg.

Problem 8. *The following figure shows the rivet with dimensions. Calculate the unit if one rivet of density is 8 gms/cc. If the rivets are manufactured from 6.5 kg of material, calculate the no. of rivets that can be manufactured. Assume that there is no. usage of material.*

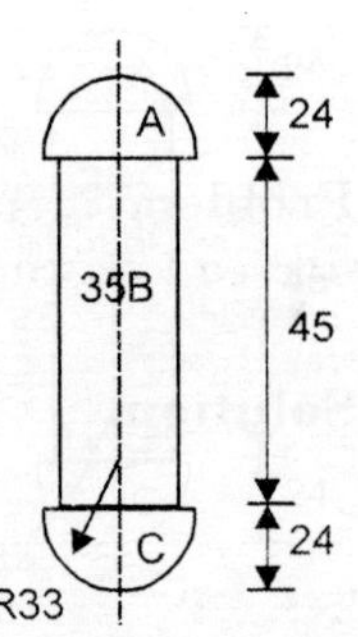

Solution. The given rivet is divided into 3 parts.

Volume of part A V_A is a segment of sphere.

$$\therefore \quad V_A = \frac{\pi}{6} h^2 (3D - 2h)$$

$$= \frac{\pi}{6} 24^2 (3 \times 66 - 2 \times 24)$$

$$V_A = \mathbf{45.238\ cm^3.}$$

$$V_B = \frac{\pi}{4} d^2 l = \frac{\pi}{4} \times 35^2 \times 45$$

$$V_B = \mathbf{43.29\ cm^3.}$$

$$V_C = V_A = \mathbf{45.238\ cm^3.}$$

$$\text{Total volume} = 43.29 \times 45.238 \times 2$$

$$= 133.77 \text{ cc.}$$

$$\text{Unit of 1 rivet} = \text{Volume} \times \text{density}$$

$$= 133.77 \times 8 \text{ gm/cc.}$$

$$= 1070.16 \text{ gm} = \mathbf{1.07\ kg.}$$

$$\text{Availability of material} = \mathbf{6.5\ kgs.}$$

$$\text{No. of rivets manf.} = \frac{6.5}{1.07} \approx 6.07$$

$$\approx \mathbf{6\ rivets.}$$

Problem 9. *As unit cost of M.S. casting shown in figure. Assume density is 7.35 gm/cc and cost of material Rs. 11/kg.*

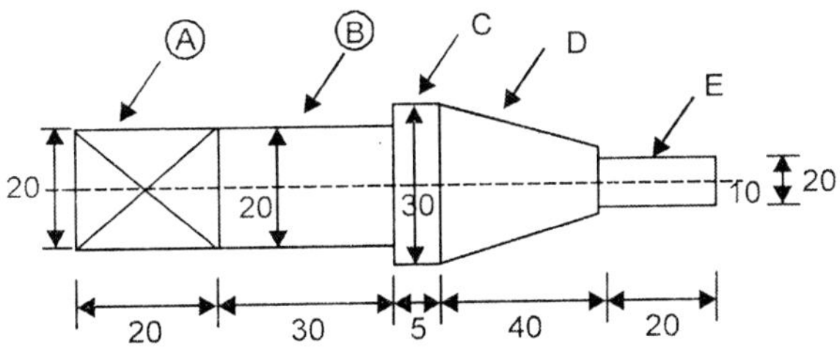

Solution.

$$V_A = \text{square}$$

$$V_A = l^3.$$

$$V_A = 20^3$$

$$= 8000 \text{ mm}^3.$$

$$\boxed{V_A = 8 \text{ cc}}$$

$$V_B = \frac{\pi}{4} d^2 l = \frac{\pi}{4} \times 20^2 \times 30$$

$$\boxed{V_B = 9.425 \text{ cc.}}$$

$$V_C = \frac{\pi}{4} d^2 l = \frac{\pi}{4} \times 30^2 \times 5$$

$$\boxed{V_C = 3.53 \text{ cc.}}$$

$$a_1 = \pi {r_1}^2$$

$$= \pi \times 2^2 = 314.1 \text{ cm}^3$$

$$a_2 = \pi {r_2}^2$$

$$= \pi \times 15^2 = 706.36 \text{ cm}^3$$

$$V_D = \frac{h}{3}\left[a_1 + a_2 \sqrt{a_1 a_2}\right]$$

$$= \frac{40}{3}\left[314.159 + 706.86 + \sqrt{314.159 \times 706.86}\right]$$

$$= \textbf{19.893 cc.}$$

$$V_E = \frac{\pi}{4} d^2 l$$

$$= \frac{\pi}{4} \times 10^2 \times 20 = \textbf{1.571 cc.}$$

$$\text{Total volume} = V_A + V_B + V_C + V_D + V_E$$

$$= 8 + 9.425 + 19.893 + 3.53 + 1.571$$

$$= \textbf{42.419 cc.}$$

$$\therefore \quad \text{Unit} = \text{vol.} \times \text{density}$$

$$= 42.418 \times 8.85$$

$$= 332.99 \text{ gms.}$$

$$= \textbf{0.332 kgs.}$$

$$\text{Cost of material} = 0.332 \times 11$$

$$\text{Cost} = \boxed{\text{Rs. } 3.6629}$$

Problem 10. *Determine the volume, weight and total cost of the following part shown in figure.*

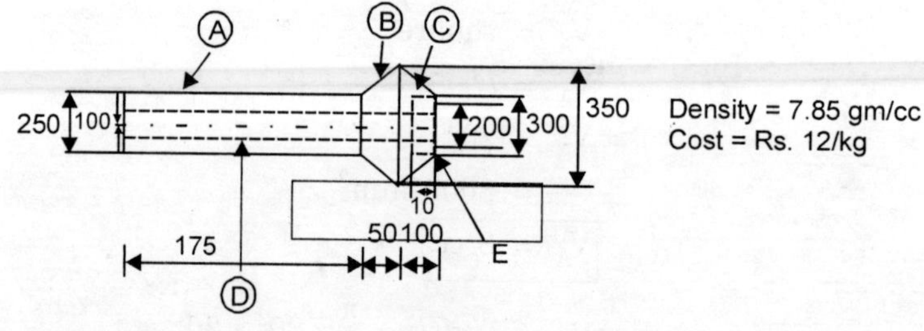

Solution.

$$V_A = \frac{\pi}{4} d^2 l$$

$$= \frac{\pi}{4} \times 250^2 \times 175 = 8590.29 \text{ cc.}$$

$$V_B = \frac{h}{3}\left[a_1 + a_2\sqrt{a_1 a_2}\right]$$

$$= \frac{50}{3}\left[96211.27 + 49087.38 + \sqrt{96211.27 \times 49087.38}\right]$$

$= \mathbf{3567.012.}$

$V_B = \mathbf{3567.01cc.}$

$a_1 = \pi {r_1}^2 \qquad a_2 = \pi {r_2}^2$

$= \pi \times 150^2 \qquad = \pi \times 125^2$

$= 70685.8 \text{ mm}^2 \quad = 49087.38 \text{ mm}^2$

$$V_C = \frac{h}{3}\left[a_1 + a_2\sqrt{a_1 a_2}\right]$$

$$= \frac{100}{3}\left[96211.27 + 706 + \sqrt{5 \times 790}\right]$$

$V_C = \mathbf{8312.13\ cc.}$

$a_1 = \pi {r_1}^2 \qquad a_2 = \pi {r_2}^2$

$= \pi \times 175^2 \qquad = \pi \times 150^2$

$= 96211.27 \text{ mm}^2 \quad = \mathbf{70685.8\ mm^2}$

$$V_D = \frac{\pi}{4} d^2 l$$

$$= \frac{\pi}{4} \times 100^2 \times (175 + 50 + 90)$$

$= \mathbf{2474.00\ cc.}$

$$V_E = \frac{\pi}{4} d^2 l = \frac{\pi}{4} \times 200^2 \times 10$$

$V_E = \mathbf{314.159\ cc.}$

Total volume $= V_A + V_B + V_C - V_D - V_E$

$= 8590.29 + 3567.01 + 8312.13 - 2474.00 - 314.159$

$V = \mathbf{17681.266\ cc.}$

$\therefore$ Unit = vol × density

$= 17681.266 \times 7.85 = \mathbf{138797.9}$

$\therefore$ Cost $= 138797.9 \times 12 = 1665.5$

Problem 11. *Determine the weight and the cost of the following component shown in figure.*

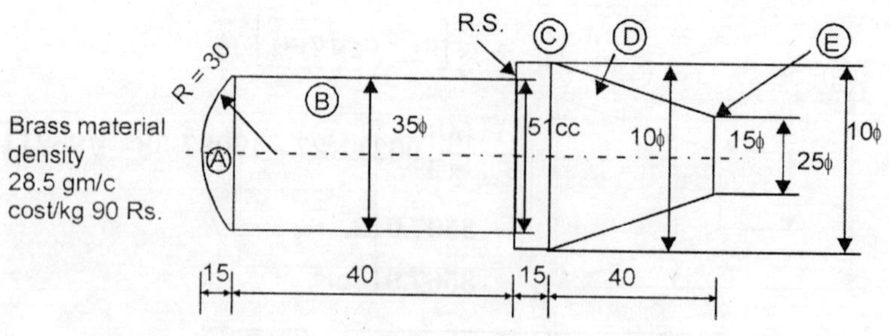

Solution.

$$V_A = \frac{\pi}{6} h^2 (3D - 2h)$$

$$= \frac{\pi}{6} \times 15^2 (3 \times (60) - 2 \times 15) = \mathbf{17.67\ cc.}$$

$$V_B = \frac{\pi}{4} d^2 l = \frac{\pi}{4} \times 35^2 \times 40$$

$$V_B = \mathbf{38.48\ cc.}$$

$$V_C = \frac{\pi}{4} d^2 l$$

$$= \frac{\pi}{4} \times 70^2 \times 15 = \mathbf{57.726\ cc.}$$

$$a_1 = \pi r_1^2 \qquad a_2 = \pi r_2^2$$

$$= \pi \times 35^2 \qquad = \pi \times 12.5^2$$

$$= 3848.4 \text{ mm}^2 \qquad = 490.87 \text{ mm}^2$$

$$V_D = \frac{h}{3}\left[a_1 + a_2 \sqrt{a_1 a_2}\right]$$

$$= \frac{40}{3}\left[3848.4 + 430.87 + \sqrt{3848.4 \times 430.87}\right]$$

$$= \mathbf{76.183\ cc.}$$

$$V_E = \frac{\pi}{4} d^2 l$$

$$= \frac{\pi}{4} \times 15^2 \times 40 = 7.068 \text{ cc.}$$

Total volume $= V_A + V_B + V_C + V_D - V_E - V_F$

$= 182.99.$

Volume of fillet $= 0.215\ R^2 \times$ mean peripheral

$= 0.215\ (15)^2 \times 52.5 = \mathbf{2.539\ cc.}$

$\therefore$ 182.99 + 2.539 = 185.529 cc.

Unit = vol. × density

Unit = 185.529 × 8.5 = **1577 gms.**

Cost/kg = 1577/gm × 80 Rs./kg

= 141.93

≈ **Rs. 142**

Problem 12. *A casting has a length of 1.5 meters and its CS is a regular hexagon. The casting opens uniformly along its length. The hexagon having a side 9cms at 1 end and 4 cms at the other end. Calculate the unit of the lasting if the material density is 7.5 gms/cc.*

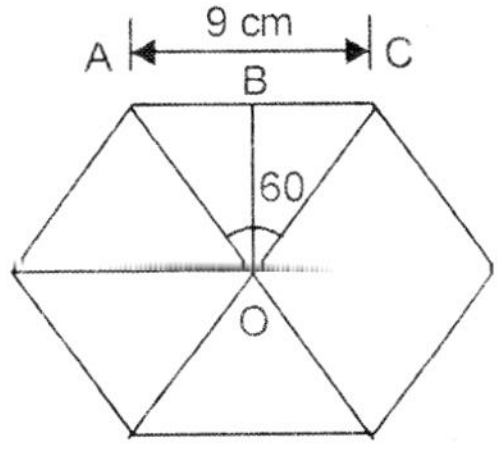

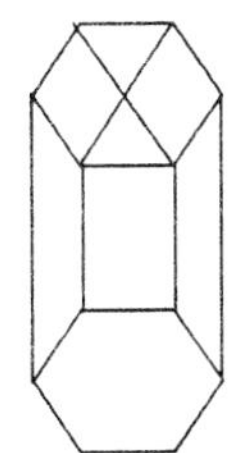

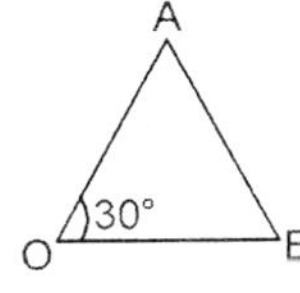

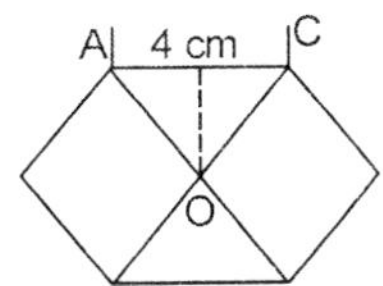

Solution.

$$\text{Tan}\theta = \frac{AB}{BO} = \frac{4.5}{\text{Tan}30} = BO = 7.79 \text{ cm.}$$

area = 1/2 × base × ht.

= 1/2 × 9 × 7.79 = **35.055 cm²**.

$\therefore$ Total area of hexagon = 6 × area of AOC

= 6 × 35.05

A_1 = **210.33 cm²**.

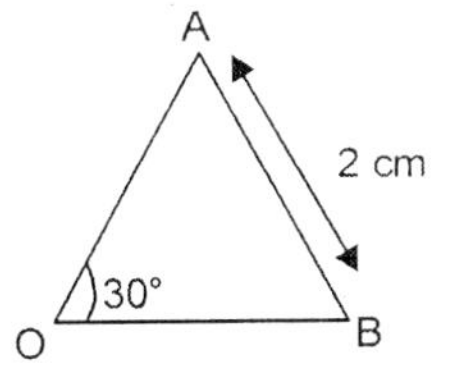

$$\text{Tan}\theta = \frac{AB}{BO} = \frac{2}{\text{Tan}30} = 3.66.$$

area = 1/2 × b × ht.

= 1/2 × 4 × 3.46 = **6.93 cm²**.

Total area of hexagon = 6 × area of AOC

= 6 × 6.93

A_2 = 41.57 cm².

The casting looks like frustrum of pyramid

$$\text{Vol. of casting} = \frac{h}{3}\left[A_1 + A_2\sqrt{A_1A_2}\right]$$

$$= \frac{1.5\times10^2}{3}\left[210.33 + 41.57 + \sqrt{210.33\times41.57}\right]$$

= **17270 cm³.**

$$\text{Unit of casting} = \text{vol.} \times \text{density}$$
$$= 17270 \times 7.5 \text{ gm/cc.}$$
$$= 129526 \text{ gm}$$
$$= \frac{129526}{1000} \text{kg.}$$
$$\text{Unit of casting} = \mathbf{129 \ kg.}$$

Problem 13. *An iron wedge is made by forging out of a 3 cm dia round bar. The length and breadth of base of wedge is 4.5 cm and 2.5 cm respectively. The length and breadth of other end of wedge is 4cm and 2.5 cm respectively. The ht. of wedge is 12cm. If the density of material remains unchanged after forging what length of bar is reqd. for making wedge.*

The wedge is like frustrum of pyramid.

Solution.

$$\text{Volume} = \frac{h}{3}\left[a_1 + a_2\sqrt{a_1 a_2}\right]$$

$$\text{Volume} = \frac{12}{3}\left[(4.5 \times 2.5) + (4 \times 2.5) + \sqrt{(4.5 \times 2.5) \times \sqrt{(4 \times 2.5)}}\right]$$

$$\boxed{V = 127.43 \text{ cm}^3.}$$

$$\text{Volume of rod} = \frac{\pi}{4} d^2 l$$

$$127.43 = \frac{\pi}{4} \times 3^2 \times l$$

$$\boxed{l = 18.02 \text{ cm.}}$$

14. *An iron wedge has dimensions as shown in figure The wedge is made of forging of 38 cm dia bar stock. Calculate length of bar reqd. to make.*

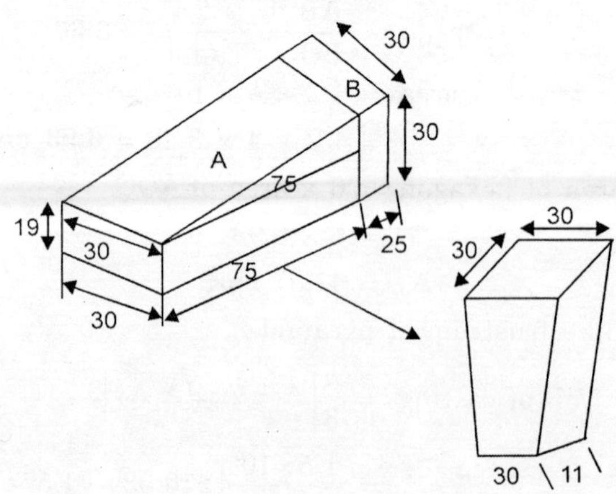

Solution. Frustrum of per $V_A = \frac{h}{3}\left[a_1 + a_2\sqrt{a_1 a_2}\right]$

$$= \frac{7.5}{3}\left[330 + 900 + \sqrt{330 \times 900}\right]$$

$$a_1 = l \times b = 30 \times 11$$

$$a_2 = l \times b = 30 \times 30$$

$$V_A = \mathbf{44.37\ cc.}$$

$$V_B = 30 \times 30 \times 25$$

$$V_B = \mathbf{22.5\ cc}$$

$$V_A + V_B = 38.44 + 22.5 = 66.87$$

$$V_C = \mathbf{61.94\ cc.}$$

$$V = \frac{\pi}{4} d^2 l$$

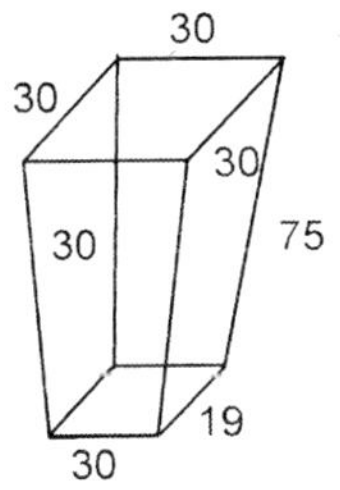

$$61.94 = \frac{\pi}{4} \times 38^2 \times l$$

$$V_A = \left(\frac{19+30}{2}\right) \times 30 \times 75 = 55.125 \text{ cc.}$$

$$V_B = \left(\frac{30+30}{2}\right) \times 25 \times 30 = 22.5 \text{ cm}^3.$$

$$\text{Total volume} = 77.625 \text{ cm}^3$$

$$= \frac{h}{3}\left[a_1 + a_2\sqrt{a_1 a_2}\right]$$

$$= \frac{75}{3}\left[30 \times 19 + 30 \times 30 + \sqrt{30 \times 30 \times 19 \times 30}\right]$$

$$= \mathbf{54.65\ cc}$$

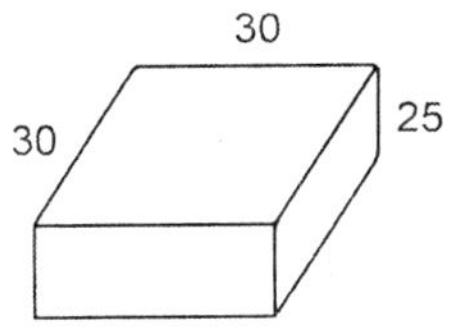

$$V_T = \frac{\pi}{4} \times 38^2 \times l$$

$$= \mathbf{11.34\ l\ cm^3.}$$

$$= 11.34l$$

$$77.625 = 11.34\ l$$

$$l = 6.8 \text{ cm.}$$

The following figure shows the bush of a gun metal. Calculate the material cost of 15 gun metal bushes assuming the density as 8.3 gm/cc. Cost of material is Rs. 120/kg. (Consider 8% material loss during the process.)

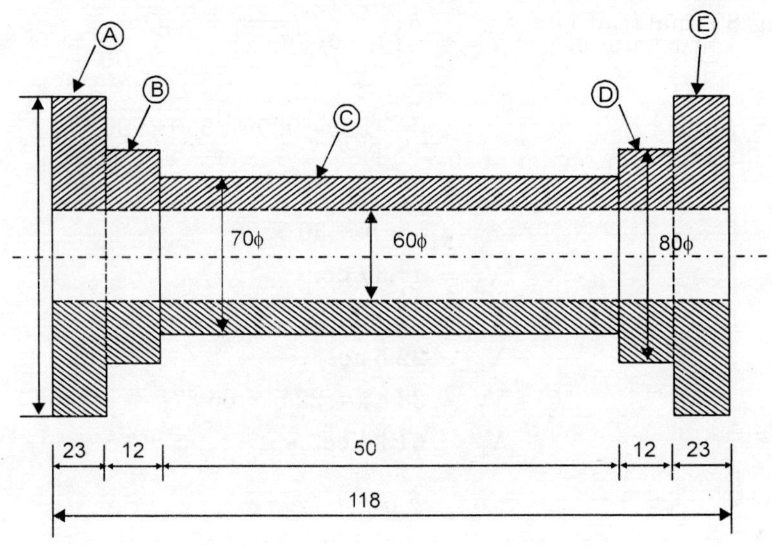

$$V_A = \frac{\pi}{4}d^2l = \frac{\pi}{4} \times 120^2 \times 22$$

$$V_A = \mathbf{248.8\ cc.}$$

$$V_B = \frac{\pi}{4}d^2l = \frac{\pi}{4} \times 80^2 \times 12$$

$$V_B = \mathbf{60.32\ cc.}$$

$$V_C = \frac{\pi}{4}d^2l = \frac{\pi}{4} \times 70^2 \times 50$$

$$V_C = \mathbf{192.42\ cc.}$$

$$V_D = \frac{\pi}{4}d^2l = \frac{\pi}{4} \times 60^2 \times 118$$

$$V_D = \mathbf{333.63\ cc.}$$

$$\text{Total volume} = 2 \times V_A + 2 \times V_B + V_C - V_D$$

$$= 2 \times 248.8 + 2 \times 60.32 + 192.42 - 333.63$$

$$\boxed{V = 477.03 \text{ cc.}}$$

$$\text{Unit} = \text{Volume} \times \text{density}$$

$$= 477.03 \text{ cc} \times 8.3 \ \frac{\text{gm}}{\text{cc}}$$

$$= 3959 \text{ gm}$$

$$= \frac{3959}{1000} \text{kg.}$$

$$\boxed{\text{Unit} = 3.96 \text{ kg.}}$$

Considering 8% material loss.

$$\therefore \quad 100 + 8\% = 1 + \frac{8}{100} = 1 + 0.08 = 1.08$$

$\therefore$ Total unit for 15 gun metal bushes

= 15 × unit of 1 gun metal bush

= 15 × 3.96

Total unit = 59.4 kg.

Total unit including actual loss = 59.4 × 1.08

Total unit + material loss = 64.152 kg.

Cost of material = 64.152 × cost/kg.

= 64.152 kg. × Rs. 120

Cost of material = Rs. 7698.24

BREAK-EVEN ANALYSIS

9.1 INTRODUCTION

Break-even analysis expresses acceptance to rejection or between competing alternatives and it is limited form of sensitivity analysis.

Many economic comparisons are a form of break-even analysis. The sensitivity studies involve an indifference level for given cash flow element at which two alternatives are equivalent the break-even point for the given element. The choice between the two rests on a judgement about which side of the break point the element will likely register.

The break-even analysis is directed to the point at which operations merely break-even, neither making nor losing money. Break-even analysis known also as cost-volume-profit-analysis is widely used for financial studies because it is simple and extracts useful insights from a modest amount of data. The studies includes an examination of production costs and operating policies. Break analysis is like a medical check-up. The physical examination reveals the current-state of health and provides clues about what should be done to become or stay healthy.

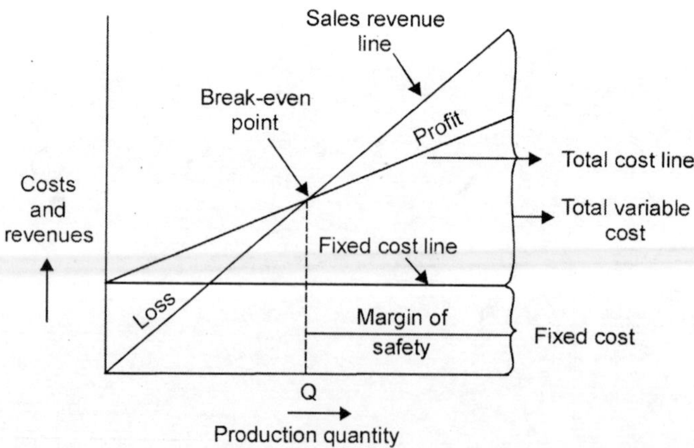

Fig. 9.1 Break-even chart.

The break-even analysis is the most widely known form of CVP analysis. Break-even analysis is a specific way of presenting and studying the inter-relationship between costs volume and profits. It also establishes a relationship between revenue and costs with result to volume. It indicates the cost and revenues are in equilibrium. The equilibrium point is commonly known as the break-even point. It is no-profit, no-loss point. The following diagram shows the fixed cost, variable cost, break-even point, loss and profit as the number quantity increases the profit, when changes it can be viewed by the Fig. 9.1. Break-even chart.

9.2 METHODS FOR LOWERING THE BREAK-EVEN POINT

A low BEP is highly desirable because it increases the safety margin of the product and it is obvious the BEP can be lowered by three methods as follows:

1. Reduce the fixed cost
2. Reduce the variable cost
3. Increases the sales volume.

1. Reduce the fixed cost from F to F′ thus lowering the BEP to $Q'_1 = Q_1\frac{F'}{F}$

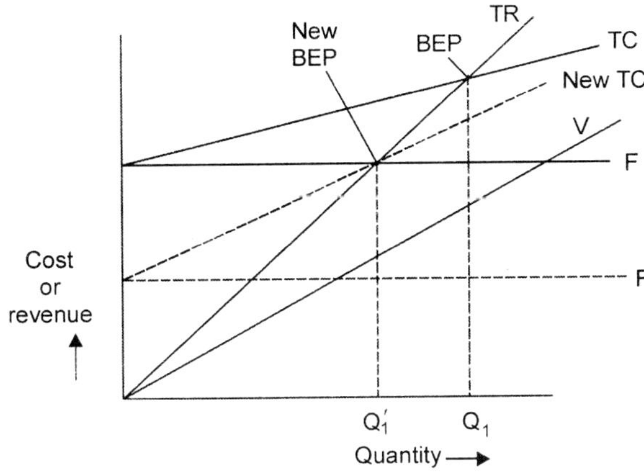

2. Reduce the variable cost coefficient a and a'. Hence $Q'_1 = Q_1\left[\frac{b-a}{b-a'}\right]$

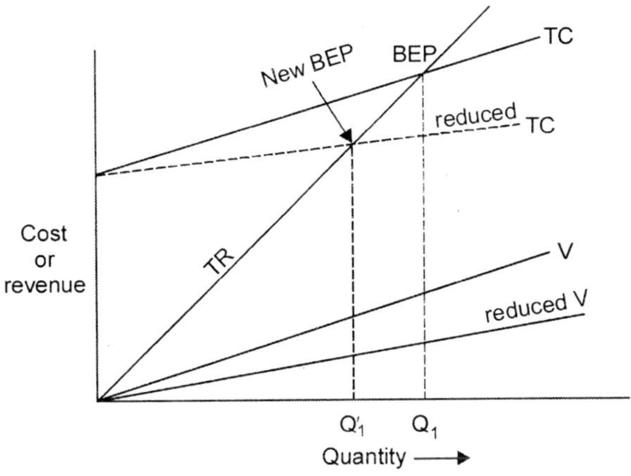

3. Increase the slope of the income line from b to b', the new BEP being

$$Q'_1 = Q_1\left[\frac{b-a}{b'-a}\right]$$

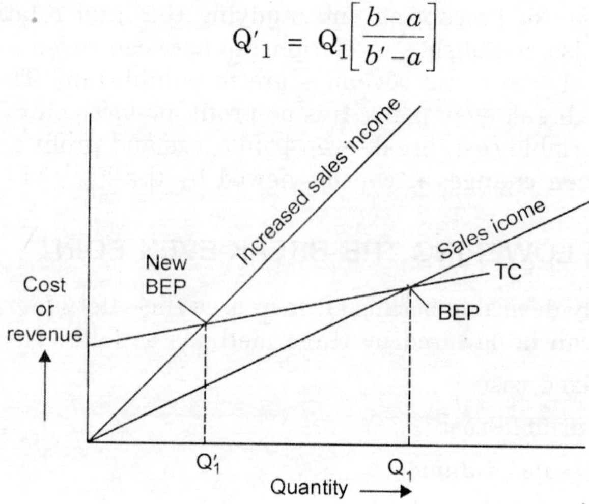

9.3 PROFIT-VOLUME CHART(P-V) CHART

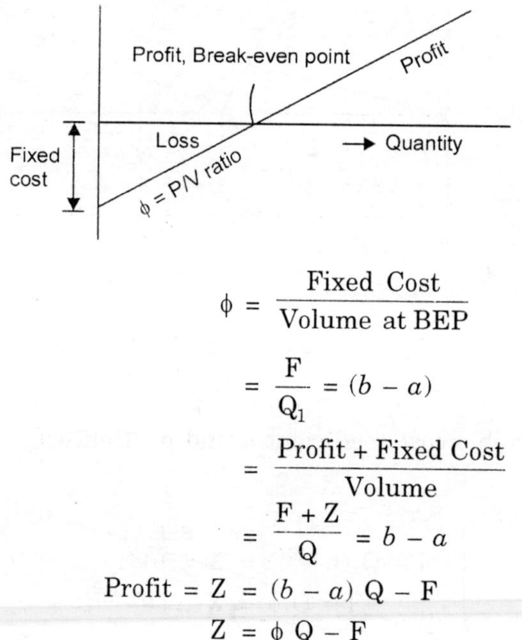

$$\phi = \frac{\text{Fixed Cost}}{\text{Volume at BEP}}$$

$$= \frac{F}{Q_1} = (b - a)$$

$$= \frac{\text{Profit + Fixed Cost}}{\text{Volume}}$$

$$= \frac{F + Z}{Q} = b - a$$

$$\text{Profit} = Z = (b - a)\,Q - F$$

$$Z = \phi\,Q - F$$

9.4 BREAK-EVEN ANALYSIS

Economic BEA acts as the relationship between cost and volume of goods to be produced by a firm. This economic model facilitates the industrial development by providing organization with a simplified profit oriented objective structure which favours innovation efficiency and growth.

The profits raise as a result of excess of revenue over total cost. Total costs includes fixed costs and variable costs. Profit can be expressed as a function of

$$P = \text{Total Revenue} - (\text{Fixed Cost} + \text{Variable Cost})$$

i.e.,

$$P = TR - (FC + VC)$$

9.5 BREAK-EVEN CHART

Break-even chart is a convenient way of graphically representing the relationship between cost and sales revenue (sales income) for different volumes of the output.

Where sales income (sales revenue) is represented by 'bQ'-line, in which Q is the quantity sold and b is the income per unit.

$$BQ = \text{Sales Revenue (Sales Income)}$$

9.6 THE TOTAL COST CONSISTS

F = Fixed Cost (This is independent of quantity produced and includes the salaries of executives, depreciation of plant and machineries etc.)

aQ = Variable Cost (This depends on the number of unit produced. All direct cost which can be identified in per unit of quantity produced are variable cost 'a' is the variable cost per unit *e.g.,* Direct material, labour and other direct cost etc.)

$$\text{Total Cost} = \text{Fixed Cost} + \text{Variable Cost}$$

$$TC = F + aQ$$

9.7 BREAK-EVEN POINT (BEP)

This point on Break Even Chart is the point of intersection of total-cost-line and sales revenue line. This point indicates that is null point at which if the corresponding quantity (QBEQ) is produced and sold the firm will remain in no profit no loss condition.

Below this quantity production and sales gives loss and these above this quantity (QBEQ) gives profit to the firm.

At BEP the

$$\text{Total cost} = \text{Total Revenue}$$

$$F + aQ_1 = bQ_1$$

or

If the plant is operating at Q_2 (greater than Q_1) it is giving profit of Z.

9.8 MARGIN OF SAFETY

Margin of safety is the ratio of profit (Z) to the corresponding fixed cost (F) on quantity produced and sold.

i.e.,

$$d = Z/F$$

or

It is the ratio of excess quantity over break-even quantity to the quantity at BEP (*i.e.,* BEQ).

9.9 PROFIT-VOLUME CHART (FOR SINGLE PRODUCT)

This chart is similar to the BEC. Here the fixed cost is marked as negative quantity on y-axis. The BEP comes on x-axis and that is the point of intersection of income line and x-axis $(b - a)Q$.

Any operation below x-axis incurs a loss and above incurs a profit.

The profitability of the product is indicated by the slope of the income line called the P/V ratio (Profit-Volume ratio) and denoted by ψ.

Solved Problems

Problem 1. *A product involves Rs. 6.000/- annum a fixed cost an yield Rs. 3.000/- profit. The sales income is Rs. 16.000/-. Draw a profit volume chart and find the Profit Volume Ratio.*

Solution.

$$\text{W.K.T. } \varnothing = \frac{P}{V} = \frac{\text{Fixed Cost + profit}}{\text{Volume}}$$

$$\varnothing = \frac{6000 + 3500}{16000}$$

$$\varnothing = 0.59$$

Problem 2. *A plant has a monthly sales income capacity of Rs. 96,000 and is producing 2 products, the data of which is as follows:*

	Product A	*Product B*
Fixed Cost	*Rs. 16,000*	*Rs. 34,000*
Break Even Point	*Rs. 43,000*	*Rs. 35,000*
Profit	*Rs. 48,000*	*Rs. 43,000*

In view of the high fixed cost and the loss incurred by the product B, it was suggested to management that product B should be eliminated and production should be concentrated on product A. Analyse the situation and comment on this suggestion.

Solution.

$$\text{W.K.T. } \varnothing_A = \frac{F}{Q_{1A}} = \frac{16000}{43000} = 0.372$$

$$\varnothing_A = \frac{F + Z}{Q_{2A}}$$

$$0.372 = \frac{16000 + 8000}{Q_{2A}}$$

$$Q_{2A} = \frac{24000}{0.372}$$

$$Q_{2A} = 64516.129$$

Volume of A now being produced.

$$\text{Similarly } \varnothing_B = \frac{F}{Q_{1B}} = \frac{34000}{35000}$$

$$\varnothing_B = 0.971 = \frac{F + Z}{Q_{2B}}$$

$$0.971 = \frac{34000 - 3000}{Q_{2B}}$$

$$Q_{2B} = 31925.84$$

Volume of B; now being produced.

Now considering the full capacity of the plant, that is, both the products are running for the monthly sales income capacity of Rs. 96,000

$$\text{W.K.T. } \varnothing = \frac{F + Z}{Q} \text{ or } \varnothing Q_1 = F + Z$$

$$Z = \varnothing Q - F$$

$$\therefore \quad Z_A = 0.372 \times 96{,}000 - 16{,}000$$

$$Z_A = \text{Rs. } 19{,}720.9$$

$$Z_B = 0.971 \times 96{,}000 - 34{,}000$$

$$Z_B = \text{Rs. } 59{,}216$$

This B should not be discontinued and product B should be manufactured at a sales volume of 96,000 using the full capacity of the plant.

The break-even points between the product A and product B can be determined as follows:

$$\varnothing_A Q - F_A = \varnothing_B Q - F_B$$

$$Q(\varnothing_A - \varnothing_B) = F_A - F_B$$

$$Q = \frac{F_A - F_B}{\varnothing_A - \varnothing_B} = \frac{16000 - 34000}{0.372 - 0.972}$$

$$Q = \frac{-18000}{-0.6}$$

$$Q = 30{,}000$$

If the plant is producing or operating at a quantity less than 30,000, it is better to purchase product A. If the plant is operating more than 30,000 we will go for B at 30,000 both product yield the same profit.

$$Z_A = 0.372 \times 30{,}000 - 16{,}000$$

$$Z_A = -4840$$

$$Z_B = 0.972 \times 30{,}000 - 34{,}000$$

$$= 29160 - 34{,}000$$

$$Z_B = -4840$$

Problem 3. *The break-even point of a product occurs at a sales income of Rs. 1,20,000 but normally the sales income is 1,80,000, the Fixed Cost being Rs, 1,00,000. A new product involved additional cost of Rs. 20,000, but P/V ratio was improved by 20% and sales income increased to 2.40,000. What net profit did the new design yield.*

Solution.

$$\varnothing_A = \frac{F}{Q} = \frac{100000}{120000} \qquad \varnothing_B = 1.2 \times 0.833$$

$$\varnothing_A = 0.833 \qquad \varnothing_B = 0.994$$

$$Z_A = \varnothing_A Q_A - F_A \qquad Z_B = 0.944 \times 2{,}40{,}000$$

$$= 0.833 \times 1{,}80{,}000 - 1{,}00{,}000 \qquad 1{,}20{,}000$$

$$Z_A = \text{Rs. } 49{,}994 \qquad Z_B = \text{Rs. } 1{,}19{,}990.4$$

$$\text{Net Profit} = 1{,}19{,}990.4 - 49{,}994$$

$$\text{Net Profit} = 69{,}996.4$$

The new design will yield a net profit of Rs. 69,996.4 over and above the previous design.

Problem 4. *A firm manufacturing some crokery products shows the following data with the help of a Profit Volume Chart. Calculate the B.E.P. and P.V. ratio for each item. Also find the equivalent B.E.P and P.V. ratio.*

Items	*Fixed Costs*	*Monthly Sales*	*Profit/ Loss*
(1) Vases	*25,000*	*60,000*	*15,000*
(2) Soap Bowls	*30,000*	*48,000*	*5,000 loss*
(3) Tea cups	45,000	78,000	7,000

Solution. Steps:

(1) Now finding the ratio ∅ for each item:

$$\varnothing_1 = \frac{25000 + 15000}{60000} = 0.666$$

$$\varnothing_1 = 0.666$$

$$\varnothing_2 = \frac{30000 - 5000}{48000} = 0.520$$

$$\varnothing_2 = 0.520$$

$$\varnothing_3 = \frac{45000 + 7000}{78000} = 0.666$$

$$\varnothing_3 = 0.666$$

(2) Now finding the individual B.E.P for all items:

$$Q_1 = \frac{F}{\varnothing_1} = \frac{25000}{0.666}$$

$$Q_1 = 37537.4$$

$$Q_2 = \frac{F}{\varnothing_2} = \frac{30000}{0.520}$$

$$Q_2 = 57692.3$$

$$Q_3 = \frac{F}{\varnothing_3} = \frac{45000}{0.666}$$

$$Q_3 = 67567.56$$

(3) Now to find the equivalent ratio:

$$\varnothing_{eq} = \frac{\text{equivalent Fixed Cost} + \text{Profit}}{\text{Equivalent Sales Income}}$$

$$= \frac{100000 + 17000}{186000}$$

$$\varnothing_{eq} = 0.629$$

(4) Equivalent B.E.P = 1,40,000

Problem 5. *A company is dealing with manufacture of nuts and bolts. The fixed costs of company is Rs. 2,00,000/- and variable costs are 50% of sales.*

(a) What sales are required to show a profit of Rs. 25,000/-

(b) What profit should be earned from sales of Rs. 5,00,000/-

(c) What sales must be achieved at the BEP.

Solution. F = Rs. 2,00,000/-

Variable Cost, aQ = 50% of Sales

(a) Z = Rs. 25,000/-

Z = Total Sales Income – (Fixed Cost + Variable Cost)

∴ Sales Income = Z + (FC + VC)

= 25,000 + (2,00,000 + 50% of Total Sales Income)

or 0.5 × Total Sales Income = 2,25,000/-

or Total Sales Income (S) = 225000,0.5 = 4,50,000/-

Hence a Sales (S) of Rs. 4,50,000/- is required to show a profit of Rs. 25,000/-

(b) S = Rs. 5,00,000/-

F = Rs. 2,00,000/-

aQ = 50% of S = 0.5 × 5,00,000 = Rs. 2,50,000/-

Z = S – (F + aQ)

= 5,00,000 – (2,00,000 + 2,50,000)

= Rs. 50,000/-

A profit of Rs. 50,000/- should be earned.

(c) At BEP,

$$F + aQ = bQ$$

Here, F = 2,00,000/-

aQ = 50% of bQ = 0.5 bQ

∴ 2,00,000 + 0.5 bQ = bQ

0.5 bQ = 8,00,000

or bQ = 4,00,000/-

A Sales of Rs. 4,00,000/- should be achieved.

Problem 6 *A company incurs expenditure of fixed costs of Rs. 16,000/- and needs a profit of Rs. 2000/-. Its annual sales income is Rs. 36,000/-. If the selling price is Rs. 8/unit, find production cost/unit and BEQ. Show the result s diagrammatically. Also calculate the profit, if the sales income increases to Rs. 50,000/-*

Solutions.

$$F = \text{Rs. } 16{,}000/-$$

$$Z = \text{Rs. } 2{,}000/-$$

$$bQ = S = \text{Rs. } 36000/-$$

$$b = \text{Rs. } 8/\text{unit}$$

$\therefore$

$$Q = \frac{S}{b} = \frac{36000}{8} = 4{,}500 \text{ units}$$

To find a

$$Z = S - (F + aQ)$$

$$-aQ = Z - S + F$$

$\therefore$

$$a = \frac{Z - S + Z}{-Q}$$

$$= \frac{2000 - 36000 + 16000}{-4500}$$

$$= \text{Rs. } 4/\text{unit.}$$

At BEQ

$$bQ = F + aQ$$

or,

$$Q = \frac{F}{b-a} = \frac{16000}{(8-4)} = 4000 \text{ units}$$

i.e.,

$$\text{BEQ} = 4000 \text{ units.}$$

It Sales Income increases by Rs. 50000/- the

$$\text{New Sales} = S_1 - (F + aQ_1)$$

$$= 50{,}000 - (16{,}000 + 4 \times 6250)$$

$$= \text{Rs. } 9000/-$$

$$\text{Hence, the new profit} = \text{Rs. } 9000/-$$

Problem 7. *A Company produces and Sales 100 units per month at Rs. 20/- each. Variable Costs per unit is Rs. 12/- and Fixed Cost is Rs. 300/- month. It is proposed to reduce sales price by 20%. Find the additional sales required to earn the same profit as before.*

Solution.

$$Q = 100$$

$$b = 120$$

$$a = 80$$

$$F = 8000$$

$$\text{Proposed reduction in sales price} = 20\%$$

Profit is retained to same

$$\text{Increment in sales} = ?$$

$$Z = bQ - (F + aQ)$$
$$= (20 \times 100) - (300 + 12 \times 100)$$
$$= 500/-$$

$$\text{New Sales price, } b_1 = 20(1 - 0.2) = \text{Rs. } 16/\text{units}$$

$$\therefore \quad Z = (b_1 - a)\, Q_1 - F$$

$$Q_1 = \frac{Z + F}{(b_1 - a)} = \frac{500 + 300}{(16 - 12)} = 200$$

$$\text{Additional Sales} = 200 - 100 = 100$$

i.e.,

$$\text{New Sales} = bQ$$
$$= 16 = 200 = \text{Rs. } 3200/-$$

$$\therefore \quad \text{Increment in Sales Income} = 3200 - 20 \times 100 = \text{Rs. } 1200/-$$

Problem 8. *A factory produces 300 units/month. The selling price is Rs. 120 and variable cost is Rs. 80/units. The fixed expenses of the factory amounts to Rs. 8000/month. Calculate.*

(*a*) *The estimated profit in a month wherein 240 units are produced.*

(*b*) *The sales to be made to earn a profit of Rs. 7000/month.*

Solution.

$$Q = 300$$
$$b = 120$$
$$a = 80$$
$$F = 8000$$

(*a*)

$$Q_1 = 240$$
$$Z = bQ_1 - (F + aQ)$$
$$= 120 \times 240 - (8000 + 80 \times 240)$$
$$= \text{Rs. } 1600/-$$

(*b*)

$$Z_1 = 7000$$
$$bQ_2 = F + aQ_2 + Z_1$$
$$120Q_2 = \frac{8000 + 7000}{(120 - 80)} = 375 \text{ units}$$

$$\text{Sales Income} = 375 \times 120 = \text{Rs. } 45000/-$$

Problem 9. *The following figures for profit and sales are obtained for the Company Ratio*

Year	*Sales (Rs.)*	*Profit(Rs.)*	*Calculate*
1985	*20,000*	*2,000*	*(a) Profit-Volume*
1986	*30,000*	*4,000*	*(b) Fixed Cost* *(c) Break Even Sales* *(d) Profit at Sales of Rs. 40,000/-* *(e) Sales to earn a profit of Rs. 5,000/-*

Solution. (*a*) Profit Volume Ratio $\psi = \dfrac{\text{Profit}}{\text{Sales revenue}}$

$$= \frac{4000-2000}{30000-20000} = \frac{2000}{10000} = 0.2$$

(*b*)
$$\psi = \frac{\text{Fixed Cost + Profit}}{\text{Volume}}$$

or,
$$\psi = \frac{F+Z}{Q}$$

$$F = \psi Q - Z$$

$$= 0.2 \times 20{,}000 - 2{,}000 = \text{Rs. } 2{,}000/-$$

or
$$F = 0.2 \times 30.000 - 4.000 = \text{Rs. } 2.000$$

(*c*)
$$\psi = \frac{F}{Q_{BEQ}}$$

∴
$$Q_{BEQ} = \frac{F}{\psi} = \frac{2000}{0.2} = 10{,}000/-$$

(*d*)
$$\psi = \frac{F+Z}{\text{Sales Volume}}$$

(or Sales Revenue)

or,
$$Z = \psi \times \text{Sales Volume} - F$$

or,
$$Z = 0.2 \times 40{,}000 - 2000 = \text{Rs. } 6{,}000/-$$

(*e*)
$$\psi = \frac{F+Z}{\text{Sales}}$$

or,
$$\text{Sales} = \frac{F+Z}{\psi} = \frac{2000+5000}{0.2}$$

$$= \text{Rs. } 35{,}000/-$$

Problem 10. *Sales = Rs. 20,00,000/-*

Variable Cost = Rs. 12,00,000/-

Fixed Cost = Rs. 6,00,000/-

(a) Find BEP

(b) How much profit Company can make

Solution. (*b*)
$$S = F + V + Z$$

∴
$$Z = S - F - V$$

$$= 20{,}00{,}00 - 6{,}00{,}000 - 12{,}00{,}000$$

$$= \text{Rs, } 2{,}00{,}000/-$$

(*a*)
$$\psi = \frac{F + Z}{S} = \frac{600000 + 200000}{2000000}$$
$$= 0.4$$

But at BEP,

$$\psi = \frac{F}{Q_{BEQ}}$$

$$\therefore \quad Q_{BEQ} = \frac{F}{\psi} = \frac{600000}{0.4} = \text{Rs. } 15{,}00{,}000/-$$

EXERCISE PROBLEMS

1.

Sales = Rs. 1,00,000/-

Variable Cost = Rs. 60,000/-

Fixed Cost = Rs. 30,000/-

Find (*a*) Profit – Volume ratio

(*b*) BEP and

(*c*) *m* origin of safety

[**Ans.** (*a*) 0.4]

(*b*) Rs. 75,000/- the BEP lies.

(*c*) 0.3333]

2. The BEP of a product occurs at a sales income of Rs. 60,000, but normally the sales income is Rs. 90,000, the fixed cost being Rs. 50,000/-. A new product involves additional costs of Rs. 10,000, but the profit volume ration is improved by 25% and sales income was increased to Rs. 1,20,000. What net profit did the new design yield.

[**Ans.** Net profit = Rs. 39,950]

3. The fixed costs for the year 75–76 are Rs. 80,000. The estimated sales for the period are valued at Rs. 2,00,000/-. The variable cost per unit for the single product made is Rs. 4/ -. If each unit is sold at Rs. 20/- and the number of units involved coincides with the expected volume of O/P.

(*a*) Construct the Break Even Chart

(*b*) Determine the BEP.

(*c*) Above how many units the company should produce in order to seek profit.

(*d*) Determine the profit earned at turnover of Rs. 1,60,000.

(*e*) Find the margin of safety

(*f*) Measure the angle of incidence

[**Ans.** (*b*) BEP = 5000 units

(*c*) From *b*, if the business is delt above the BEP at which volume is = 5000, the business will earn profit.

(*d*) Profit Z_1 = Rs. 48,000/-

(*e*) Margin of Safety = 1]

4. A lathe manufacturer buys a component at Rs. 10/- each. If he decides to make himself, his fixed and variable cost will be Rs. 10,000/- and Rs. 5/- per component respectively. Decide whether he has to make or buy the component.

Ans. BEQ = 2000 units

5. A company produces 3 products A, B and C. The following are the results for the year 87:

Product	Sales (Rs.)	Variable Cost	Profit
A	10,000	2,000	2,000
B	5,000	3,000	–4000
C	5,000	6,000	–6000

Fixed profit volume ratio for each of the 3 products. Plot the P/V chart and also find the equivalent P/V ratio.

6. There are three alternatives available to meet the demand of a particular product. They are as follows:

(*a*) Manufacuring the product by using process A.

(*b*) Manufactering the product by using procss B

(*c*) Buying the product.

The details are as given in the following table:

Cost-elements	Manufacturing the product by Process A	Manufacturing the product by Process B	Buy
Fixed cost/year (Rs.)	500,000	600,000	
Variable/unit (Rs.)	175	150	
Purchage Price/unit (Rs.)			125

INTRODUCTION, SCOPE OF FINANCE, FINANCE FUNCTIONS

10.1 DEFINITION

Management accounting is the application of professional knowledge and skill in the preparation and presentation of the accounting information is such a way to assist the management in the formulation of the policies, planning and control of the operations undertaking.

10.2 FUNDAMENTALS OF BOOK KEEPING

Definition

It is the system of recording the business transactions in the books of accounts in accordance with the principles of accountancy, in order to ascertain the net profit or net loss of the business and the financial position of the business.

OBJECTIVES OR IMPORTANCE OF BOOK KEEPING

1. To have a permanent record for all the business transactions for future reference.
2. To ascertain the net results of the business for a particular trading period.
3. To know the exact reasons leading to the net profit or net loss of the business.
4. To know the progress of the business from year to year.
5. To know what amounts are due to the business and from whom the amounts are due.
6. To know the exact financial position of the business as on particular date.
7. To have a valuable information for legal and tax purposes.

10.3 SYSTEMS OF BOOK KEEPING

Single Entry System of Book Keeping

It is the system under which a complete record of each and every transaction is not kept. For most of the transactions only one aspect is recorded. Hence it is called a single entry system of book keeping. This is simple and economical but unscientific and incomplete.

Double Entry System of Book Keeping

A system of making two entries in two accounts in opposite directions in each of the parties book for recording a transaction completely is called a double entry system of book keeping.

Advantages of Double Entry System

1. If gives the complete record of all transactions of the business as it records both aspects of every transaction.
2. As both aspects of every transaction are recorded, it is possible to prepare the trial balance and to check arithmetical accuracy in books of accounts.
3. As the nominal accounts are maintained it is possible to determine the net profit or net loss of a business using profit and loss account for any particular year.
4. As the information of assets, liabilities and capital are maintained under double entry system of book keeping, it is possible to determine the financial position of the business.

Single entry	*Double entry*
1. Only one aspect is recorded (maybe debit or credit)	1. Both aspects are recorded
2. It won't give a detailed information	2. It gives a detailed information
3. Simple and economical	3. Not economical
4. Unscientific	4. Scientific
5. Impractical	5. Practical
6. No final accounts	6. Final accounts can be prepared

10.4 TYPE OF ACCOUNTS

1. **Personal Account :** These are the accounts of the party's with whom the business is carried on

   Example: *x a/c, Bank a/c*

2. **Real or Asset Account:** These are the accounts of the assets or the properties with which the business is carried on.

   Example: *Machinery account, furniture account, cash account etc.*

3. **Nominal Account :** These are the expenses and losses which a concern incurs and the incomes and gains which a concern earns in the course of business.

   Example: *Salary account, Rent account, commission account.*

Rules to be followed while debiting and crediting the various accounts.

1. **Personal Account**

   Debit the receiver, credit the giver.

2. **Real Account**

   Debit what all comes in, credit what all goes out.

3. **Nominal Account**

   Debit the expenses and losses, credit the incomes and gains.

10.5 JOURNAL

1. A journal contains a record of various transactions that take place every day together.
2. As the transactions are recorded in the journal in the order of the date, it provides the complete record of each transaction in one place as both the debit and credit aspects are entered together in one entry in the journal.
3. The process of entering the debit and credit aspects of each business transaction in the journal is known as journalising.
4. After the transactions are entered in the journal they become journal entries.

Problem 1. *Journalise the following transactions in the books of Ram.*

1.1.99	Ram commenced the business with cash	Rs. 10,000.
3.1.99	Paid into bank	Rs. 2000
4.1.99	Bought goods for cash	Rs. 2000
5.1.99	Bought office furniture for cash	Rs. 400
6.1.99	Sold goods for cash	Rs. 800
7.1.99	Bought goods from Raman on credit	Rs. 600
10.1.99	Paid rent to the landlord	Rs. 400
12.1.99	Paid salary to the manager	Rs. 200
15.1.99	Sold office furniture for cash	Rs. 400
16.1.99	Received commission	Rs. 100
18.1.99	Bought goods	Rs. 400
20.1.99	Sold goods	Rs. 500
22.1.99	Sold goods to Gopal	Rs. 500
23.1.99	Bought goods from Ramu	Rs. 500
24.1.99	Bought goods from Rai for cash	Rs. 600
25.1.99	Paid postage	Rs. 50
26.1.99	Sold goods to Mohan for cash	Rs. 1000
27.1.99	Paid carriage	Rs. 100
31.1.99	Withdrew cash from the office for the personal use	Rs. 200

Journal entries of ram

Date	*Particulars*	*L.F. No.*	*Dr.* *Rs.*	*Cr.* *Rs.*
1.1.99	Cash account	Dr.	10,000	
	To capital account (being cash introduced in the business by proprietor)			10,000
3.1.99	Bank account	Dr.	2000	
	To cash account (being paid into bank)			2000
4.1.99	Purchase account	Dr.	2000	
	To cash account (being goods bought for cash)			2000
5.1.99	Office furniture a/c	Dr.	400	
	To cash account (being office furniture bought for cash)			400
6.1.99	Cash account	Dr.	800	
	To sales account (being goods sold for cash)			800
7.1.99	Purchase account	Dr.	600	
	To Raman's account (being goods bought from Raman on credit)			600
10.1.99	Rent account	Dr.	400	
	To cash account (being rent paid to the landlord)			400
12.1.99	Salary account	Dr.	200	
	To cash account (being salary paid to the manager)			200
15.1.99	Cash account	Dr.	400	
	To office furniture a/c (being office furniture sold for cash)			400
16.1.99	Cash account	Dr.	100	
	To commission account			100

	(being received commission			
18.1.99	Purchase account	Dr.	400	
	To cash account (being goods bought for cash)			400
20.1.99	cash account	Dr.	500	
	To sales account (being goods sold for cash)			500
22.1.99	Gopal's account	Dr.	500	
	To sales account (being goods sold to Gopal on credit)			500
23.1.99	Purchase account	Dr.	500	
	To Ramu's account (being goods bought from Ramu on credit)			500
24.1.99	Postage account	Dr.	600	
	To cash account (being bought goods from Rai for cash)			600
25.1.99	Postage account	Dr.	50	
	To cash account (being paid for postage)			50
26.1.99	Cash account	Dr.	1000	
	To sales account (being goods sold to Mohan for cash)			1000
27.1.99	carriage account	Dr.	100	
	To cash account (being paid for carriage)			100
31.1.99	Drawings account	Dr.	200	
	To cash account (being withdrawn cash from the office for the personal use)			200

Problem 2. *Journalise the following transactions in the books of Mr.* A:

1.1.98	Mr. A commenced the business with cash	Rs. 50,000
3.1.98	Paid into bank	Rs. 10,000
4.1.98	Bought goods for cash	Rs. 5000
5.1.98	Bought office furniture for	Rs. 5000
6.1.98	Sold goods for cash	Rs. 6000
8.1.98	Sold goods to Murthy on credit	Rs. 4000
8.1.99	Bought goods from Narayan on credit	Rs. 5000
10.1.98	Paid rent to the land!ord	Rs. 3000
16.1.98	Received commission from Suresh	Rs. 200
18.1.98	Bought goods	Rs. 4000

Journal entries or Mr. A

Date	*Particulars*	*L.F. No.*	*Dr. Rs.*	*Cr. Rs.*
1.1.98	Cash account To capital account (being cash introduced into the business by the proprietor)	Dr.	50,000	 50,000
3.1.98	Bank account To cash account (being cash paid into bank)	Dr.	10,000	 10,000
4.1.98	Purchase account To cash account (being goods bought for cash)	Dr.	5000	 5000
5.1.98	Office furniture a/c To cash account (being office furniture bought for cash)	Dr.	5000	 5000
6.1.98	Cash account To sales a/c (being goods sold for cash)	Dr.	6000	 6000
8.1.98	Murthy's account To sales a/c (being goods sold to Murthy on credit)	Dr.	4000	 4000

8.1.98	Purchase account To Narayan's a/c (being goods bought from Narayan on credit)	Dr.	5000	5000
10.1.98	Rent account To cash account (being rent paid to the landlord)	Dr.	3000	3000
16.1.98	Cash account To commission account (being received commission from Suresh)	Dr.	200	200
18.1.98	Purchase account To cash account (being goods bought for cash)	Dr.	4000	4000

10.6 LEDGER

A ledger is a book which contains various accounts. As the transactions are recorded in the journal in the order of the date, the transactions of similar nature may be found on different pages of a journal. As a result one cannot know the net effect of similar transaction on any particular date with the help of the journal.

To know easily the exact position of each and every account as on particular date, the ledger has been introduced.

In the ledger the transactions of similar nature are grouped together in one place in the form of an account through the process called the *posting*.

Points to be noted while preparing a ledger:

1. In a ledger, a separate account should be opened for every account that is found in the journal.
2. The journal entry should be posted to the ledger account in the order of the date.
3. While posting the debit side of a account, the name of the a/c which is credited in the journal should be written in the particulars column (debit side). On the other hand while posting the credit side of an account, the name of the a/c which is debited in the journal should be written in the particulars column (credit side)

Problem 1. *Prepare the various ledger account for prob. 1 of the journal.*

Capital Account

Date	Particulars	L.F. No.	Amount	Date	Particulars	L.F. No.	Amount
31.1.99	To balance c/d		10,000	1.1.99	By cash a/c		10,000
			10,000				10,000
				1.2.99	By balance b/d		10,000

Cash Account

Date	Particulars	L.F. No.	Amount	Date	Particulars	L.F. No.	Amount
1.1.99	To capital a/c		10,000	3.1.99	By Bank account		2000
6.1.99	To sales a/c		800	4.1.99	By purchase account		2000
15.1.99	To furniture a/c		400	5.1.99	By office furniture a/c		400
16.1.99	To commission a/c		100	10.1.99	By Rent account		400
20.1.99	To sales account		500	12.1.99	By salary account		200
26.1.99	To sales account		1000	18.1.99	By purchase account		400
				24.1.99	By purchase account		600
				25.1.99	By postage account		50
				27.1.99	By carriage account		100
				31.1.99	By drawing's account		200
				31.1.99	By balance c/d		6450
			12,800				12,800
1.2.99	To balance b/d		6450				

Purchase Account

Date	Particulars	L.F. No.	Amount	Date	Particulars	L.F. No.	Amount
4.1.99	To cash account		2000	31.1.99	By balance c/d		4100
7.1.99	To Ramains account		600				
18.1.99	To cash account		400				
23.1.99	To Ramu's account		500				
24.1.99	To cash account		600				
			4100				4100
1.2.99	To balance b/d		4100				

Dr **Bank Account** Cr

Date	Particulars	L.F. No.	Amount	Date	Particulars	L.F. No.	Amount
3.1.99	To cash account		2000	31.1.99	By balance c/d		2000
			2000				2000
1.2.99	To balance c/d		2000				

Sales Account

Date	Particulars	L.F. No.	Amount	Date	Particulars	L.F. No.	Amount
				6.1.99	By cash account		800
31.1.99	To balance b/d		2800	20.1.99	By cash account		500
				22.1.99	By Gopal's account		500
				26.1.99	By cash account		1000
			2800	1.2.99	By balance b/d		2800
							2800

Office Furinuture Account

Date	Particulars	L.F. No.	Amount	Date	Particulars	L.F. No.	Amount
5.1.99	To cash account		400	15.1.99	By cash account		400
			400				400

Raman's Account

Date	Particulars	L.F. No.	Amount	Date	Particulars	L.F. No.	Amount
31.1.99	To balance c/d		600	7.1.99	By purchase account		600
			600				600
				1.2.99	By balance b/d		600

Rent Account

Date	Particulars	L.F. No.	Amount	Date	Particulars	L.F. No.	Amount
10.1.99	To cash account		400	31.1.99	By balance c/d		400
			400				400
1.2.99	To balance b/d		400				

Salary Account

Date	Particulars	L.F. No.	Amount	Date	Particulars	L.F. No.	Amount
12.1.99	To cash account		200	31.1.99	By balance c/d		200
			200				200
1.2.99	To balance b/d		200				

Commission Account

Date	Particulars	L.F. No.	Amount	Date	Particulars	L.F. No.	Amount
31.1.99	To balance c/d		100	16.1.99	By cash account		100
			100				100
				1.2.99	By balance b/d		100

Gopal's Account

Date	Particulars	L.F. No.	Amount	Date	Particulars	L.F. No.	Amount
22.1.99	To sales account		500	31.1.99	By balance c/d		500
			500				500
1.2.99	To balance b/d		500				

Ramu's Account

Date	Particulars	L.F. No.	Amount	Date	Particulars	L.F. No.	Amount
31.1.99	To balance c/d		500	23.1.99	By purchase account		500
			500				500
			1.2.99		By balance b/d		500

Postage Account

Date	Particulars	L.F. No.	Amount	Date	Particulars	L.F. No.	Amount
25.1.99	To cash account		50	31.1.99	By balance c/d		50
			50				50
1.2.99	To balance b/d		50				

Carriage Account

Date	*Particulars*	*L.F. No.*	*Amount*	*Date*	*Particulars*	*L.F. No.*	*Amount*
27.1.99	To cash account		100	31.1.99	By balance c/d		100
			100				100
1.2.99	To balance b/d		100				

Drawings Account

Date	*Particulars*	*L.F. No.*	*Amount*	*Date*	*Particulars*	*L.F. No.*	*Amount*
31.1.99	To cash account		200	31.1.99	By balance c/d		200
			200				200
1.2.99	To balance b/d		200				

Problem 2

Capital Account

Date	*Particulars*	*L.F. No.*	*Amount*	*Date*	*Particulars*	*L.F. No.*	*Amount*
31.1.98	To balance c/d		50,000	1.1.98	By cash account		50,000
			50,000				50,000
				1.2.98	By balance b/d		50,000

Cash Account

Date	*Particulars*	*L.F. No.*	*Amount*	*Date*	*Particulars*	*L.F. No.*	*Amount*
1.1.98	To capital account		50,000	3.1.98	By bank account		10,000
6.1.98	To sales account		6,000	4.1.98	By purchase account		5,000
16.1.98	To commission account		200	5.1.98	By office furniture a/c		5,000
				10.1.98	By rent account		3,000
				18.1.98	By purchase account		4,000
				31.1.98	By balance c/d		29,200
			56,200				56,200
1.2.98	To balance b/d		29,200				

Bank Account

Date	Particulars	L.F. No.	Amount	Date	Particulars	L.F. No.	Amount
3.1.98	To cash account		10,000	31.1.98	By balance c/d		10,000
			10,000				10,000
1.2.98	To balance b/d		10,000				

Office Furniture Account

Date	Particulars	L.F. No.	Amount	Date	Particulars	L.F. No.	Amount
5.1.98	To cash account		5,000	31.1.98	By balance c/d		5,000
			5,000				5,000
1.2.98	To balance b/d		5,000				

Purchase Account

Date	Particulars	L.F. No.	Amount	Date	Particulars	L.F. No.	Amount
4.1.98	To cash account		5,000	31.1.98	By balance c/d		14,000
8.1.98	To Narayan's account		5,000				
18.1.98	To cash account		4,000				
			14,000				14,000
1.2.98	To balance c/d		14,000				

Sales Account

Date	Particulars	L.F. No.	Amount	Date	Particulars	L.F. No.	Amount
31.1.98	To balance c/d		10,000	6.1.98	By cash account		6,000
				8.1.98	By Murthy's account		4,000
			10,000				10,000
				1.2.98	By balance b/d		10,000

Murthy's Account

Date	Particulars	L.F. No.	Amount	Date	Particulars	L.F. No.	Amount
8.1.98	To sales account		4,000	31.1.98	By balance c/d		4,000
			4,000				4,000
1.2.98	To balance b/d		4,000				

Narayan's Account

Date	Particulars	L.F. No.	Amount	Date	Particulars	L.F. No.	Amount
31.1.98	To balance c/d		5,000	8.1.98	By purchase account		5,000
			5,000				5,000
				1.2.98	By balance b/d		5,000

Rent Account

Date	Particulars	L.F. No.	Amount	Date	Particulars	L.F. No.	Amount
10.1.98	To cash account		3,000	31.1.98	By balance c/d		3,000
			3,000				3,000
1.2.98	To balance b/d		3,000				

Commission Account

Date	Particulars	L.F. No.	Amount	Date	Particulars	L.F. No.	Amount
31.1.98	To balance c/d		200	16.1.98	By cash account		200
			200				200
				1.2.98	By balance b/d		200

Trial balance is a list of debit and credit balance of ledger account prepared on any particular date to verify whether the entries in the books of accounts are arithmetically correct.

Advantages

1. It gives us all the balances of all the accounts in ledger in one place.
2. Balance of any ledger account can be found out easily from the trial balance without going through the ledger.
3. It is a check on accuracy of the books of accounts.

Points to be remembered while preparing a trial balance.

1. Capital account – Credit balance
2. Debtors – Debit balance
3. Creditors – Credit balance
4. Drawings a/c – Debit balance
5. Bank a/c – Debit or credit

If the bank a/c shows debit balance, it means cash at bank so it should appear on debit side of trial balance, on the other hand if the bank a/c shows credit balance, it means bank overdraft so it should appear on credit side of trial balance

6. Cash a/c – Debit
7. Purchase a/c – Debit
8. Sales a/c – Credit balance
9. Sales returns a/c – Debit balance
10. Purchase returns a/c – Credit
11. Opening stock a/c – Debit
12. Investment a/c – Debit balance
13. Fixed assets a/c – Debit
14. Bills receivables a/c – Debit
15. Bills payable a/c – Credit
16. Accounts of expenses & losses – Debit
17. Accounts of incomes & gains – Credit
18. Closing stock account – Usually it will not appear in trial balance.

Problem 1. *Prepare a trial balance for the following ledger balances:*

	Rs.
Capital account	10,000
Debtors	2,700
Purchases	9,500
Wages	5,000
Goods in trade	2,000

(Contd...)

Sales	14,500
Carriage	150
Machinery	3,500
Drawings	900
Creditors	1,400
Bank	1,500
Rent	450
Sundry expenses	200

Trial Balance

Sl. No.	*Name of account*	*Dr.*	*Cr.*
1.	capital account		10,000
2.	Debtors	2,700	
3.	Purchase account	9,500	
4.	Wages account	5,000	
5.	Goods in trade (os)	2,000	
6.	Sales		14,500
7.	Carriage	150	
8.	Machinery	3,500	
9.	Drawings	900	
10.	Creditors		1,400
11.	Bank	1,500	
12.	Rent	450	
13.	Sundry expenses	200	
	Total	25,900	25,900

Problem 2. *Prepare a trial balance from the following ledger balance:*

1.	Capital a/c (28,000)		28,000
2.	Stock of goods (10,000)	10,000	
3.	Motor car (8000)	8,000	
4.	Discount received (400)		400
5.	Bad debts (400)	400	
6.	Sales (40,000)		40,000
7.	Cash at bank (4000)	4000	
8.	Returns inwards (2000)	2000	
9.	Cash in hand (6000)	6000	
10.	Rent (3500)	3500	
11.	Discount allowed (300)	300	

(Contd...)

12.	Carriage (1500)	1500	
13.	Purchases (15,000)	15,000	
14.	Furniture (5,000)	5,000	
15.	Wages (8,200)	8,200	
16.	Sundry creditors (6,500)		6,500
17.	Commission received (600)		600
18.	Returns outwards (1,000)		1,000
19.	Debtors (200)	200	
20.	Sundry expenses (300)	300	
21.	Interest received (200)		200
22.	Advertisement (500)	500	
23.	Salaries (2,800)	2800	
24.	Plant and machinery (9,000)	9,000	
		76,700	76,700

Problem 3. *Prepare a trial balance of ABC company as on 31st Dec. 1997 for the following ledger balances:*

Sl. No.	*Name of the account*	*Dr.*	*Cr.*
1.	Cash in hand (115)	115	
2.	Bank overdraft (5,025)		5025
	Capital a/c (84,551)		84,551
	Loan (20,000)		20,000
	Wilson and Gray (Cr) (5,535)		5,535
	Leasehold premises (18,000)	18,000	
	Machinery (27,300)	27,300	
	Furniture (1,500)	1,500	
	Stock (51,900)	51,900	
	Mohan and Com. (Dr-5,160)	5,160	
	Purchases (14,558)	14,558	
	Sales (10,760)		10,760
	Returns outwards (448)		448
	Sharp & Brothers Com. (Cr-1,100)		1,100
	John & Brown Com. (Cr-3,360)		3,360
	Dale & Sons Comp. (Cr-5,215)		5,215
	Salaries outstanding (2,000)		2,000
	Blunt & Company (Dr-4,500)	4,500	
	Bad debts (1760)	1,760	
	Wages (1435)	1,435	

(Contd...)

	Customer duty (6860)	6,860	
	Repairs to machinery (250)	250	
	Advertisement (280)	280	
	Repairs to furniture (30)	30	
	Gound rent (600)	600	
	Drawings (410)	410	
	Interest (386)	386	
	Advertising posters (1100)	1,100	
	Salaries (2000)	2,000	
	Discount received (150)		150
		1,38,194	1,38,144

Since the debit and credit amounts are not balanced, the entries in the account is not arithmetically correct.

Prepare a trial balance for the following ledger balances:

Opening stock (500)	500	
Bills receivables (2250)	2,250	
Purchases (19,500)	19,500	
Wages (1,400)	1,400	
Insurance (550)	550	
Sundry debtors (15,000)	15,000	
Carriage inwards (400)	400	
Commission paid (400)	400	
Allow interest on capital (350)	350	
Stationery (225)	225	
Returns inwards (250)	250	
Trade expenses (100)	100	
Office furniture (500)	500	
Cash in hand (250)	250	
Cash at bank (2375)	2375	
Rent & taxes (550)	550	
Carriage outwards (750)	750	
Sales (25,000)		25,000
Bills payables (1500)		1,500
Creditors (9,825)		9,825
Capital (8950)		8,950

10.7 FINAL ACCOUNTS

The trader has to ascertain the trading results (profit or loss) of a business for a particular trading period and the financial position of the business as on particular date in order to ascertain the profit or loss of a business, he has to prepare an account called profit and loss account. The profit and loss a/c is divided into to sections

(*a*) Trading account

(*b*) Profit and loss a/c

(*a*) This is an account, which shows the results of trading i.e., buying and selling goods in order to ascertain the gross profit or gross loss. It is prepared for only those translations which are directly connected with goods.

PROFORMA OF A TRADING ACCOUNT
TRADING ACCOUNT FOR THE PERIOD ENDING.....

Particulars	*Amount (Rs.)*	*Particulars*	*Amount (Rs.)*
To opening stock	....	By sales	....
To purchases	....	*Less* sales returns	....
Less purchase returns	....	....	....
To expenses incurred in acquiring and bringing the materials to the business premises		By closing stock	....
		By profit and loss account	....
• Cartage	....	(Gross loss transferred)	....
• Carriage	....		
• Excise duty	....		
• Octroi	....		
• Import duty	....		
• Clearing charges	....		
To expenses incurred in making goods ready for sale			
• Wages	....		
• Manufacturing expenses	....		
• Power and fuel	....		
• Coal and gas	....		
To profit and loss account (gross profit transferred)	....		
Total	....		....

PROFORMA OF A PROFIT AND LOSS ACCOUNT
PROFIT AND LOSS ACCOUNT FOR THE PERIOD ENDING.....

Dr. *Particulars*	*Amount (Rs.)*	*Particulars*	*Cr.* *Amount (Rs.)*
To trading account	....	By trading account	....

(Gross loss transfered)		(gross profit transferred)	
To salaries	....	By rent received	....
To rent & taxes	....	By commission received	....
To office lighting & heating	....	By interest received	....
To insurance	....	By discount received	....
To printing & stationery	....	By interest on drawing	....
To postage & telegrams	....	By capital a/c	....
To General expenses	....	(Net loss transferred)	
To telephone charges	....		
To discount allowed	....		
To bad debts	....		
To repairs	....		
To depreciation	....		
To commission paid	....		
To carriage inwards	....		
To travelling expenses	....		
To allow interest on capital	....		
To capital account (Net profit transferred)	....		
Total	....		....

Balance Sheet

Liabilities	*Amount*	*Assets*	*Amount*
I Current liabilities		I. Current assets	
* Short-term loans	....	* Cash in hand	....
* Bank overdraft	....	* Cash at bank	....
* Outstanding expenses	....	* Bills receivables	....
* Income received in advance but not earned	....	* Sundry debtors	....
* Creditors	....	* Closing stock	....
* Bills payables	....		
II. Fixed liabilities		* Prepaid expenses	....
* Long-term loans taken	....	* Outstanding incomes	....
		* Short-term investment	

(Contd...)

III. To capital		II. Long-term investment Fixed deposit in bank III. Fixed assets * Tools * Fixtures * Additions to fixtures * Furniture * Plant and machinery * Land & building * Goodwill	
Total	....		....

Problem 1. The following balances are extracted from the books of ABC and company on 31st December 1995. You are required to prepare the trading account, profit and loss account and balance sheet as on that day

Debit balances

Plant & machinery	10,000
Land & building	12,000
Furniture & other equipment	5,000
Sundry debtors	10,000
Trade expenses	1,000
Depreciation	2,000
Cash in hand	10,000
Cash at bank	15,000
Wages	6,000
Repair	1,000
Purchases	60,000
Opening stock	20,000
Discount (allowed)	800
Drawings	1,000
Bill receivables	4,000
Bad debts	400

Credit Balances

Capital	50,000
Sales	90,000
Sundry creditors	12,000
Rent	1,200

(*Contd...*)

Purchase returns	1,000
Bills payables	3,000
Interest	1,000

Trading account for the period ending 31.12.95

Dr. *Particulars*	*Amount (Rs)*	*Particulars*	*Cr.* *Amount (Rs)*
To opening stock	20,000	By sales	90,000
To purchases 60,000			
Less purchase returns 1,000		By closing stock	14,000
	59,000		
To wages	6,000		
To profit and loss a/c			
(Gross profit transferred)	19,000		
	1,04,000		1,04,000

Profit and loss account for period ending 31.12.95

Dr. *Particulars*	*Amount*	*Particulars*	*Cr.* *Amount*
To Trade expenses	1,000	By trading a/c	19,000
To depreciation	2,000	(Gross profit transferred)	
To Repairs	1,000	By rent received	1,200
To discount allowed	800	By interest	1,000
To Bad debts	400		
To capital a/c	16,000		
(Net profit transferred)			
	21,200		21,200

Balance sheet as on 31st Dec. 1995

Liabilities	*A*	*Assets*	
I. Current liabilities		Current assets	
Sundry creditors	12,000	Cash in hand	10,000
Bills payables	3,000	Cash at bank	15,000
		Bills receivables	4,000
II. Fixed liabilities	NIL	Sundry debtors	10,000
		Closing stock	14,000
III. Capital	50,000		
Add net profit	16,000	II. long-term investment	NIL
	66,000		

(Contd...)

Less drawings	1,000 65,000	III. Fixed assets Furniture Plant & machinery Land & building	 5000 10,000 12,000
Total	8,0000		80,000

Problem 2. The following are the balances extracted from the books of XYZ company as 31 Dec. 1998. You are required to prepare trading a/c, profit and loss.

Stock on 1st Jan. 1998	500
Bills receivables	2,250
Purchases	19,500
Wages	1,400
Insurance	550
Sundry debtors	15,000
Carriage inwards	400
Commission (Dr.)	400
Allow interest on capital	350
Stationery	225
Returns inwards	650
Commission (Cr.)	200
Returns outwards	250
Trade expenses	100
Office furniture	500
Cash in hand	250
Cash at bank	2375
Rent and taxes	550
Carriage outwards	725
Sales	25,000
Bills payables	1500
Creditors	9825
Capital	8950

Note: Closing stock was valued at Rs. 12,500

Solutions. Trading account for the period ending 31.12.98.

Dr. *Particulars*	*Amount*	*Particulars*	*Cr.* *Amount*
To opening stock To purchases 19,500	500	By sales 25,000 *Less* sales returns 650	

(*Contd...*)

Less purchase returns 250			24,350
	19,250		
		By closing stock	12,500
To wages	1,400		
To carriage inwards	400		
To profit and loss a/c	15,300		
(Gross profit transferred)			
	36,850		36,850

Profit and loss account for period ending 31.12.1998

Dr. *Particulars*	*Amount*	*Particulars*	*Cr.* *Amount*
To trade expenses	100	By trading a/c	15,300
To insurance	550	(Gross profit transferred)	
To commission paid	400	By commission	200
To interest on capital	350		
To stationery	225		
To rent and taxes	550		
To carriage outwards	725		
To capital account	12,600		
(Net profit transferred)	15,500		15,500

Balance sheet for period ending 31.12.98

Liabilities	*Amount*	*Assets*	*Amount*
I. Current liabilities		I. Current assets	
Sundry creditors	9825	Cash in hand	250
Bills payables	1500	Cash at bank	2375
		Bills receivables	2250
II. Fixed liabilities	NIL	Sundry debtors	15,000
		Closing stock	12,500
III. Capital	8950		
Add Net profit	12600	II. Long-term Investment	NIL
	21,550		
		III. Fixed assets	
		Office furniture	550
	32,875		32,875

Problem 3. The following is the trial balance of Mr. Haribabu associates for year ending 31st March 1996. Prepare trading a/c, Profit and loss a/c, Balance sheet.

	Dr.	Cr.
Sales		2,15,000
Purchases	1,35,500	
Sales returns		3000
Purchase returns	2000	
Sundry debtors	30,500	
Sundry creditors		20,600
Opening stock	20,400	
Salaries and wages	27,500	
Furniture	6,600	
Repairs to shop	3,200	
Postage & telegrams	2,800	
Power and electricity	500	
Trade expenses	1,200	
Rent and taxes	4800	
Bad debts	750	
Fixed deposit in bank	13,500	
Interest on deposit	750	
Insurance		600
Prepaid insurance	200 (250)	
Cash in hand	550	
Bank balance	2,300	
Outstanding salary		2,200
Depreciation on furniture	1,000	
Drawings		4,000
Capital	18,350	

Closing stock was valued at Rs. 19,500

	Particulars	Dr.	Cr.
1.	Sales		2,15,000
2.	Purchases	1,35,500	
3.	Sales returns	3,000	
4.	Purchase returns		2000
5.	Sundry debtors	30,500	
6.	Sundry creditors		20,600
7.	Opening stock	20,400	
8.	Salaries and wages	27,500	
9.	Furniture	6,600	

(Contd...)

10.	Repairs to shop	3,200	
11.	Postage and telegrams	2,800	
12.	Power and electricity	500	
13.	Trade expenses	1,200	
14.	Rent and taxes	4,800	
15.	Bad debts	750	
16.	Fixed deposit in bank	13,500	
17.	Interest on deposit		750
18.	Insurance	600	
19.	Pre-paid insurance	200	
20.	Cash in hand	550	
21.	Bank balance	2,300	
22.	Outstanding salaries		2,200
23.	Depreciation on furniture	1000	
24.	Drawings	4000	
25.	Capital		18,350
		258900	258900

Trading account for the period ending 31st March 1996

Dr. *Particulars*	*Amount*	*Particulars*	*Cr.* *Amount*
To opening stock	20,400	By sales 2,15,000	
To Purchases	1,35,500	Less sales returns 3000	
Less Purchase returns 2,000			2,12,000
	1,33,500		
		By closing stock	19,500
To power and electricity	500		
To profit and loss a/c	77100		
(Gross profit transferred)			
	2,31,500		2,31,500

Profit and loss account for period ending 31st March 1996

Dr. *Particulars*	*Amount*	*Particulars*	*Cr.* *Amount*
To salaries and wages	27,000	By trading account	77,100
To repairs to shop	3,200	(Gross profit transferred)	
To postage and telegrams	2,800		

(Contd...)

To trade expenses	1,200	By interest on deposit	750
To Rent and taxes	4,800		
To Bad debts	750		
To Insurance	600		
To Depreciation on furniture	1000		
To capital account (Net profit transferred)	36,000		
	77,850		77,850

Balance sheet as on 31st March 1996

Liabilities	*Amount*	*Assets*	*Amount*
I. Current liabilities		I. current assets	
Sundry creditors	20,600	Cash in hand	550
Outstanding salaries	2,200	Bank balance	2,300
		Sundry debtors	30,500
II. Fixed liabilities	NIL	Prepaid Insurance	200
		closing stock	19,500
III. Capital	18,350	II. Long-term Invest.	
Add Net Profit	36,000		
	54,350	Fixed deposit in bank	13,500
Less drawings	4,000		
	50,350	III Fixed assets	
		Furniture	6,600
Total	73,150		73,150

1. From the following trial balance of Karthik Enterprise prepare Trading a/c, Profit & Loss a/c for the year ending 30 June 1985 and a balance sheet as on that date.

Debit balances

Opening stock	750
Purchases	1,490
Returns inwards	40
Duty on import goods	260
Carriage on purchases	140
Carriage on sales	200
Office salaries	240
Drawings	400
Rent paid	180

(*Contd...*)

General Expenses	150
Bank balance	300
Cash in hand	100
Sundry debtors	1000
Buildings	2000
Machinery	1000
Bills receivables	250
Depreciation	200
Horses and carts	150
Interest	90
Discount	10

Credit balances

Capital	2625	
Returns outwards	45	
Sales	3810	
Commission earned	200	
Bills payables	1500	
Sundry creditors	770	
Adjustments		
Closing stock	985	and
Rent due but not paid	30	
(outstanding expenses)		

Trading account for period ending 30th June 1985

Dr. *Particulars*	*Amount*	*Particulars*	*Cr.* *Amount*
To opening stock	750	By sales 3,810	
To purchase 1,490		*Less* Sales returns 40	
Less Purchase returns 45			3770
	1445		
		By closing stock	985
To duty on import goods	260		
To carriage purchases	140		
To profit & loss a/c	2160		
(Gross profit transferred)			
	4755		4755

Profit and loss account for the period ending 30th June 1985

Dr. *Particulars*	*Amount*	*Particulars*	*Cr.* *Amount*
		By trading a/c	2160
To carriage on sales	200	(Gross profit transferred)	
To office salaries	240		
To Rand	180	By commission earned	200
Rent outstanding 30			
	210		
To General Expenses	150		
To depreciation	200		
To Interest	90		
To discount	10		
To capital a/c	1260		
Net profit transferred)			
	2360		2360

Balance sheet as on 30th June 1985

Liabilities	*Amount*	*Assets*	*Amount*
I. **Current liabilities**		I. **Current assets**	
Sundry creditors	770	Bank balance	300
Bills payables	1500	cash in hand	100
outstanding Rent	30	Sundry debtors	1000
		Bills receivables	250
II. **Fixed liabilities**	NIL	Closing stock	985
Capital	2625	II. Long-term Invest.	NIL
+Add profit	1260		
	3885	III. **Fixed assets**	
Less Drawings	400	Buildings	2000
	3485	Machinery	1000
		Horses and carts	150
	5,785		5,785

FINANCIAL RATIO ANALYSIS

11.1 INTRODUCTION

A basic limitation of the traditional financial statement comprising the balance sheet and the profit & loss account which will not give all the information regarding the financial operation of a firm. The financial statements provide a summarised view of the financial position and operation of the firm. The analysis of financial statements is an important aid to financial analysis. The first task of the financial analyst is to select the information relevant to the decision under consideration from the total information contained in the financial statement. The second step involved in financial analysis is to arrange the information in a way to highlight significant relationships. The final step is interpretation and drawings of inferences and conclusions.

In brief, financial analysis is the process of selection, relation and evaluation.

This chapter focuses on ratio analysis as the most widely used technique of financial statement analysis.

11.2 NATURE OF RATIO ANALYSIS

Ratio analysis is a widely used tool of financial analysis. It is defined as the systematic use of ratio to interpret the financial statements so that strengths and weaknesses of a firm as well as its historical, performance and current financial condition can be determined. The term ratio refers to the numerical or quantitative relationship between two items/variables. Comparison is relationship between two items variables, with related facts. There are four types of comparisons involved:

(1) Trend ratios

(2) Inter-firm comparison

(3) Comparison of items within a single year's financial statement of a firm.

(4) Comparison with standard or plans.

11.3 TYPES OF RATIOS

Ratios can be classified in four broad groups:

(*i*) Liquidity ratios

(*ii*) Capital structure/leverage ratios

(*iii*) Profitability ratios

(*iv*) Activity ratios

11.3.1 Liquidity ratios

The importance of adequate liquidity in the sense of the ability of a firm to meet current short-term obligations when they become due for payment can hardly be over-stressed. Liquidity is a prerequisite for the very survival of a firm. The short-term creditors of the firm are interested in the short-term solvency or liquidity of a firm.

A proper balance between the two contradictory requirements i.e., Liquidity and Profitability, is required for efficient financial management. The ratios which indicate the liquidity of a firm are :

(*i*) Net working capital

(*ii*) Current ratios

(*iii*) Acid test/quick ratios

(*iv*) Super quick ratios

(*v*) Turnover ratios.

Net Working Capital (NWC)

Net working capital represents the excess of current assets over current liabilities. Current assets refers assets which in the normal course of business get converted into cash over a short period. Current liabilities are those liabilities which are required to be paid in short period (1 year). An enterprise should have sufficient NWC in order to meet the claims of the creditors and meeting the day-to-day needs of business.

Current Ratio

The current ratio is the ratio of total current assets to total current liabilities. It is calculated by dividing current assets by current liabilities.

$$\text{Current ratio} = \frac{\text{Current assets}}{\text{Current liabilities}}$$

The current assets of a firm are those assets which can be ordinary course of business, converted into cash within a short period of time (1 year) which includes cash, bank balance, securities, inventory of raw materials, semi-finished and finished goods. The current liabilities which are short-term maturing obligations to be met (1 year) consist of trade creditors, bills payable, bank credit, provision for taxation, dividend payable and outstanding expenses.

Acid Test or Quick Ratio

Any defect in current ratio fails to convey any information on the composition of the current assets of a firm.

The acid test ratio is a measure of liquidity designed to overcome this defect of the current ratio. It is also referred to as quick ratio, because it is a measurement of a firm's ability to convert its current assets quickly into cash in order to meet its current liabilities. It is a measure of quick or acid liquidity.

$$\text{Acid test ratio} = \frac{\text{Quick Assets}}{\text{Current liabilities}}$$

Turn-over Ratio

Another, way of examining the liquidity is to determine how quickly certain current assets are converted into cash. The ratios to measure these are referred to as turnover ratios. These are three turnover ratios that are relevant:

(1) Inventory turnover ratio

(2) Debtors' turnover ratio

(3) Creditors' turnover ratio

$$\text{Inventory turnover ratios} = \frac{\text{Cost of goods sold}}{\text{Average inventory}}$$

if high ratio is good from the viewpoint of liquidity and vice versa. A low ratio would signify that inventory does not sell fast and stays on the sheet or in the warehouse for a long time.

$$\text{Debtors' turnover ratios} = \frac{\text{Net credit sales}}{\text{Average debtors}}$$

The ratio measures how rapidly debts are collected. A high ratio indicates shorter time lag between credit sales and cash collection. A low ratio shows that debts are not being collected rapidly.

Creditors' Turnover Ratio

It is a ratio between net credit purchases and the average amount of creditors outstanding during the year.

$$\text{Creditors' turnover ratio} = \frac{\text{Net credit purchases}}{\text{Average creditors}}$$

A low turnover ratio reflects liberal credit terms granted by suppliers, while a high ratio shows that accounts are to be settled rapidly. The creditors turnover ratio is an important tool of analysis as a firm can reduce its requirement of current assets relying on supplier's credit.

11.3.2 Leverage/Capital Structure Ratios

The second category of financial ratios is leverage or capital structure ratios. The leverage or capital structure ratios may be defined as financial ratios which throw light on the long-term solvency of a firm reflected in its ability to assure the long-term creditors with regard to (*i*) Periodic payment of interest during the period of the loan (*ii*) Repayment of principal on maturity or in predetermined instalments on due dates.

There are two aspects of the long-term solvency of a firm: (*i*) ability to repay the principal when due and (*ii*) regular payment of the interest.

Accordingly, there are two different types of leverage ratios. First, ratios which are based on the relationship between borrowed funds and owner's capital. These ratios are computed from the balance sheet and have many variations, such as (*a*) debt-equity ratios, (*b*) debt-assets ratio, (*c*) equity-assets ratio.

The second type of capital structure ratios, popularly called coverage ratios are calculated from the profit and loss account. Included in this category are: (*a*) interest coverage ratio, (*b*) dividends coverages (*c*) total fixed charges coverage ratio, (*d*) cash flow coverage ratio and (*e*) debt service coverage ratio.

Debt-Equity Ratios

The relationship between borrowed funds and owner's capital is a popular measure of the long-term financial solvency of a firm. This ratio reflects the relative claims of creditors and shareholders against the assets of the firm. The relationship between outsiders' claims and owner's capital can be shown in different ways and accordingly there are many variants of the debt equity (D/E) ratio.

$$\text{D/E ratio} = \frac{\text{Long-term debt}}{\text{Shareholder's equity}}$$

or

$$= \frac{\text{Total debt}}{\text{Shareholder's equity.}}$$

Debt to Total Capital Ratio

The relationship between creditors' funds and owner's capital can also be expressed in terms of leverage ratios. It is called Debt to Total Capital Ratio.

$$\text{Debt to total capital ratio} = \frac{\text{Long-term debt}}{\text{Permanent capital}}$$

Another approach to calculating the debt to capital ratio is to relate the total debt to the total assets of the firm. The total debt of the firm comprises long-term debt plus current liabilities. The total assets consist of permanent capital plus current liabilities.

$$\text{Debt to total assets/capital ratio} = \frac{\text{Total debt}}{\text{Permanent capital} + \text{Current liabilities}}$$

Coverage Ratios

The second category of leverage ratios are coverage ratios. There ratios are computed from information available in the profit and loss account. The obligations of a firm are normally met out of the earnings or operating profits.

Its claims consist of i) interest on loans ii) preference dividend and iii) amortisation of principal or repayment of the instalment of loans or redemption of preference capital on maturity.

Interest Coverage Ratios

This ratio measures the debt servicing capacity of a firm, as fixed interest on long-term loan is concerned. It is determined by dividing the operating profits or earning before interest and taxes (EBIT) by fixed interest changes on loans.

$$\text{Interest coverage} = \frac{\text{EBIT}}{\text{Interest}}$$

Dividend Coverage Ratio

It measures the ability of a firm to pay dividend on preference shares which carry a stated rate of return. This ratio is the ratio of net profits after taxes (EAT) and the amount of preference dividend.

$$\text{Dividend Coverage} = \frac{\text{EAT}}{\text{Presence dividend}}$$

Total Coverage Ratio

The total coverage ratio has wider scope and takes into account all the fixed obligations of a firm *i.e.*, (*i*) interest on loan, (*ii*) preference dividend, (*iii*) lease payments and (*iv*) repayment of principal.

$$\text{Total Coverage} = \frac{\text{EBIT} + \text{lease payment}}{\text{Interest} + \text{Lease payment} + (\text{Preference dividend} + \text{Instalment of principal})/(1-t)}$$

The above coverage ratio suffers from one major limitation, i.e., relates the firm's ability to meet its various financial obligations to its earnings. The ratio is determined

Total cash flow coverage ratio

$$= \frac{\text{EBIT} + \text{lease payment} + \text{Depreciation} + \text{Non-cash expenses}}{\text{Lease payment} + \text{Interest} + \left(\frac{\text{Principal repayment}}{(1-t)}\right)}, \quad \frac{\text{Preference dividend}}{(1-t)}$$

11.3.3 Profitability Ratios

Profitability is a measure of efficiency and the search for it provides an incentive to achieve efficiency. Profitability also indicates public acceptance of the product and shows that the firm can produce competitively. The profitability ratios are designed to provide answers to questions such as

(*i*) Is the profit earned by the terms adequate?

(*ii*) What rate of return does it present?

(*iii*) What is the rate of profit for various divisions and segments of the firm?

(*iv*) What is the earnings per share ?

(*v*) What amount was paid in dividends?

(*vi*) What is the rate of return to equity holders?

Profitability ratios can be determined on the basis of either sales or investments. In relation to sales are

(*a*) Profit margin

(*b*) Expenses ratio or operating ratio.

In relation to investments are

(*a*) return on assets

(*b*) return on capital employed

(*c*) return on shareholder's equity

Profit Margin

Profit margin measures the relationship between profit and sales. There are two types of profit margin: (1) Gross profit margin, (2) Net profit margin.

$$\text{Gross profit margin} = \frac{\text{Gross Profits}}{\text{Sales}} \times 100$$

Since, gross profit = sales – cost of goods sold, the gross profit margin can also be calculated as follows :

$$\text{Gross profit margin} = \frac{\text{Sales} - \text{Cost of goods sold}}{\text{Sales}} \times 100$$

A high ratio of gross profits to sales is a sign of good management as it implies that the cost of production of the firm is relatively low.

A relatively low gross margin is definitely a danger signal, warranting a careful and detailed analysis of the factors responsible for it. The factors contributing are (*i*) a high cost of production, (*ii*) a low selling price resulting from severe competition.

Net Profit Margin

The measure of the relationship between net profits and sales of a firm depends on the concepts of net profit employed. This ratio can be computed in two ways:

(*i*) $\text{Operating profit ratio} = \dfrac{\text{Earnings before interest and taxes (EBIT)}}{\text{Sales}}$

(*ii*) $\text{Net profit ratio} = \dfrac{\text{Earnings after interest and taxes (EAT)}}{\text{Sales}}$

A high net profit margin would ensure adequate return to the owners as well as enable a firm to withstand adverse economic conditions when selling price is declining, cost of production is rising and demand for the product is falling.

A low net profit margin has the opposite implications.

Expense Ratio

The term expenses includes

(*i*) cost of goods sold

(*ii*) administrative expenses

(*iii*) selling and distribution expenses.

There are different variants of expense ratios:

(*i*) $\text{Cost of goods sold ratio} = \dfrac{\text{Cost of goods sold}}{\text{Net sales}} \times 100$

(*ii*) Operating expenses ratio = $\dfrac{\text{Administrative expenses} + \text{Selling expenses}}{\text{Net sales}}$

(*iii*) Administrative expenses ratio = $\dfrac{\text{Administrative expenses}}{\text{Net sales}} \times 100$

(*iv*) Selling expenses ratio = $\dfrac{\text{Selling expenses}}{\text{Net sales}} \times 100$

(*v*) Operating ratio = $\dfrac{\text{Cost of goods sold} + \text{Operating expenses}}{\text{Net sales}} \times 100$

(*vi*) Financial expenses ratio = $\dfrac{\text{Financial expenses}}{\text{Net sales}} \times 100$

Return on Assets (RoA)

The profitability ratio is measured in terms of the relationship between net profits and assets. The RoA may also be called profit-to-assets ratio.

The assets may be defined as (*i*) total assets (*ii*) fixed assets and (*iii*) tangible assets.

$$\text{Return on Assets} = \frac{\text{Net profit after taxes}}{\text{Average total assets}}$$

Return on Capital Employed

A comparison of this ratio with similar firms, with the industry average and overtime would provide sufficient insight into how efficiently the long-term funds of owners and creditors are being used.

$$\text{Return on capital employed} = \frac{\text{Net profit after taxes}}{\text{Average total capital employed}}$$

$$\text{Return on shareholders equity} = \frac{\text{Net profit after taxes}}{\text{Average total shareholder's equity}}$$

$$\text{Return on equity funds} = \frac{\text{Net profit after taxes} - \text{Preference dividend}}{\text{Average ordinary shareholder's equity or net worth.}}$$

$$\text{Earnings per share} = \frac{\text{Net profit available to equity holders}}{\text{Number of ordinary shares outstanding}}$$

$$\text{Dividend per share} = \frac{\text{Net profit after interest \& preference dividend paid to ordinary shareholders}}{\text{Number of ordinary shareholders}}$$

$$\text{Earning power} = \text{Net profit margin} \times \text{Investment turnover.}$$

$$\text{Net profit margin} = \frac{\text{Net profit after taxes}}{\text{Sales}}$$

$$\text{Investment turnover} = \frac{\text{Sales}}{\text{Average total investment}}$$

11.3.4 Activity Ratios

Activity ratios are concerned with measuring the efficiency in asset management. It is also called efficiency ratios or asset utilisation ratios. An activity ratio may be defined as a test of the relationship between sales and the various assets of a firm. Depending upon the various type of assets, there are various types of activity ratios.

Inventory Turnover Ratios

It measures the relationship between the cost of goods sold and the inventory level.

$$\text{Inventory turnover} = \frac{\text{Cost of goods sold}}{\text{Average inventory}}$$

$$\text{Cost of goods sold} = \text{Opening stock} - \text{Manufacturing cost} - \text{Closing stock.}$$

$$\text{Inventory turnover} = \frac{\text{Sales}}{\text{Closing inventory}}$$

$$\text{Raw material turnover} = \frac{\text{Cost of raw material}}{\text{Average raw material inventory}}$$

$$\text{Work-in-progress turnover} = \frac{\text{Cost of goods manufactured}}{\text{Average work-in-progress inventory}}$$

Receivables Turnover Ratio and Collection Period

$$\text{Debtors' turnover} = \frac{\text{Credit sales}}{\text{Average debtors} + \text{Average bills receivable}}$$

$$\text{Average collection period} = \frac{\text{Months in a year}}{\text{Debtors' turnover}}$$

$$\text{Total assets turnover} = \frac{\text{Cost of goods sold}}{\text{Average total assets}}$$

$$\text{Fixed assets turnover} = \frac{\text{Cost of goods sold}}{\text{Average fixed assets}}$$

$$\text{Capital turnover} = \frac{\text{Cost of goods sold}}{\text{Average capital employed}}$$

$$\text{Current assets turnover} = \frac{\text{Cost of goods sold}}{\text{Average current assets}}$$

$$\text{Working capital turnover ratio} = \frac{\text{Cost of goods sold}}{\text{Net working capital}}$$

Evaluation of a Firm's Earning Power

RONA or ROE is the measure of the firm's operating performance. It indicates the firm's earning power. It is product of the asset turnover, gross profit margin and operating leverage.

$$\text{RONA} = \frac{\text{EBIT}}{\text{NA}} = \frac{\text{Sales}}{\text{NA}} \times \frac{\text{GP}}{\text{Sales}} \times \frac{\text{EBIT}}{\text{GP}}$$

All firms would like to improve their RONA. A firm can convert it's RONA into an impressive ROE through financial efficiency. Financial leverage and debt equity ratios affect ROE and reflect efficiency.

ROE is a product of RONA and financial leverage ratios.

$$\text{ROE} = \text{Operating performance} \times \text{leverage factor.}$$

$$\text{ROE} = \frac{\text{PAT}}{\text{NW}} = \frac{\text{EBIT}}{\text{NA}} \times \frac{\text{PAT}}{\text{EBIT}} \times \frac{\text{NA}}{\text{NW}}$$

Comparative Statement Analysis

A simple method of tracking periodic changes in the financial performance of a company is to prepare comparative statements. Comparative financial statements will contain items at least for two periods.

Changes increase and decrease—in income statement and balance sheet over period can be shown in two ways:

(1) aggregate changes and

(2) proportional changes.

Aggregate changes can be indicated by drawing special columns for aggregate amount

Proportional changes are indicated by recording percentage calculated in relation to a common base in special columns.

Problems and Dangers of Budgeting

Budgeting is a systematic approach to the solution of problems. It suffers from certain problems and limitations. The major problems in developing a budgeting system are:

* Seeking the support and involvement of all levels of management.
* Developing meaningful forecast and plans.
* Educating all individuals to be involved in the budgeting process and gaining their full participation.
* Establishing realistic objectives, policies, procedures and standards of desired performance.

* Applying the budgeting system in a flexible manner.
* Maintaining effective follow-up procedure and adapting the budgeting system whenever the circumstance changes.

Management must consider the following limitations in using the budgeting system to solve managerial problems:

(1) Management judgement
(2) Continuous adaptation
(3) Implementation
(4) Management complacency
(5) Unnecessary details
(6) Goal conflict
(7) Evaluation deficiencies
(8) Unrealistic targets

11.4 RATIO ANALYSIS OR ACCOUNTING RATIOS

A ratio is a mathematical expression used as a technique of analysis and interpretation of financial statement. It is a relationship expressed between two figures in the final accounts. The ratio analysis is required from the point of view of creditors, shareholders, management and government. The creditors are interested in solving liquidity and solvency ratios. Shareholders are interested in solvency ratios and capital gearing ratios. The government is interested in profit earning capacity and solvency ratios. The management is interested in all the above ratios.

Advantages of Ratio Analysis

1. It helps to analyse the probable casual relation among the different items of analysing and scrutinising the past results.
2. This result helps the management to prepare the budget and to prepare the future plan of action.
3. It helps to take the time dimension into account by trend analysis *i.e.*, whether the firm is improving or deteriorating over the no. of years.
4. It throws light on degree of efficiency of management and the utilization of the assets.

Disadvantages or Limitations

1. The ratios are calculated on the basis of past results, so it does not help properly to predict the future and to prepare the budget.
2. It is very difficult to ascertain the standard ratios in order to make a proper comparison.
3. The ratio analysis depends on figures appeared in the financial statements. But in most of the cases these figures are not exactly correct. So, as a result, the correct picture cannot be drawn from the ratio analysis.

11.5 CLASSIFICATION OF RATIOS

I. Balance Sheet Ratios

(a) Current ratio $= \dfrac{\text{Current assets}}{\text{Current liabilities}}$

Current assets are cash in hand, cash at bank, short-term investments. bills receivable, stock in trade, debtors, inventory, prepaid expenses etc.

The current liabilities are bank overdraft, bills payable, provision for taxation, outstanding expenses, creditors etc.

Generally 2:1 is considered as ideal ratio

(b) Acid test Ratio or Liquidity Ratio or Quick Ratio $= \dfrac{\text{Liquid assets}}{\text{Current liabilities}}$

Liquid assets = Current assets – Stock.

(c) Absolute liquidity ratio $= \dfrac{\text{Cash in hand + bank balances + Short-term loans}}{\text{Current liabilities}}$

(d) Ratio of Inventory to the working capital $= \dfrac{\text{Inventory}}{\text{working capital}}$

Inventory means stock at the end.

Working capital = Current assets – Current liabilities

Ideal ratio = 1 : 1

(e) Ratio of current assets tõ fixed assets $= \dfrac{\text{Current assets}}{\text{Net fixed assets}}$

Net fixed assets = Fixed – Dep.

(f) Debt equity ratio $= \dfrac{\text{External equities}}{\text{Internal equities}}$

or $= \dfrac{\text{Outsider's fund}}{\text{Shareholder's fund}}$

The external equities are long-term debts like debentures, long-term loans taken etc.

Internal equities are preferential shares, equity share, profit and loss a/c balances, capital reserve etc.

Ideal ratio is 2 : 1

(g) Proprietory Ratio $= \dfrac{\text{Shareholder's fund}}{\text{Total assets}}$

Total assets = Fixed assets + Current assets

(*h*) Capital Gearing Ratio = $\dfrac{\text{Equity share capital}}{\text{Fixed interest bearing securities}}$

Fixed interest bearing securities means preferential shares and debentures.

II. Profit and Loss a/c Ratios

(*a*) Gross Profit Ratio = $\dfrac{\text{Gross profit}}{\text{Net sales}} \times 100$

Net sales = Sales – Sales returns.

(*b*) Operating ratio = $\dfrac{[\text{Cost of goods sold + Operating expenses}]}{\text{Net sales}}$

Cost of goods sold = Opening stock + Purchases + Manufacturing expenses – closing stock

or Cost of goods sold = Sales – Gross profit

The operating expenses are like office, administrative, selling and distribution expenses.

(*c*) Net profit ratio = $\dfrac{\text{Net profit after taxes}}{\text{Net sales}}$

Net profit = Gross profit + operating and non operating income – operating and non operating expenses.

(*d*) Operating expenses ratio = $\dfrac{\text{Operating expenses}}{\text{Net sales}}$

(*e*) Operating profit ratio = $\dfrac{\text{Operating profit}}{\text{Net sales}}$

Pperating profit = Net profit – Nonoperating expenses + Non-operating incomes

III. Combined Ratios

(*a*) Stock turnover ratio or inventory turnover ratio

$$= \frac{\text{Cost of goods sold}}{\text{Average stock or inventory}}$$

$$\text{Avg. stock} = \frac{\text{Opening stock + Closing stock}}{2}$$

(*b*) Debtors turnover ratio or Avg. collection period

$$= \frac{\text{Debtors + Bills receivables}}{\text{Credit sales}} \times 365$$

(*c*) Working capital turnover ratio $= \dfrac{\text{Net sales}}{\text{Net working capital}}$

(*d*) Fixed assets turnover ratio $= \dfrac{\text{Net sales}}{\text{Net fixed assets}}$

(*e*) Capital turnover ratio $= \dfrac{\text{Net sales}}{\text{Capital employed}}$

Capital employed = Total assets – Current liabilities

(*f*) Total assets turnover ratio $= \dfrac{\text{Net sales}}{\text{Total assets}}$

(*g*) Return on shareholders investment or Return on proprietors fund

$$= \frac{\text{Net profit}}{\text{Shareholders fund}}$$

(*h*) Return on total resources $= \dfrac{\text{Net profit}}{\text{Total assets}}$

(*i*) Return on capital employed $= \dfrac{\text{Net profit}}{\text{Capital employed}}$

(*j*) Creditors turnover ratio $= \dfrac{\text{Creditors + Bills payables}}{\text{Credit purchases}} \times 365$

Problem 1. *Following is the summarised profit and loss a/c for XYZ Ltd. for the year ending 31st Dec. 1987 and a balance sheet as on that date.*

Trading A/c and Profit and Loss A/c

Particulars	*Amount*	*Particulars*	*Amount*
To opening stock	99,500	By sales	8,50,000
To purchases	5,45,250		
To manufacturing expenses	14,250	By closing stock	1,49,000
To gross profit	3,40,000		
	9,99,000		9,99,000
To operating expenses		By gross profit	3,40,000
• Selling and distribution	30,000		
• Administrative expenses	1,50,000	By non-operating income	
• Finance	15,000		
To Nonoperating expenses		Int	3,000
• Loss on sale of assets		Profit on sale of shares	6,000
To Net profit	4,000		
	1,50,000		
	3,49,000		3,49,000

BALANCE SHEET

Liabilities	*Amount*	*Assets*	*Amount*
2000 Equity shares at Rs. 100 each	2,00,000	Land and Building	1,50,000
		Plant and m/c	80,000
Capital reserve	90,000	Stock in trade	1,49,000
		Sundry debtors	71,000
Current liabilities	1,30,000	Cash and bank balances	30,000
Profit and loss a/c balance	60,000		
	4,80,000		4,80,000

From the above statement you are required to calculate the following ratios:

Current ratio
Operating ratio
Stock turnover ratio
Return on total resources
Turnover of fixed assets
Turnover of total assets
Return on capital employed
Working capital turnover ratio

Solution. $$\text{Current ratio} = \frac{\text{Current assets}}{\text{Current liabilities}}$$

Current assets are stock in trade, sundry debtors, cash and bank balances.

$$\text{Current assets} = 1,49,000 + 71,000 + 30,000$$
$$= 2,50,000$$
$$\text{Current ratio} = \frac{2,50,000}{1,30,000}$$
$$= 1.92 : 1$$

• $$\text{Operating ratio} = \frac{\text{Cost of goods sold + Operating expenses}}{\text{Net sales}}$$

$$\text{Cost of goods sold} = \text{Sales} - \text{Gross profit}$$
$$= 8,50,000 - 3,40,000$$
$$= 5,10,000$$

Operating expenses are selling and distribution + administrative + finance

$$= 30,000 + 1,65,000$$
$$= 1,95,000$$

• $$\text{Operating ratio} = \frac{5,10,000 + 1,95,000}{8,50,000}$$
$$= 0.829 : 1$$

- Stock turnover ratio $= \dfrac{\text{Cost of goods sold}}{\text{Avg. stock}}$

Avg. stock $= \dfrac{\text{Opening stock + Closing stock}}{2}$

$= \dfrac{99,500+1,49,000}{2} = 124250$

Ratio $= \dfrac{5,10,000}{1,24,250} = 4.1 : 1$

- Return on total resources $= \dfrac{\text{Net profit}}{\text{Total assets}}$

$= \dfrac{1,50,000}{4,80,000} = 0.3125 : 1$

- Turnover of fixed assets ratio $= \dfrac{\text{Net sales}}{\text{Net fixed assets}}$

$= \dfrac{8,50,000}{2,30,000} = 3.69 : 1$

Net fixed assets = 1,50,000 + 80,000

= Land and B + P and m/c

- Turnover of total assets ratio $= \dfrac{\text{Net sales}}{\text{Net total assets}}$

$= \dfrac{8,50,000}{4,80,000} = 1.77 : 1$

- Return on capital employed $= \dfrac{\text{Net profit}}{\text{Capital employed}}$

Capital employed = Total assets – Current liabilities

= 4,80,000 – 1,30,000

= 3,50,000

$= \dfrac{1,50,000}{3,50,000} = 0.428 : 1$

- Working capital turnover ratio $= \dfrac{\text{Net sales}}{\text{Net working capital}}$

$= \dfrac{8,50,000}{\text{Current assets} - \text{Current liabilities}}$

$= \dfrac{8,50,000}{1,20,000} = 7.08 : 1$

Problem 2. *The following is the balance sheet as on 31st March 1995.*

Liabilities		*Amt.*	*Assets*	*Amt.*
Share Capital	5,000	5,00,000	Land and building	5,00,000
Shares at Rs. 100			P and m/c	2,00,000
each			Stock	1,50,000
			Debtors	2,50,000
Capital reserve		4,00,000	Cash at bank	1,50,000
Profit and loss a/c balance		1,50,000		
Creditors		2,00,000		
		12,50,000		12,50,000

Trading and profit and loss a/c

O.S.	2,50,000	By sales	18,00,000
Purchases	10,50,000	By closing stock	1,50,000
To gross profit	6,50,000		
	19,50,000		19,50,000
To selling and dis.	1,00,000	By Gross profit	6,50,000
To admini. exp.	2,30,000	By profit on sale of	
Finance	20,000	fixed assets	50,000
Net profit	3,50,000		
	7,00,000		7,00,000

You are required to calculate

Current ratio

Quick ratio.

Operating ratio

Stock turnover ratio

Turnover of FA assets

Return on resources

Return on proprietor's fund

Problem 3. *Make an assessment of comparative position of the firms A, B and C after taking the following data. Calculate the relevant ratios and comment on it.*

Particulars	*Firm A*	*Firm B*	*Firm C*
Avg. Inventory	10,00,000	15,00,000	20,00,000
Sales	66,00,000	83,59,000	89,60,000
Cost of goods sold	60,00,000	75,00,000	80,00,000
Expenses of managements	5,00,000	7,60,000	10,00,000
Receivables	13,20,000	24,97,500	35,84,000

Solution:

Name of ratio	*A*	*B*	*C*	*Remark*
Inventory turnover ratio $= \dfrac{\text{Cost of goods sold}}{\text{Avg. Inventory}}$	6 : 1	5 : 1	4 : 1	Firm A is better
Debtors turnover Ratio $= \dfrac{\text{Debtors + Bills receivables}}{\text{Credit sales}} \times 365$	73 days	109 days	146 days	Firm A is better
Gross profit Ratio $= \dfrac{\text{Gross profit}}{\text{Net sales}} \times 100$ Sales Cost of goods sold =	9.09%	10.21	10.71	Firm C is better
Operating Ratio $= \dfrac{\text{Cost of goods sold + Operating expenses}}{\text{Net sales}}$	0.985 : 1	0.98 : 1	1.004 : 1	Firm A is better
Net profit Ratio $= \dfrac{\text{Net profit}}{\text{Net sales}}$ Gross profit – Expenses	0.015 : 1	0.011 : 1	–0.004 : 1	Firm A is better
Operating expense ratio	0.075 : 1	0.091 : 1	0.112 : 1	

Problem 4. *Calculate the current assets of the company with the following information.*

Stock turnover ratio = 5 times
Stock at the end = 5000 more than the stock in the beginning
Sales (cr) = 200,000
Gross Profit Ratio = 20%
Current liabilities = Rs. 60,000
Quick Ratio = 0.75

Solution:

$$\text{Gross Profit Ratio} = \frac{\text{Gross profit}}{\text{Net sales}}$$

$$\text{Quick Ratio} = \frac{\text{Liquid assets}}{\text{Current liabilities}}$$

$$0.75 = \frac{\text{L.A.}}{60,000}$$

$$\boxed{\text{Liquid assets} = 45,000}$$

$$\text{Stock turnover Ratio} = \frac{\text{Cost of goods sold}}{\text{Avg. stock}}$$

$$\text{Cost of goods sold} = \text{Sales} - \text{Gross profit}$$

$$= 2,00,000 - 40,000 = 1,60,000$$

$$\text{Avg. stock} = \frac{\text{Opening stock + Closing stock}}{2}$$

$$\therefore \quad 5 = \frac{1,60,000}{\text{Avg. stock}}$$

$$\text{Avg. stock} = 32,000$$

$$32000 = \frac{x + x + 5000}{2}$$

$$2x + 5000 = 64,000$$

$$x = 29,500$$

$$\text{Opening stock} = 29,500$$

$$\text{Closing stock} = 34,500$$

$$\text{Current assets} = \text{Liquid assets + closing stock}$$

$$= 45,000 + 34,500 = 79,500.$$

Problem 5. *The following is the balance sheet of Quick Sales Ltd. as on 31st Dec. 1997.*

Liabilities	*Amt.*	*Assets*	*Amt.*
Share capital		Goodwill	80000
80000 equity shares	8,00,000	Land and buildings	200000
at Rs. 10 each		P and m/c	500000
		patents	20000
Reserve for capital	2,10,000		
6% mortgage debentures	4,00,000	**Current assets**	
Sundry creditors	2,00,000	Stock of raw m/c	100000
		Finished goods	300000
		Sundry debtors	150000
		Booked debts	
		Cash at bank	210000
		B/R	50000
	16,10,000		16,10000

The gross profit ratio for the year is $33\frac{1}{3}\%$.

The Net sales for the year were Rs. 30,00,000.

Opening stocks of raw m/c and finished goods are Rs. 140000, Rs. 340000 respectively. Work out the various turnover ratios.

Solution. Capital turnover ratio $= \dfrac{\text{Sales}}{\text{Capital employed}}$

$= \text{Total assets} - \text{Current}$

$= \dfrac{30,00,000}{16,10,000-2,00,000}$

• Stock of turnover ratio $= \dfrac{\text{Cost of goods sold}}{\text{Avg. stock}}$

$= \dfrac{30,00,000 - 10,00,000}{44,0000} = 4.5 : 1$

Avg. stock $= \dfrac{1,40,000 + 3,40,000 + 1,00,000 + 3,00,000}{2}$

$= 4,40,000$

• Fixed assets turnover ratio $= \dfrac{\text{Net sales}}{\text{Net Fixed assets}}$

$= \dfrac{30,00,000}{80,0000} = 3.75 : 1$

• Net working capital turnover ratio

$= \dfrac{\text{Net sales}}{\text{Net working capital}}$

$= \dfrac{3000000}{810000 - 200000} = 4.918 : 1$

• Turnover of land and buildings $= \dfrac{\text{Net sales}}{\text{Land \& buildings}}$

$= \dfrac{30,00,000}{2,00,000}$

• Turnover of plant and m/c $= \dfrac{\text{Net sales}}{\text{Plant and m/c}}$

$= 3,000$

• Debtors turnover ratio $= \dfrac{\text{Debtors + Bills receivables}}{\text{Credit sales}} \times 365$

$= \dfrac{1,50,000+50,000}{30,00,000} \times 365$

$= 24.3$ days.

• Current ratio $= \dfrac{\text{Current assets}}{\text{Current liabilities}}$

Current assets are stock, debtors, cash at bank

$$= 1{,}50{,}000 + 2{,}50{,}000 + 1{,}50{,}000 = 5{,}50{,}000$$

$$\text{Current liabilities, creditors} = 2{,}00{,}000$$

$$\text{current ratio} = \frac{55{,}0000}{2{,}00000}$$

$$= 2.75 : 1$$

- $$\text{Quick ratio} = \frac{\text{Liquid assets}}{\text{Current liabilities}}$$

$$\text{Liquid assets} = \text{current assets} - \text{Stock}$$

$$= 550000 - 150000 = 4{,}00{,}000$$

$$\text{Quick ratio} = \frac{4{,}00{,}000}{2{,}00{,}000} = 2 : 1$$

- $$\text{Operating ratio} = \frac{\text{Cost of goods sold} + \text{Operating expenses}}{\text{Net sales}}$$

$$\text{Cost of goods sold} = \text{Sales} - \text{Gross profit}$$

$$= 18{,}00{,}000 - 6{,}50{,}000$$

$$= 11{,}50{,}000$$

- $$\text{Operating ratio} = \frac{11{,}50{,}000 + 3{,}50{,}000}{18{,}00{,}000}$$

$$= 0.833 : 1$$

Operating expenses are selling and distribution, administrative expenses and finance

$$= 100000 + 230000 + 20000 = 350000$$

- $$\text{Stock turnover ratio} = \frac{\text{Cost of goods sold}}{\text{Avg. stock}}$$

$$\text{Avg. stock} = \frac{\text{O.S} + \text{C.S}}{2}$$

$$= \frac{2{,}50{,}000 + 1{,}50{,}000}{2}$$

$$= 2{,}00{,}000.$$

$$\text{Stock turnover ratio} = \frac{11{,}50{,}000}{2{,}00{,}000}$$

$$= 5.75 : 1.$$

- $$\text{Turnover of fixed assets} = \frac{\text{Net sales}}{\text{Net fixed assets}}$$

$$= \frac{18{,}00{,}000}{7{,}00{,}000} = 2.57 : 1$$

Net fixed assets are land and building, plant and machinery

$$= 5{,}00{,}000 + 2{,}00{,}000 = 7{,}00{,}000$$

* Return on Resources $= \dfrac{\text{Net profit}}{\text{Total assets}}$

$$= \frac{3{,}50{,}000}{12{,}50{,}000} = 0.28 : 1$$

• Return on proprietor's fund $= \dfrac{\text{Net profit}}{\text{Shareholder's fund}}$

Proprietor's fund = Shareholder's fund

= Share capital + Capital reserve + Profit and loss a/c balance**

$$= 5{,}00{,}000 + 4{,}00{,}000 + 1{,}50{,}000$$

$$= 10{,}50{,}000.$$

• RPF $= \dfrac{3{,}50{,}000}{10{,}50{,}000} = 0.33 : 1.$

Problem 6. *Following are the figures related to trading activity of Jaitina Ltd. for year ended 30th June 2000.*

Particular	*Amt.*
Sales	5,20,000
Purchase	3,22,250
Opening stock	76,250
Sales return	20,000
Closing stock	98,500
Sellers and distributors	
Salaries	15,300
Advt.	4,700
Travelling expenses	2,000
Administrative expenses	
Salaries	27,000
Rent and postage	5,200
Depreciation	9,300
Other changes	16,500
Provision for taxation	40,000
Non-operating income	
Dividend of shares	9,000
Profit on the loss on sale and shares	3,000
Non-operating stock	
Loss on the sales of shares	4,000

You are required to rearrange the above figure in a suitable form for analysis. Show separately relevant profit and loss account ratio.

Solution.

Particulars	*Amount*	*Amount*	*Amount*
Sales		5,20,000	
Less sales returns		20,000	
Net sales			5,00,000
Opening stock	76,250		
Add purchases	3,22,250		
		3,98,500	
Less closing stock		98,500	
Cost of goods sold.			3,00,000
Gross profit			2,00,000
Operating expenses :			
Selling and distribution			
Salaries	15,300		
Advertisement	4,700		
Travelling expenses	2000		
		22,000	
Administration expenses			
Rent and postage	5200		
Depreciation	9300		
Other charges	16500		
Provision for taxation	40,000		
Salaries	27000		
		98,000	
Total operating expenses			1,20,000

Net operating profit = Gross profit – Total operating expenses

= 80,000.

Add Non-operating incomes			
Divident on shares		9,000	
Profit on sale of shares		3,000	
			12,000
Less Non-operating			
expenses			92,000
Loss on sale of shares			4,000
Net profit			88,000

Profit and Loss account Ratio

1. Gross profit ratio $= \frac{\text{Gross profit}}{\text{Net sales}} \times 100$

$= \frac{2,00,000}{5,00,000} \times 100 = 40\%.$

2. Operating ratio $= \frac{\text{Cost of goods sold + Operating expenses}}{\text{Net sales}}$

$= \frac{3,00,000 + 1,20,000}{5,00,000} = 0.84 : 1$

3. Net profit ratio $= \frac{\text{Net profit after taxes}}{\text{Net sales}} \times 100$

$= \frac{88,000}{5,00,000} \times 100 = 17.6\%$

4. Operating expenses ratio $= \frac{\text{Total operating expenses}}{\text{Net sales}}$

$= \frac{1,20,000}{5,00,000} = 0.24 : 1$

5. Net operating profit ratio $= \frac{\text{Net operating profit}}{\text{Net sales}}$

$= \frac{80,000}{5,00,000} = 0.16 : 1$

Problem 7. *The following is the summarised profit and loss a/c of ABC Com. Ltd. for the year ending 1997. Work out the profitability ratios from the given profit and loss a/c. Rearrange the following figures suitably before calculating the ratios. The capital employed during the year is 15 lakhs.*

Particulars	*Amt.*	*Particulars*	*Amt.*
Opening stock of raw m/l	99,500	Sales	8,50,000
Purchase of materials	3,20,000	Closing stock of m/l	89,000
Direct wages	2,25,250	Closing stock of	60,000
Manuf. expenses	14,250	Finished goods	
Selling and distribution	30,000	Nonoperating income	9,000
Administrative expenses	1,50,000		
Finance charges	15,000		
Non-operating expenses	4,000		
	1,50,000		
Net profit	1,08,000		1,08,000

Solution: **Income statement for the year ending 1997**

Particulars	*Amt.*	*Amt.*	*Amt.*
Sales			8,50,000
Actual m/ls consumed			
opening stock of raw m/l	99,500		
Add purchases of m/l	3,20,000		
		4,19,500	
Less C.S. materials		89,000	
Actual materials consumed		3,30,500	
Wages	2,25,250		
Manufacturing expenses	14,250		
	2,39,500		
Less closing stock of	60,000		
finished goods.		1,79,500	
Cost of goods sold			5,10,000
Gross profit			3,40,000

Operating expenses:			
Selling and distribution		30,000	
Administrative expenses		1,50,000	
Finance charges		15,000	
Total operating expenses			1,95,000
Net operating profit			1,45,000
Add Non-operating income			9,000
			1,54,000
Less non-operating expenses			4,000
Net profit			1,50,000

* 1.

$$\text{Gross profit ratio} = \frac{\text{Gross profit}}{\text{Net sales}} \times 100$$

$$= \frac{3,40,000}{8,50,000} \times 100 = 40\%$$

2.

$$\text{Operating ratio} = \frac{\text{Cost of goods sold} + \text{Total operating expenses}}{\text{Net sales}}$$

$$= \frac{5,10,000 + 1,95,000}{8,50,000} = 0.829 : 1$$

3. Net profit ratio $= \frac{\text{Net profit}}{\text{Net sales}}$

$$= \frac{1,50,000}{8,50,000} = 0.1765 : 1$$

4. Operating expenses ratio $= \frac{\text{Total op. expenses}}{\text{Net sales}}$

$$= \frac{1,95,000}{8,50,000} = 0.229 : 1$$

5. Operating profit ratio $= \frac{\text{Net op. profit}}{\text{Net sales}}$

6. Profitability ratio $= \frac{\text{Net profit}}{\text{Capital employed}}$

$$= \frac{1,50,000}{15,00,000} = 0.1:1$$

SUB RATIOS

1. Selling and distribution expenses ratio

$$= \frac{\text{S \& D expens}}{\text{Net sales}}$$

$$= \frac{30,000}{8,50,000}$$

2. Wages to sales ration $= \frac{2,25,250}{8,50,000} = 0.265 : 1$

3. Manufacturing expenses to net sales

$$= \frac{14.250}{8,50,000} = 0.016$$

Problem 8. *Using the information and the form below compute the balance sheet items for a firm having a sale of 36,00,000.*

Sales : Total assets = 3
Sales : Fixed assets = 5
Sales : Current assets = 7.5
Sales : Inventory = 20
Sales : Debtors = 15

$$\text{Current ratio} = 2$$

$$\text{Total assets : Net worth} = 2.5$$

$$\text{Debt : Equity ratio} = 1$$

Solution:

BALANCE SHEET

Liabilities	*Amount*	*Assets*	*Amount*
Net worth	4,80,000	Fixed assets	7,20,000
Long-term debt	4,80,000	Inventories	1,80,000
Current liabilities	2,40,000	Debtors	2,40,000
		Liquid assets	60,000
		Total current assets	4,80,000
Total liabilities	12,00,000	Total assets	12,00,000

Working notes:

$$\text{Sales} = 36{,}00{,}000$$

$$\frac{\text{Sales}}{\text{Total assets}} = 3 \Rightarrow \text{Total assets} = 12{,}00{,}000$$

$$\frac{\text{Sales}}{\text{Fixed assets}} = 5 \Rightarrow \text{Fixed assets} = 7{,}20{,}000$$

$$\frac{\text{Sales}}{\text{Current assets}} = 7.5 \Rightarrow \text{Current assets} = 4{,}80{,}000$$

$$\frac{\text{Sales}}{\text{Inventory}} = 20 \Rightarrow \text{Inventory} = 1{,}80{,}000$$

$$\frac{\text{Sales}}{\text{Debtors}} = 15 \Rightarrow \text{Debtors} = 2{,}40{,}000$$

$$\text{Current ratio} = \frac{\text{Current assets}}{\text{Current liabilities}}$$

$$= 2.$$

$$\therefore \quad \text{Current liabilities} = 2{,}40{,}000.$$

$$\frac{\text{Total assets}}{\text{Net worth}} = 2.5$$

$$\text{Net worth} = 48{,}000$$

Problem 9. *From the following details prepare the balance sheet of Mr. Raghu and Company.*

$$\textit{Stock velocity} = 6$$

$$\textit{Capital turnover ratio} = 2$$

$$\textit{Fixed assets to Ratio} = 4$$

$$\textit{Gross profit} = 20\%$$

Debt collection period = 2 months

Creditors payments period = 73 days

Gross profit was Rs. 60,000. Closing stock was Rs. 5,000 in excess of opening stock.

Solution. • GPR $= \frac{\text{G.P.}}{\text{Net sales}} \times 100$

Net gross sales = Rs. 3,00,000

Stock turnover ratio $= \frac{\text{Cost of goods sold}}{\text{Avg. stock}}$

Cost of goods sold = Sales – G.P.

= 3,00,000 – 60,000 = 2,40,000

Avg. Inventory = 40,000

$$\frac{x + x + 5,000}{2} = 40,000$$

$$2x + 5000 = 40,000$$

$$x = 37,500 = \text{opening stock}$$

Closing stock = 42,500

FA to ratio $= \frac{\text{Cost of goods sold}}{\text{Fixed assets}}$

$$4 = \frac{2,40,000}{\text{FA}}$$

Fixed assets = 60,000

• Debt collection period or debtors velocity

$$= \frac{\text{Debtors}}{\text{Net sales}} \times 12$$

$$2 = \frac{\text{Debtors}}{3,00,000} \times 12$$

Debtors = 50,000

• Creditors turnover ratio $= \frac{\text{Creditors}}{\text{Purchases}} \times 365$

$$73 = \frac{\text{Creditors}}{2,40,000 + 5,000} \times 365$$

Creditors = 49000

Capital turnover ratio $= \frac{\text{Net sales}}{\text{Capital employed}}$

$$\text{Proprietor's fund} = \frac{\text{Cost of sales}}{\text{Capital turnover ratio}}$$

$$\text{Proprietor's fund} = 1{,}20{,}000$$

BALANCE SHEET

Liabilities	*Amount*	*Assets*	*Amount*
Creditors	49,000	Stock	42,500
Proprietor's fund	1,20,000	Debtors	50,000
		Fixed assets	60,000
		Cash in hand	16,500
	1,69,000		1,69,000

FINANCIAL AND PROFIT PLANNING

12.1 INTRODUCTION

The strategic plan of a firm spells out its corporate purpose, corporate scope, corporate objectives and corporate strategies.

As a key member of the top management group that formulates the strategic plan, the financial manager must (*i*) sensitise the strategic planning group to the financial implications of various choices (*ii*) ensure that the chosen strategic plan is financially feasible (*iii*) translate the strategic plan that is finally adopted and co-ordinate the development of the budget.

There are various facts of financial planning and budgeting as what and why of financial planning, sales forecast, proforma profit and loss account, proforma balance sheet, computerised financial planning systems, growth and external financing requirement, sustainable growth rate, budgeting, responsibility budgeting, budgeting in India.

12.2 FINANCIAL PLANNING

A long-term financial plan represents a blueprint of what a firm proposes to do in the future. It covers a period of three to ten years. Planning over such an extended time horizon tends to be in fairly aggregative terms. There is considerable variation in the scope, degree of formality and level of sophistication in financial planning across firms. Most corporate financial plans have certain common elements.

1. Economic assumption is based on certain assumptions about the economic environment. (Interest rate, inflation rate, growth rate, exchange rate.)
2. Sales forecast : The sales forecast is typically the starting point of the financial forecasting exercise.
3. Proforma statements. The heart of a financial plan are the proforma (forecast) profit and loss account and balance sheets.
4. Asset requirements: Firms need to invest in plant and equipment and working capital. The financial plan spells out the projected capital investments and working capital requirements over time.
5. Financial Plan : Suitable sources of financing have to be thought of for supporting the investment in capital expenditures and working capital.
6. Cash budget: The cash budget shows the cash inflows and outflows expected in the budget period.

Companies spend considerable time and resources in financial planning. It is reasonable to ask what are the benefits of financial planning.

- Identifies advance actions to be taken in various areas.
- Seeks to develop a number of options in various areas that can be exercised under different conditions.
- Facilitates a systematic interaction between investment and financing decisions.
- Classifies the links between present and future decisions.
- Forecasts what is likely to happen in future and hence, helps in avoiding surprises.
- Ensures that the strategic plan of the firm is financially viable.
- Provides bench-marks against which future performance may be measured.

12.3 PROFIT PLANNING

Profit plan is a short-term financial plan. It is an action plan to guide managers in achieving the objectives of a firm. A profit plan is a comprehensive and coordinated plan for the operations and resources of an enterprises for some specific period in the future.

The basic elements of a profit plan are

(1) It is a comprehensive and co-ordinated plan.

(2) It is expressed in financial terms.

(3) It is a plan for the firm's operation and resources.

(4) It is a future plan for a specified period.

The profit planning should have the following qualities to make management active to influence the environment in the interest of the enterprise.

(*i*) Integrated plan

(*ii*) Financial quantification

(*iii*) Operation and resources

(*iv*) Time element

12.4 OBJECTIVES OF PROFIT PLANNING

The process of preparing and using budgets to achieve management objectives is called budgeting. A comprehensive profit planning and controlling of budgeting is a systematic and formalised approach for stating and communicating the firm's expectations and accomplishing the planning, coordinating and control responsibilities of management in such a way as to maximise the use of given resources.

The major purposes of profit planning are:

- To state the firm's expectation (goals) in clear and formal terms to avoid confusion and to facilitate their attainability.
- To communicate expectations to all concerned with the management of the firm so that they are understood, supported and implemented.

- To provide detailed plan of action for reducing uncertainty and for the proper direction of individual and group efforts to achieve goals.
- To co-ordinate the activities and efforts in such a way that the use of resources is maximised.
- To provide a means of measuring and controlling the performance of individuals and units and to supply information on the basis of which the necessary corrective action can be taken.

12.5 STATEMENT OF EXPECTATIONS

The long-run objectives are pursued in successive, short-run steps in the future periods of time. A budget is a means of expressing goals to be achieved in short-run in formal terms. The target of expected performance are laid down when a budget is prepared. These targets are directional and motivational in nature. They direct individual and group efforts towards a common goal. They motivate individuals to evaluate performance.

12.5.1 Communication

Managers of a firm should know what the goals are, they should understand and support them. It is the function of top management to inform people at lower-levels of management about the performance expected of them. Top management uses budgeting as a vehicle to communicate goals or expectation to employees. A clear written communication will help employees to understand, support and accomplish goals through a proper coordination of goals and means.

12.5.2 Planning

Planning is essential to accomplish goals. Formalised planning indicates the responsibility of management and provides an alternative to grouping without direction. Budgeting or planning involves the determination of what should be done, how the goals may be reached and what individuals or units are to assume responsibility and be held accountable.

12.5.3 Coordination

Coordination is a major function of budgeting. Coordination implies striking a proper balance between labour, material and other resources so that the goals are attained at a minimum cost. Coordination between the purchase department and the production department is also required. The point to be emphasised is that the activities of all departments must mesh. Budgeting requires each manager to establish a proper rapport between the activities of his department and that of other departments.

12.5.4 Control

A budget may be used to serve as an index for measuring employees' performance. The actual performance of employees is compared with the budgeted performance to provide a feedback. Once the causes for the difference between the actual and budgeted performance have been identified a corrective action should be initiated. It should be remembered that budget is just a method of control, it is not a control system in itself.

12.6 ESSENTIALS OF PROFIT PLANNING

A successful and sound planning and budgeting system is based upon certain prerequisites. These prerequisites represent management attitude, organisation structure and managerial approaches necessary for the effective and efficient application of the budget system.

The following are some of the important essentials or fundamentals of a successful profit planning or budgeting.

- Top management support
- Clear and realistic goals
- Assignment of authority and responsibility
- Creation of responsibility centres.
- Adaptation of the accounting system
- Full participation
- Effective communication
- Budget education
- Flexibility

12.6.1 Top Management Support

A budgeting system will be an utter failure if it is not initiated and supported by top management. Top management must realise that budgeting is not merely an accounting device, but it is an important management tool. Top management must (*i*) understand the nature and characteristics of budgeting (*ii*) be convinced that this particular approach to managing is preferable for their situation (*iii*) be willing to devote the effort required to make it operative (*iv*) support the programme in all its ramification and (*v*) view the result of the planning process as performance commitments. The support of top management for the budgeting system implies that it is confident about its capability to plan the future course of action and run the enterprise successfully.

Top management should not only have a positive attitude towards budgeting but should also devote necessary time and resources to the preparation and implementation of budgets. Top management should also initiate a follow-up procedure to see that there is effective implementation of budgets.

12.6.2 Clear and Realistic Goals

Budgeting means to achieve goals and objectives. Budgeting will not succeed if the goals to be achieved are not clear, then budget implementation will not be systematic.

The enterprise objectives and budget goals to be accomplished through budgeting should be reasonable and realistic and it should be capable of attainment. Goals set at a very high level are impossible to attain and as a result have a depression effect on the employees' morale. Goals set at a very low level do not provide any challenge to employees'. The enterprise objectives and budget goals must provide a real challenge and should be capable of motivating employees. Goals set realistically provide better motivation to employees in the long run.

12.6.3 Assignment of Authority and Responsibility

A sound organisational structure is essential for the success of the budgetary system. Authorities and responsibilities of each manager should be clearly identified and established. The budgetary system should be established in terms of the assigned authorities and responsibilities, the performance of each manager should be evaluated accordingly.

The type of organisational structure for an enterprise will depend upon the leadership style of top management. Whatever the organisational structure, the budgetary system should be tuned in accordance with such structure.

12.6.4 Creation of Responsibility Centres

A small firm can be managed by an individual or a small group of individuals. But, a large firm cannot be supervised by an individual or a few individuals. For effective control of all activities a large firm is divided into meaningful segments, departments or divisions. Each subunit has certain activities to be performed by its manager who is assigned specific authority and responsibility to carry out those activities and held responsible for his decision affecting those activities. The sub-units of an enterprise for the purpose of control are called responsibility centres or decision centres. The subunits are:

Cost Centre: A cost centre is a responsibility centre where the manager is responsible only for costs incurred in the sub-unit. He is not responsible for profit or investment in the centre.

In a cost centre the consequences of decision are measured by costs alone, the accomplishments of the cost centre are not measured in financial terms. A cost centre spending the least is considered as the best.

Profit Centre: A profit centre is a responsibility centre where the manager is responsible for both costs and revenues and profit. A profit centre provides more effective assessment of performance as both costs and revenues are measured in financial terms.

To ensure effective control through the profit centre control system the controllable and noncontrollable activities should be identified. If the allocation of indirect costs is avoided, one may think in terms of contribution margin centres rather than profit centres. Contribution margin is the difference between variable costs and revenues of the centres.

Investment Centre: An investment centre is a responsibility centre where the manager is responsible for costs and revenues as well as for investments in assets used by the centre. In an investment centre performance is assessed not only by profit, but by relating profit to investment. In a sense, investment centres are treated as separate firms where the manager is responsible for the overall activities affecting costs, revenues and investment.

12.6.5 Adaptation of the Accounting System

The accounting system catering only to the needs of external users is not adequate for the purpose of profit planning and control and internal management. A responsibility accounting system is primarily oriented towards the organisational responsibilities and is a means to achieving effective control. An accounting system tailored to the responsibility structure of the enterprise generates data that are relevant to the planning and control system.

A cost accounting system has two primary aims: (*i*) to measured the cost of production and (*ii*) to furnish data for planning and control. In summary, it may be stated that for effective and successful budgeting the accounting system must be structured around the planning and control needs of management.

12.6.6 Full Participation

Full participation of managers and their subordinates at all levels should be sought in developing the budgeting system. The participation should be meaningful and real. A meaningful participation creates a positive motivation. Participation tends to increase commitment, commitment tends to heighten motivation, motivation which is job-oriented tends to make managers work harder and more productively, and harder and more productive work by managers tends to enhance the company's prosperity, participation good.

To seek meaningful participation of all persons in the enterprise is a difficult task. All individuals should be motivated to participate in the budgeting process. It should be understood well that the responsibility to prepare budgets lies on the line managers. It is essential that those line executives who have to implement budgets must also participate in proving information for preparing their respective departmental budgets.

12.6.7 Effective Communication

Communication is the process of transmitting ideas or information from one person to another. The basic purpose of communication is to instil mutual trust between two or more persons by creating similar understanding of ideas or thoughts. Communication is fundamental and vital to all decisions, it should not be taken for granted for an adequate flow of information. It must be ensured that communication is effective. Effective communication implies transmission of information as well as understanding. Budgeting is a formal way of communicating plans, objectives and budget goals to various responsibility centres. Budget developed through full participation tailored around the organisational structure of authorities and responsibilities provides for better understanding of goals and plans.

12.6.8 Budget Education

Participation can be meaningful only when people at all levels of management are convinced of the usefulness of budgets, understand the nature and characteristics of budgets and role which they have to play in profit planning and control. The line executives who prepare the budgets should not only be confident of their ability to plan for the future with reasonable precision but should understand the technicalities of budgeting. This requires a continuous budget education. The employees of an enterprise must be educated about the nature, characteristics, value and methods of budgeting. They should also be taught how to interpret the budget results and how the performances are evaluated through budgets. Seminars, conferences, lectures, discussions, executives' development programme etc. can be organised. Written material can be distributed.

12.6.9 Flexibility

The budgeting system should be flexible enough to take advantage of all opportunities that arise from time to time.

A rigidly administered budgeting programme causes tension and anxiety and imposes 'Strait Jackets' in implementing the budgets. Top management would exercise a light control over lower levels of management and would put restrictions on them to make decisions in the absence of a sophisticated budgeting system. A flexible and comprehensive budgeting permits the management to readjust plans when a new situation arises.

12.7 BUDGETING AND BUDGETING CONTROL

Budgeting control can be defined as establishment of budgets relating to the responsibilities of executives to the requirement of policies and a continuous comparison of the actual results with the budgeted results.

In short, budgeting control means laying down the monetary and quantitative terms i.e. what actually has to be done and how exactly it has to be done over the coming period. Then to ensure the actual results do not diverge from the planned course more than the necessary.

12.8 OBJECTIVES OF BUDGETING

1. To forecast the future and plan to avoid the loss, but more positively to maximise profit.
2. To bring out the coordination between the different functions of an enterprise which is essential for the success of an enterprise.
3. To create a definite idea in terms of the no. i.e., quantity about the long-term or the short-term of an organisation.
4. To motivate the related departments and persons for achieving the desired goal.
5. To ensure the action in tune with the target (to take a suitable corrective action to meet the target).

12.9 ADVANTAGES OF BUDGETING

1. Budgeting leads to the maximum utilisation of the resources with a view to ensure maximum returns.
2. It creates a sense of awareness at all levels of management in the process of fulfillment of the targets.
3. Budgeting leads to a coordination and have the understanding between different functions of the organisation.
4. It is a process of self examination and subcriticism which is essential for the success of an enterprise.
5. Budgeting enlists the support and active participation of the top management.

12.10 DISADVANTAGES

1. The installation of the budgeting system is an elaborate process and takes time.
2. The budgeting should be followed up by an effective control action. This is often lacking in many organisations which fails the very purpose of budgeting.
3. The basic requirement for the success of budgeting is absolute support and enthusiasm provided by the top management. If it is lacking at any time, the whole system will collapse.
4. Budgeting is not an exact science. It is only a certain amount of judgement.

12.11 CLASSIFICATION OF BUDGET

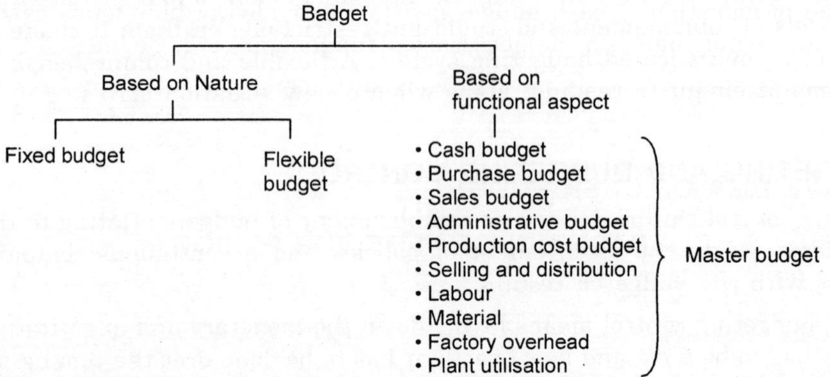

12.12 ORGANISATION CHART FOR BUDGETING

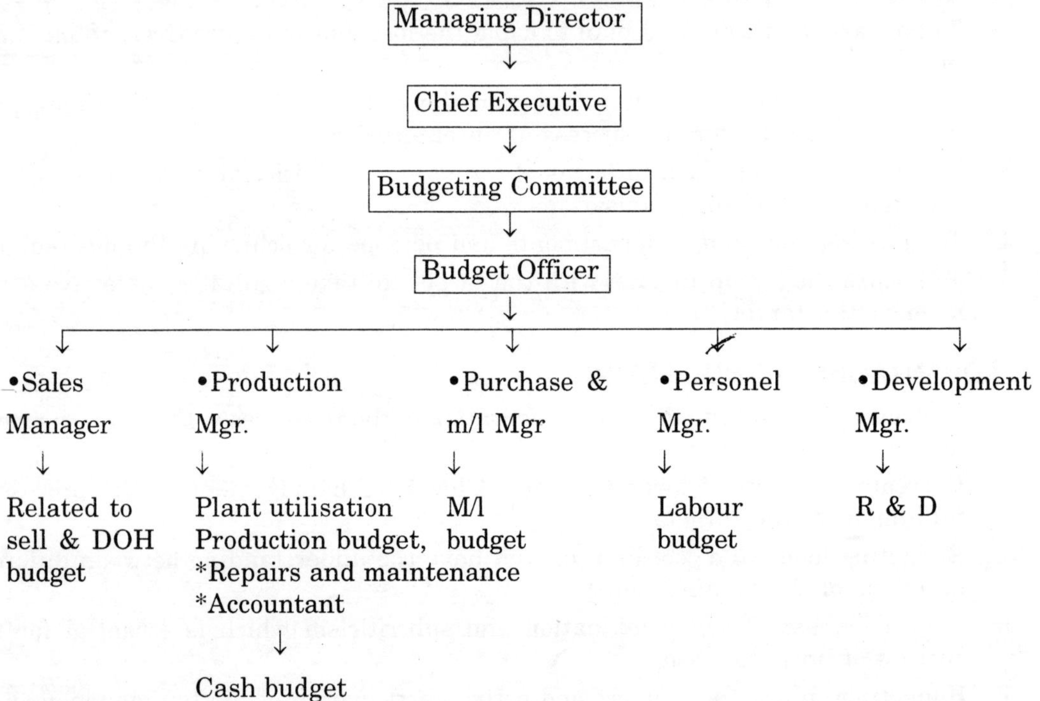

12.13 CASH BUDGET

It is an estimate of receipts and payments for each month or for a particular budgeted period. It is essential to allow a certain time lag between certain transactions. A cash budget is prepared by any one of the following methods.

(*a*) Receipt and payment method

(*b*) Profit and loss a/c

(*c*) Balance sheet method

The receipt & payment method is very useful for the preparation of cash budget.

12.14 ADVANTAGES

1. The preparation of cash budget gives a clear idea of how much cash is required and at what time and the necessary arrangements may be made for that purpose.
2. Cash budget expresses the deficit or surplus of cash. The surplus of cash if any will be invested properly.

12.15 PROFORMA OF CASH BUDGET

CASH BUDGET FOR THE PERIOD

Particulars	*Jan.*	*Feb.*	*Mar.*	*April*
(*a*) Opening balance				
(*b*) *Receipts* :				
Cash sales				
Credit sales (Debtors)				
Sale of assets				
Other receipts				
Total (*a* + *b*)				
(*c*) *Payments* :				
Purchases (Creditors)				
Overheads				
Wages				
Sales commission				
Other payments				
Total				
Balance *c*/*d*[(*a* + *b*) – *c*]				

Problems 1. *ABC Company Ltd. wishes to prepare a cash budget from Jan. to June. The following are the estimated revenues and expenses.*

Month	*Total sales Rs.*	*Materials (purchases)*	*Wages*	*POH*	*SDOH*
Jan.	20,000	20,000	4000	3200	800
Feb.	22,000	14,000	4400	3300	900
Mar.	24,000	14,000	4600	3300	800
Apr.	26,000	12,000	4600	3400	900
May	28,000	12,000	4800	3500	900
June	30,000	12,000	4800	3600	1000

The following information are given :

(1) Cash balance on 1st Jan. Rs. 10,000.

(2) A new m/c is to be installed at Rs. 30,000 on credit to be repaid by two equal installments *i.e.*, March and April.

(3) Sales commission at 5% of total sales to be paid in the month time.

(4) Rs. 10,000 being the amount of second call may be received in March.

(5) Share premium amounting to Rs. 2000 is also obtainable with second call.

(6) Period of credit by suppliers is two months.

(7) Period of credit to customers is one month.

(8) Delay in payment of OH 1 month.

Delay in the payment of wages ½ month. Assume cash sales are 50% of total sales. Cash balance at the end of June = 15,200.

Solution.

CASH BUDGET FOR ABC COMPANY FOR THE PERIOD JAN.-JUNE

Particulars	*Jan.*	*Feb.*	*Mar.*	*Apr.*	*May*	*June*
(*a*) Opening balance	10,000	18,000	29,800	20,000	6,100	8,800
(*b*) *Receipts:*						
Cash sales (50%)	10,000	11,000	12,000	13,000	14,000	15,000
Credit sales						
(Debtors 50%)	–	10,000	11,000	12,000	13,000	14,000
Share premium	–	–	2,000	–	–	–
Capital	–	–	10,000	–	–	–
Total (*a* + *b*)	20,000	39,000	64,800	45,000	33,100	37,800
(*c*) *Payments:*						
Purchases	–	–	20,000	14,000	14,000	12,000
(Creditors)						
Wages	2,000	4,200	4,500	4,600	4,700	4,800
POH	–	3,200	3,300	3,300	3,400	3,500
SDOH	–	800	900	800	900	900
Sales commission	–	1,000	1,100	1,200	1,300	1,400
Asset purchased	–	–	15000	15000	–	–
(*c*)	2,000	9,200	44,800	38,900	24,300	22,600
Bal c/d (*a* + *b* – *c*)	18,000	29,800	20,000	6,100	8,800	15,200

Problem 2. *From the following information and assumptions the balance on hand on 1st Jan. 1996 is 72,500.*

Month	*Sales*	*M/l*	*Wages*	*SDOH*	*POH*	*AOH*
Jan.	72,000	25,000	10,000	4,000	6,000	1,500
Feb.	97,000	31,000	12,100	5,000	6,300	1,700
Mar.	86,000	25,500	10,600	5,500	6,000	2,000
Apr.	88,600	30,600	25,000	6,700	6,500	2,200
May	102,500	37,000	22,000	8,500	8,000	2,500
June	108,500	38,800	23,000	9,000	8,200	2,500

The following information:

1. Assume 50% cash sales.
2. Assets have to be acquired in Feb. and March and provision should be made for payment of 8,000 and Rs. 25000.
3. An application has been made for the grant of a loan of Rs. 30,000 and hoped that it will be received in May.
4. It is anticipated that a dividend of 35,000 Rs. will be paid in June.
5. Debtors allowed 1 month credit.
6. Sales commission at 3% of total sales has to be paid.
7. Creditors (m/l & OH) grant one month credit.

Solution.

CASH BUDGET FOR THE PERIOD JAN.-JUNE

Particulars	Jan.	Feb.	Mar.	April	May	June
(a) Opening balance	72,500	96,340	1,21,330	1,30,650	1,51,292	2,05,767
(b) *Receipts* :						
Cash sales (50%)	36,000	48,500	43,000	44,300	51,250	54,250
Credit sales (Debtors)	–	36,000	48,500	43,000	44,300	51,250
Loan	–	–	–	–	30,000	–
	1,08,500	1,80,840	2,12,830	2,17,950	2,76,842	3,11,267
(c) *Payments* :						
Purchases (creditors)	–	25,000	31,000	25,500	30,600	37,000
Wages	10,000	12,100	10,600	25,000	22,000	23,000
SDOH	–	4,000	5,000	5,500	6,700	8,500
POH	–	6,000	6,300	6,000	6,500	8,000
AOH	–	1,500	1,700	2,000	2,200	2,500
Purchase of assets	–	8,000	25,000	–	–	–
Dividend	–	–	–	–	–	35,000
Sales commission	2,160	2,910	2,580	2,658	3,075	3,255
	12,160	59,510	82,180	66,658	71,075	1,17,255
	96,240	1,21,330	1,30,650	1,51,292	2,05,767	1,94,012

Problem 3. *From the following forecast of income and expenses for half-yearly ended with 30th June 1998.*

Month	Credit sales	Credit purchases	Wages	Manuf. expenses	Adv. expenses	S.D. expenses
1997 Nov.	25,000	10,000	2,500	1,100	1,000	600
Dec.	30,000	15,000	2,800	1,200	975	650

(Contd...)

1998 Jan.	20,000	10,000	2,000	1,250	1,060	550
Feb.	25,000	15,000	2,200	1,150	1,040	650
Mar.	30,000	17,000	2,400	1,300	1,105	750
April.	35,000	20,000	2,600	1,350	1,120	800
May.	40,000	22,000	2,800	1,450	1,180	825
June.	45,000	25,000	3,000	1,500	1,185	875

1. Sales commission of 5% of total sales to be paid in two months.
2. Plant purchased on 1st Jan. for Rs. 10,000 payment immediately.
3. New building purchased in Feb. for Rs. 80,000 payable in two half-yearly instalments. The first one in Feb.
4. Dividend of Rs. 5000 is paid in April.
5. Period of credit allowed by debtors and creditors is 2 months.
6. Lag in the payment of wages is 1/8th of month.
7. Lag in the payment of expenses 1 month.
8. Cash balance on 1st Jan. is 37500

Solution.

CASH BUDGET FOR HALF YEARLY ENDED 30TH JUNE 1998

Particulars	*Jan.*	*Feb.*	*March.*	*April.*	*May.*	*June*
(*a*) Opening balance	37,500	36,325	4,790	8575	6595	12,050
(*b*) *Receipts :*						
Credit sales (Debtors)	25,000	30,000	30,000	25,000	30,000	35,000
Total (*a* + *b*)	62,500	66,325	24,740	33,575	36,595	47,050
(*c*) *Payments.*						
Purchases (Creditors)	10,000	15,000	10,000	15,000	17,000	20,000
Wages	2,100	2,175	2,375	2,575	2,775	2,975
Manufacturing expenses	1,200	1,250	1,150	1,300	1,350	1,450
Administrative expenses	975	1,060	1,040	1,105	1,120	1,180
S & D expenses	650	550	650	750	800	825
Sales commission	1,250	1,500	1,000	1,250	1,500	1,750
Purchase of plant	10,000	–	–	–	–	–
Purchase of new building	–	40,000	–	–	–	–
Dividend	–	–	–	5000	–	–
(*c*)	26,175	61,535	16,215	26,980	24,545	28,180
Cash c/d (*a* + *b* – *c*)	36,325	4,790	8,575	6,595	2,050	18,870

Problem 4. *From the following information prepare a cash budget ending 31st Dec.*

Expected sales

Sep. – 50,000

Oct. – 60,000

Nov. – 45,000

Dec. – 80,000

Expected purchases

Sep. – 32,000

Oct. – 60,000

Nov. – 70,000

Dec. – 45,000

1. Wages paid to the worker Rs. 6,000 each month.
2. Dividends from the investment amounting to Rs. 1,000 are expected on 31st Dec.
3. Income tax to be paid in advance in Dec. Rs. 2,000.
4. Preferential share dividend Rs. 5000 to be paid in 30th Nov.
5. Bank balance on 1st Sep. is Rs. 6,000.

Solution.

CASH BUDGET FOR THE PERIOD SEP.–DEC.

Particulars	*Sep.*	*Oct.*	*Nov.*	*Dec.*
(*a*) Opening balance	6,000	18,000	12,000	–24,000
(*b*) *Receipts:*				
Sales	50,000	60,000	45,000	80,000
Dividends	–	–	–	1000
Total = (*a* + *b*)	56,000	78,000	57,000	57,000
(*c*) *Payments :*				
Purchases	32,000	60,000	70,000	45,000
Wages	6,000	6,000	6,000	6,000
Income tax paid	–	–	–	2000
P-share dividends.	–	–	5000	–
(c)	38,000	66,000	81,000	53,000
Total = (*a* + *b* – *c*)	18,000	12,000	–24,000	4,000

Cash balance at the end of Dec. is Rs. 4000.

Problem 5. *Prepare a cash budget for the month of May, June and July on the basis of the following information.*

Month	*Credit sales*	*Credit purchases*	*Wages*	*Man. expn.*	*Office expn.*	*Selling*
Mar.	60,000	36,000	9,000	4,000	2,000	4,000
April	62,000	38,000	8,000	3,000	1,500	5,000
May	64,000	33,000	10,000	9,500	2,500	4,500
June	58,000	35,000	8,500	3,500	2,000	3,500
July	56,000	39,000	9,500	4,000	1,000	4,500
Aug.	60,000	34,000	8,000	3,000	1,500	4,500

1. Cash balance on 1st May is Rs. 8000.
2. Plant costing Rs. 16,000 is due for delivery in July payable 10% on delivery and balance after 3 months.
3. Advance tax Rs. 8,000 is payable in May and June.
4. Period of credit allowed by suppliers is 2 months and for customers is one month.
5. Lag in the payment of manuf. expenses is ½ month.
6. Lag in the payment of office and selling expenses is 1 month.

Solution.

CASH BUDGET FOR THE MONTH OF MAY-JULY

	Particulars	*May*	*June*	*July*
(*a*)	Opening balance	8,000	7,250	7,250
(*b*)	*Receipts* :			
	Credit sales	62,000	64,000	58,000
	Total $(a + b)$	70,000	71,250	65,250
(*c*)	*Payments* :			
	Credit purchases	36,000	38,000	33,000
	Wages	10,000	8,500	9,500
	Manufacturing expense	6,250	6,500	3,750
	Office expenses	1,500	2,500	2,000
	Selling expenses	5,000	4,500	3,500
	Purchase of plant	–	–	1,600
	Tax payment	4,000	4,000	–
	(c)	62,750	64,000	53,350
	Bal. = *c*/*d* $(a + b - c)$	3,250	7,250	

12.16 PRODUCTION BUDGET

This budget is prepared after the preparation of sales budget to determine the quality of goods which should be produced to meet the budgeted sales. It is expressed in physical terms such as (*a*) Units of output

(*b*) Labour hours

(*c*) Material requirement

The production budget is prepared by the production manager and is submitted to the budget committee for its approval. The following points are to be noted while its preparation.

1. To determine the quantity of each product which will be produced during the budget period.
2. To prepare the production plan on the basis of the sales budget.
3. To consider the production plant capacity and production planning.
4. To consider the volume of production.
5. To consider the availability of m/l, labour etc.

$$\text{AMC} = \text{OS} + \text{Purchase} - \text{CS}$$

Problem 1. *Prepare a purchase budget from the following particulars when the estimated price per kg of material is X – Rs. 2, Y – Rs. 3, Z – Rs. 4.*

Materials	*Estimated consumption of m/ls in kgs.*
X	*1,00,000*
Y	*2,00,000*
Z	*2,50,000*

Materials	*Stock in the beginning*	*Estimated stock at the end*
X	*30,000*	*15,000*
Y	*40,000*	*20,000*
Z	*45,000*	*50,000*

Solution.

PURCHASE BUDGET

Particulars	M_X	M_Y	M_Z
Estimated consumption of m/l	1,00,000	2,00,000	2,50,000
Add stock at the end	15,000	20,000	1,50,000
	1,15,000	2,20,000	3,00,000
Less opening stock	30,000	40,000	45,000
Actual Qty. purchased	85,000	1,80,000	2,55,000
Price per kg.	2	3	4
Total cost	1,70,000	5,40,000	1,02,000

Problem 2. *From the following particulars presented by ABC com., prepare a production budget for the year 1998.*

Product	*Sales as per the budgeted units*	*Estimated units in the beginning*	*Estimated units at the end*
X	*4,00,000*	*20,000*	*30,000*
Y	*3,00,000*	*25,000*	*40,000*
Z	*8,00,000*	*60,000*	*45,000*

Solution. **PRODUCTION BUDGET FOR THE YEAR 1998**

Particulars	*P–X*	*P–Y*	*P–Z*
Sales as per budgeted units	4,00,000	3,00,000	8,00,000
Add C.S.	30,000	40,000	45,000
	4,30,000	3,40,000	8,45,000
Less OP.	20,000	25,000	60,000
Actual Qty.	4,10,000	3,15,000	7,85,000

Problem 3. *XYZ Com. manufactures products C & G. During Jan. it expects to sell 5000 kg. of C and 20000 kg. of G at Rs. 20, Rs. 10 respectively. Direct materials A, B and E are mixed in equal proportion to produce product C. materials D, B and E are mixed in the proportion of 5 : 3 : 2 to produce product G. These is no loss of weight in the production. Actual and budgeted inventory in Quantities and the cost for the month are as follow.*

M/l	*Opening Inventory (kgs)*	*Closing Inventory*	*Anticipated Cost in Rs/kg.*
A	*1500*	*1000*	5.5
B	*1000*	*2000*	5
D	*10000*	*3000*	1
E	*5000*	*6000*	3.5

Product	*Opening Inventory kg.*	*Closing Inventory kg.*
C	*1,000*	*500*
G	*5,000*	*6,000*

You are required (a) to prepare the product budget.

(b) Materials purchase budget indicating the expenditure for the materials for the month of Jan.

Solution. **Production budget for the month of Jan.**

Particulars	*Product C*	*Product G*
Sales	5,000	20,000
Add CS	500	6,000
	5,500	26,000
Less OS	1,000	5,000
Actual Materials purchased	4,500	21,000

Materials Purchase Budget for Jan.

Particulars	M_A	M_B	M_D	M_E
M/ls required for the production of products				
C	1,500	1,500	–	1,500
G	–	6,300	10,500	4,200
(Ratio 5 : 3 : 2)				
	1,500	7,800	10,500	5,700
Add C.I.	1,000	2,000	3,000	6,000
	2,500	9,800	13,500	11,700
Less O.I.	1,500	1,000	10,000	5,000
Actual m/l purchased	1,000	8,800	3,500	6,700
Cost/kg	5.5	5	1	3.5

Problem 4. *From the given particulars presented by Luchi Ltd. company you are required to prepare a material budget.*

Estimated sales = *50,000 units (each unit of product requires 2 units of M-X & 4 units of M-Y)*

Estimated opening balance:

Finished goods – *10,000 units*
M-X – *15,000 units*
M-Y – *25,000 units*

Estimated closing balance:

Finished goods – *6,000 units*
M-X – *16,000 units*
M-Y – *30,000 units*

Estimated m/ls opening on order:

M-X – *8,000 units*
M-Y – *12,000 units*

Estimated m/ls on order closing

M-X – *6,000 units*

M-Y – *8,000 units*

Solution.

Estimated production = E. sales – E.O.S. of finishes goods + C.S. of finished goods.

= 50,000 + 6,000 – 10,000

= 46,000

Material Budget

Particulars	*M-X*	*M-Y*
Requirement of m/ls M-X for 46000 units.	92,000	1,84,000
Add closing balance	16,000	30,000
Add C.S. on order	6,000	8,000
	1,14,000	2,22,000
Less opening balance	15,000	25,000
Less O. on order	8,000	12,000
Actual Qty. purchased	91,000	1,85,000

Problem 5. A fine product ltd. has a daily capacity to produce the following:

Product	*No. of units*	*Std. O/P per hr.*
A	10,000	100
B	12,000	120
C	9,000	150

For a budgeted period, the production targets per day are as follows: 8000 units of A, 9000 units of B, 9000 units of C.

Determine the idle capacity.

Solution. The std. hrs. required for std. o/p. of A

$$= \frac{10,000}{100} = 100$$

$$\text{For B} = \frac{12,000}{120} = 100 \text{ hrs.}$$

$$\text{For C} = \frac{9,000}{150} = 60$$

Total Std. hrs = 260.

Std. hrs. required for actual production of A.

$$= \frac{8,000}{100} = 80 \text{ hrs.}$$

$$B = \frac{9,000}{120} = 75 \text{ hrs.}$$

$$C = \frac{9,000}{150} = 60 \text{ hrs.}$$

Total Std. hrs. required for actual production

= 215

∴ Idle hrs = 260 – 215

= 45 hrs.

= 17.3%.

Problem 6. *The following are the budgeted and the actual costs for a firm of a particular production:*

Particulars	*Budgeted (50,000 units)*	*Actual (40,000 units)*
Variable cost	*5,00,000*	*4,40,000*
Semi-fixed cost	*3,00,000*	*2,50,000*
Fixed cost	*2,00,000*	*2,10,000*

$33\frac{1}{3}\%$ *of the semifixed cost is fixed. Prepare a revised budget and compare the actual against the revised budget.*

Solution.

Total cost = 5,00,000 + 1,00,000

VC/unit = 5,00,000, 50,000 = 10 Rs.

Actual production is 40,000 = Rs. 4,00,000 variable cost.

Fixed cost/unit = Rs. 4

Actual prod^n F.C. = Rs. 1,60,000

Semi-variable Cost = Rs. 2,00,000

Semi-fixed Cost = Rs. 1,00,000

S.VC/unit = Rs. 4.

For actual production = 4 × 40,000 + 1,00,000

= 2,60,000

Actual	*Revised*	*Variance*
440000	400000	40000
250000	260000	10000
210000	200000	10000

12.17 FLEXIBLE BUDGET

It is the one which is designed to change in accordance with the level of activity attained. This budget is defined as a budget which by recognising the difference in the behaviour between the fixed and variable cost in relation to the fluctuation of output. It is based on the assumption that the forecast of the business activity will not be correct and as such adjustments will have to be made in the budget. Since the actual level of activity is known the main classification under this budget is fixed cost, variable cost and semi variable cost which is used while preparing the flexible budget.

Advantages

1. The element of control is maintained even though there is a decrease or increase in the level of production. It is used by those firms whose level of activity frequently changes.

Problems 1. *The expenses for the budgeted production of 10000 units in a factory are given below:*

Particulars	*Cost/unit Rs.*
Materials	70
Labour	25
Variable OH	20
Fixed OH (100000) &	10
Variable expenses (Direct)	5
Selling expenses (10% fixed)	13
Distribution expenses (20% fixed)	7
Administrative expenses (50,000)	5
Total cost of sales/unit	155

Prepare a flexible budget for the production of 8000 and 6000 units. Assume the administrative expenses are rigid for all levels of production.

Solution. **FLEXIBLE BUDGET**

Particulars	*10,000 (100%)*		*8000 (80%)*		*6000 (60%)*	
	Amt.	*Amt.*	*A*			
Direct materials	7,00,000		5,60,000		4,20,000	
Direct Labour	2,50,000		2,00,000		1,50,000	
Direct Expenses	50,000		40,000		30,000	
Prime cost		10,00,000		800000		6,00,000
Variable OH : General	2,00,000		1,60,000		1,20,000	
(@ Rs. 11.7) Selling expenses	1,17,000		93,600		70,200	
Distributing (@ 5.6)	56,000		44,800		33,600	
		3,73,000		298400		2,23,800

Fixed OH : Selling @ 1.3	13,000		13,000		130,00	
Distribution @ 1.4	14,000		14,000		14,000	
Administrative	50,000		50,000		50,000	
FOH	1,00,000		1,00,000		1,00,000	
		1,77,000		1,77,000		1,77,000
Cost per unit	155	159.42	166.8			

Problem 2. *A manufacturing company has a production capacity of 20,000 units of product A. The expenses for the production of 10,000 units for a period is as follows.*

Particulars	*Cost / unit Rs.*
Materials	*40*
Wages	*10*
VOH	*10*
Manuf./expenses (40% fixed)	*10*
Administrative (all fixed)	*5*
Selling & distr. exp. (50% fixed)	*5*
Profit	*20*
Selling	*100*

Prepare a flexible budget to show 70 and 100% level of activity. It is expected that the present unit selling price will remain constant upto 60% beyond which 5% reduction is contemplated up to 100% level of activity. Also give your opinion on which level of activity should be selected.

Solution.

Particulars	*10,000 (50%)*		*60% – 12,000*		*70% (14,000)*		*100% (20,000)*	
	Amt.	*Amt.*	*Amt.*	*Amt.*	*Amt.*	*Amt.*	*Amt.*	*Amt.*
DM	4,00,000		4,8,0000		5,60,000		8,00,000	
Wages	1,00,000		1,20,000		1,40,000		2,00,000	
Prime cost		5,00,000		6,00,000		7,00,000		10,00,000
VOH .								
General	1,00,000		1,20,000		1,40,000		2,00,000	
Manuff/exp.	60,000		72,000		84,000		1,20,000	
Selling	25,000		30,000		35,000		50,000	
		1,85,000		2,22,000		2,59,000		3,70,000
Fixed OH								
Manf/exp.	40,000		40,000		40,000		40,000	
Selling	25,000		25,000		25,000		25,000	
Admi.	50,000		50,000		50,000		50,000	
		1,15,000		1,15,000		1,15,000		1,15,000
Total cost		8,00,000		9,37,000		10,74,000		1,48,500

Selling price		10,00,000		12,00,000		13,30,000		19,000
Profit		2,00,000		2,63,000		2,56,000		41,500
Profit/unit		20		21.92		18.28		20.79
Cost/unit		80		78.08		76.71		74.26

Problem 3. *Draw a flexible budget for the OH expenses on the basis of the following data and determine the OH rate at 70%, 90% and 80% plant capacity.*

Particulars	*Plant capacities (80%)*
VOH:	
Indirect labour	Rs. 12,000
Stores including spares	4,000
Semi VOH:	
Power (30% fixed)	20,000
Repairs & mnce (60% fixed)	2,000
Fixed OH:	
Depreciation	11,000
Insurance	3,000
Salaries	10,000
Total amount	62000

Estimated direct labour hrs. – 124000

Solution. **FLEXIBLE BUDGET**

Particulars	*70%*		*80%*		*90%*	
	Amt.	*Amt.*	*Amt.*	*Amt.*	*Amt.*	*Amt.*
VOH :						
Indirect labour	10,500		12,000		13,500	
Stores inclu	3,500		4,000		4,500	
Power	12,250		14,000		15,750	
Repairs & maintence	700		800		900	
		26,950		30800		34,650
FOH :						
Depreciation	11,000		11,000		11,000	
Salaries	10,000		10,000		10,000	
Insurance	3,000		3,000		3,000	
Power	6,000		6,000		6,000	
Repairs & mnce	1,200		1,200		1,200	
	s	31,200		31,200		31,200
Total cost		58,150		62,000		65,850
Estimated labour hours	1,08,500		1,24,000		1,39,500	
Over head rate/hr.	0.53		0.5		0.47	

As production plant capacity increases the OH recovery rate per hour decreases.

Problem 4. *Prepare a flexible budget for the OH expenses and ascertain the OH rate at 50% and 70%. The following particulars are given for 60%.*

Particulars	*Amount*
VOH	
Indirect m/l	*6,000*
Indirect labour	*18,000*
Semi VOH	
Electricity (40% fixed)	*30,000*
Repairs & mnce	*3,000*
Fixed OH	
Depreciation	*16,500*
Insurance	*4,500*
Salaries	*15,000*
Estimated direct labour hrs. is 1,86,000	

Flexible Budget

Particulars	*50%*		*60%*		*70%*	
	Amt.	*Amt.*	*Amt.*	*Amt.*	*Amt.*	*Amt.*
VOH						
Indirect m/l	5,000		6,000		7,000	
Indirect labour	15,000		18,000		21,000	
Electricity	15,000		18,000		21,000	
Repairs & mnce	1,250		1,500		1,750	
		36,250		43500		50,750
FOH						
Depreciation	16,500		16,500		16,500	
Insurance	4,500		4500		4500	
Salaries	15,000		15,000		15,000	
Electricity	12,000		12000		12,000	
Repairs & mnce	1,500		1500.		1500	
		49,500		49,500		49,500
Total cost		85,750		93,000		1,00,250
Estimated labour hrs.		1,55,000		1,86,000		2,17,000
OH recovery rate per hr.		0.55		0.5		0.46

12.18 FLEXIBLE BUDGET WITH STANDARD COSTING FORMULAS

1. $$\text{Std. capacity usage ratio} = \frac{\text{Actual no. of working hrs.}}{\text{Budgeted no. of working hrs.}}$$

2. $$\text{Actual capacity usage ratio} = \frac{\text{Actual no. of working hrs.}}{\text{Budgeted no. of working hrs. in the period}}$$

3. $$\text{Activity ratio} = \frac{\text{Actual production in terms of std. hrs.}}{\text{Budgeted production in terms of std. hrs.}}$$

4. $$\text{Efficiency ratio} = \frac{\text{Actual production in terms of std. hrs.}}{\text{Actual hrs. worked}}$$

5. $$\text{Calendar ratio} = \frac{\text{No. of actual working days in a period}}{\text{No. of working days in a budgeted period}}$$

Problems 1. *From the following information calculate the important ratios used for the control purposes.*

(a) *Max. possible working hrs. in a budgeted period = 5000*

(b) *Budgeted no. of working hrs. = 4,000*

(c) *Actual no. of working hrs. = 3,600*

(d) *Budgeted prod*^{n} *in terms of std. hrs. = 6,000*

(e) *Actual prod*^{n} *in terms of std. hrs. = 5,400*

(f) *Actual working days during the period = 24*

(g) *Budgeted working days during the period = 25*

Solution.

1. $$\text{Std. cap. usage ratio} = \frac{3{,}600}{4{,}000} = 0.9$$

2. $$\text{Act. cap. usage ratio} = \frac{3{,}600}{5{,}000} = 0.72$$

3. $$\text{Activity ratio} = \frac{5{,}400}{6{,}000} = 0.9$$

Problem 2. *Two articles x and y are manufactured in a department. Their specification shows that 2 units of x or 8 units of y can be produced in an hour. The budgeted production for June 98 is 200 units of x, 400 units of y. The actual production at the end of the month was 250 units of x, 480 units of y and the actual hrs. spent for this production was 160. Find out the capacity, activity & efficiency ratios for June 98 and also calculate the calendar ratio of the actual working days during the month be 27 corresponding to 25 days budgeted.*

Solution.

$$\text{Cap. ratio} = \frac{\text{Actual prod}^{\text{n}} \text{ in terms of std. hrs.}}{\text{Budgeted prod}^{\text{n}} \text{ in terms of std. hrs.}}$$

For 1 hr. no. of units x produced in 2 units.

For 1 unit ½ hrs.

For y = 1/8 hrs.

$$= \frac{\frac{1}{2} \times 250 + \frac{1}{8} \times 480}{\frac{1}{2} \times 200 + \frac{1}{8} \times 400} = 1.23$$

$$\text{Efficiency ratio} = \frac{\text{Actual prod}^{n} \text{ in terms of std. hrs.}}{\text{Actual hrs. worked}}$$

$$= \frac{250/2 + 480/8}{160}$$

$$= 1.15$$

$$\text{Capacity ratio} = \frac{\text{Actual no. of working hrs.}}{\text{Budgeted no. of working hrs.}}$$

$$= \frac{160}{\frac{200}{2} + \frac{400}{8}} = 1.06$$

$$\text{Calendar ratio} = \frac{\text{No. of actual working days}}{\text{No. of working days}}$$

$$= \frac{27}{25} = 1.08$$

12.19 LABOUR BUDGET

This budget indicates the requirement a total direct labour cost for a given level of prod^{n} to be shown in the prod^{n} budget. The various types and grades of labour required for the purpose are ascertained on the basis of labour hrs. and estimate the rate of wages per hr. Therefore, the total direct labour cost can be estimated by multiplying the total labour hrs. consumed by the rate of wages per hr. This budget helps the financial manager to determine the estimated requirement of cash for the total direct labour cost. The purpose of preparing this budget is

1. It assists the mgmt. in estimating the total outlay on the total direct cost.
2. Defining the total requirement of labour force.

Problem 1. *The direct labour hrs. requirement of the three products manufacturing in a factory, each involving more than one labour operations are estimated as follows.*

Direct labour hrs/unit. in min.				
		Product		
O/P^n		*1*	*2*	*3*
	1	*18*	*42*	*30*
	2	–	*12*	*24*
	3	*9*	*6*	–

The factory works 8 hrs. a day, and 6 days a week. The budget quarter is taken as 13 weeks. and during the quarter lost hrs. due to leave and hot days and other causes are estimated to be 124 hrs.

The budgeted hourly rates for the workers manning the operation 1, 2 and 3 are Rs. 2, Rs 2.5 and Rs. 3 respectively. The budgeted sales of the product during the quarter are as follows:

Product	*Sales in units*
1	*9,000*
2	*15,000*
3	*12,000*

There is a carry over of 5,000 units of product 2 and 4,000 units of product 3 and is proposed to build up a stock at the end of the budget quarter as

Product 1 = 1,000

Product 3 = 2,000 units.

Prepare a manpower budget showing for each operation:

(a) *Direct labour hrs.*

(b) *Direct cost*

(c) *No. of man power.*

Solution. **PRODUCTION BUDGET**

Particulars	*P1*	*P2*	*P3*
Sales	9,000	15,000	12,000
Add C.S.	1,000	–	2,000
	10,000	15,000	14,000
Less O.S.	–	5,000	4,000
Actual Qty. Budgeted prod^n	10,000	10,000	10,000

SAMPLE CALCULATIONS FOR LABOURS HRS.

Labour cost for product 1 & operation 1.

$$\text{Direct labour hrs.} = \frac{10,000 \times 18}{60}$$

$$= 3,000 \text{ hrs.}$$

$$\text{Labour cost} = 3,000 \times 2 = \text{Rs. } 6,000.$$

The sample calculation for the manpower requirement.

$$\text{The no. of workers required} = \frac{\text{Total Direct Labour hrs.}}{\text{Total hrs. available for one worker}}$$

No. of hrs. available for a worker in a quarter

$$= 13 \times 6 \times 8 - 124 = 500$$

$$\text{No. of workers required} = \frac{3,000}{500} = 6$$

Table for labour hrs.

Operation	*Rate / hr.*	*P1*		*P2*		*P3*		*Total*		*No. of workers.*
		L-hrs.	*Cost*	*L-hrs.*	*Cost*	*L-hrs.*	*Cost*	*L-hrs.*	*Cost*	
1	2	3,000	6,000	7,000	14,000	5,000	10,000	15,000	30,000	30
2	2.5	–	–	2,000	5,000	4,000	10,000	6,000	5,000	12
3	3	1,500	4,500	1,000	3,000	–	–	2,500	7,500	5
Total		4,500	10,500	10,000	22,000	9,000	20,000	23,500	52,500	47

12.20 ADMINISTRATIVE COST BUDGET AND SDOH COST BUDGET

The administrative budget represents the estimated expenditure of administration i.e., expenditure in framing the policies, controlling the business operations etc. Since most of the expenses on administration are fixed in nature, this budget is there easy to prepare in comparison with the other functional budget.

SDOH budget is the forecast of the cost relating to the selling and distribution of the product for the budget period. It is related to the sales budget and is prepared by the sales manager with the help of advertising manager, distribution manager, sales office manager and the accountant. However, the SD cost budget may be prepared on the elements of cost which are

(*a*) Direct selling expenses.

Ex : Salary, commission, expenses of the salesmen etc.

(*b*) Distribution expenses.

Ex : Rent rates, wages, insurance etc. of the warehouse.

(*c*) Cost of the sales office expenses.

Ex : Salaries, rent, lighting of the sales office etc.

(*d*) Publicity expenses.

Ex. : Posters, TV, Radio etc.

Problem 1. *You are required to prepare the sales OH budget from the estimates given below*

Particulars	*Amount*
Advertisement	*2,500*
Salaries of salesmen	*5,000*
Expenses of sales dept.	*1,500*
Counter salesmen salaries & DA	*6,000*

Commission to the counter sales at 1% on their sales.

Travelling salesmen commission at 10% on their sales

Expenses at 5% on their sales.

The sales during the period were estimated as follows :

Area	*Counter sales (Rs.)*	*Travelling salesmen (Rs.)*
1	*80,000*	*10,000*
2	*1,20,000*	*15,000*
3	*1,40,000*	*20,000*

Solution. **SALES OVERHEAD BUDGET**

Particulars	*Area 1*		*A2*		*A3*	
VOH						
Commission						
Counter sales at 1%	800		1,200		1,400	
Travelling salesmen at 10%	1,000		1,500		2,000	
Expenses at 5%	500		750		1,000	
		2,300		3450		4,400
FOH						
Advertisement	2,500		2500		2,500	
Salaries of salesmen	5,000		5000		5,000	
Expenses of sales dept.	1,500		1500		1,500	
Counter sales salaries	6,000		6000		6,000	
		15,000		15,000		15,000
Total SDOH		17,300		18,450		19,400

Problem 2. *From the following expenditure regarding SD of ABC Company in the last budgeted period, you are required to prepare an SD expense budget for the given period. Expenses are as follows:*

Particulars	*West Bengal*	*Bihar*	*Assam*	*Total*
Commission on sales	*8000*	*12,000*	*16,000*	*36,000*
Salaries	*6000*	*8000*	*10,000*	
Selling expenses	*3000*	*2000*	*1000*	
Advertisement	*4000*	*5000*	*6000*	
Rent & taxes	*2000*	*4000*	*3000*	
Warehouse expenses	*5000*	*5000*	*6000*	

During the budget period, the following changes are to be made:

1. *Commission on sales will be increased by 5% in all the states.*
2. *Salaries will be increased by Rs. 2000, Rs. 3000, Rs. 4000 respectively.*
3. *Rent will be increased by 5% in all the states.*
4. *Warehouse expenses increased by 1% in West Bengal and Bihar.*
5. *In Assam, advertisement expenses increased by 2000.*
6. *Selling expenses will increase by 10% in West Bengal and Bihar.*

Solution.

SELLING AND DISTRIBUTION EXPENSES BUDGET

Particulars	*West Bengal*	*Bihar*	*Assam*	*Total*
Commission on	8400	12,600	16,800	37,800
Salaries	8000	11,000	14000	33,000
Rent	2100	4200	3150	9450
Warehouse expense	5050	5050	6000	16,100
Advertisement	4000	5000	8000	17,000

TABLE FOR DISCRETE SERIES, COMPOUNDING INTEREST FACTORS

	Single Payment		Uniform Series					
	Compound Amount Factor	Present Worth Factor	Capital Recovery Factor	Present Worth Factor	Sinking Fund Factor	Compound Amount Factor	Gradient Factor	
N	(F/P, ½, N)	(P/F, ½, N)	(A/P, ½, N)	(P/A, ½, N)	(A/F, ½, N)	(F/A, ½, N)	(A/G, ½, N)	N
1	1.00500	0.99502	1.00600	0.99502	1.00000	1.00000	0.00000	1
2	1.01003	0.99007	0.50375	1.98510	0.49875	2.00500	0.49875	2
3	1.01508	0.98515	0.33667	297025	0.33167	3.01503	0.99667	3
4	1.02015	0.98025	0.25313	3.95050	0.24813	4.03010	1.49377	4
5	1.02625	0.97537	0.20301	4.92587	0.19801	5.05025	1.99003	5
6.	1.03038	0.97062	0.16960	5.89638	0.16460	6.07550	2.48545	6
7	1.03553	0.96569	0.14573	6.86207	0.14073	7.10588	2.98005	7
8	1.04071	0.96089	0.12783	7.82296	0.12283	8.14141	3.47382	8
9	1.04591	0.95610	0.11391	8.77906	0.10891	9.18212	3.96675	9
10	1.05114	0.96135	0.10277	9.73041	0.09777	10.22803	4.45885	10
11	1.05640	0.94661	0.09366	10.67703	0.08866	11.27917	4.95013	11
12	1.06168	0.94191	0.08607	11.61893	0.08107	12.23556	5.44057	12
13	1.06699	0.93722	0.07964	12.55615	0.07464	13.39724	5.93018	13
14	1.07232	0.93256	0.07414	13.48871	0.06914	14.46423	6.41896	14
15	1.07768	0.92792	0.06936	14.41662	0.06436	15.53655	6.90691	15
16	1.08307	0.92330	0.06519	15.33993	0.06019	16.61423	7.35403	16
17	1.08849	0.91871	0.06151	16.25863	0.05651	17.69730	7.88031	17
18	1.09393	0.91414	0.05823	17.17277	0.05323	18.78579	8.36577	18
19	1.09940	0.90969	0.05530	18.08236	0.05030	19.87972	8.85040	19
20	1.10490	0.90506	0.05267	18.98742	0.04767	20.97912	9.33419	20

(Contd...)

21	1.11042	0.90056	0.05028	19.88798	0.04528	22.08401	9.81716	21
22	1.11597	0.89608	0.04811	20.78406	0.04311	23.19443	10.29929	22
23	1.12155	0.89162	0.04613	21.67568	0.04113	24.31040	10.78060	23
24	1.12716	0.88719	0.04432	22.56287	0.03932	25.43196	11.26107	24
25	1.13280	0.88277	0.04265	23.44564	0.03765	26.55912	11.74072	25
26	1.13846	0.87838	0.04111	24.32402	0.03611	27.69191	12.21963	26
27	1.14415	0.87401	0.03969	25.19803	0.03469	28.83037	12.69751	27
28	1.14987	0.86966	0.03836	26.06769	0.03336	29.97452	13.17467	28
29	1.15562	0.86533	0.03713	26.93302	0.03213	31.12439	13.65099	29
30	1.16140	0.86103	0.03598	27.79405	0.03098	32.28002	14.12649	30
31	1.16721	0.85675	0.03490	28.65080	0.02990	33.44142	14.60116	31
32	1.17304	0.85248	0.03389	29.50328	0.02889	34.60862	15.07499	32
33	1.17891	0.84824	0.03295	30.35153	0.02795	35.78167	15.54800	33
34	1.18480	0.84402	0.03206	31.19555	0.02706	36.96058	16.02018	34
35	1.19073	0.83982	0.03122	32.03537	0.02622	38.14538	16.49153	35
40	1.22079	0.81914	0.02765	36.17223	0.02265	44.15885	18.83585	40
45	1.25162	0.79896	0.02487	40.20720	0.01987	50.32416	21.15947	45
50	1.28323	0.77929	0.02265	44.14279	0.01765	56.64516	23.46242	50
55	1.31563	0.76009	0.02084	47.98145	0.01584	63.12577	25.74471	55
60	1.34885	0.74137	0.01933	51.72556	0.01433	69.77003	28.00638	60
65	1.38291	0.72311	0.01806	55.37746	0.01306	76.58206	30.24745	65
70	1.41783	0.70530	0.01697	58.93942	0.01197	83.56611	34.66796	70
75	1.45363	0.68793	0.01602	62.41365	0.01102	90.72650	34.66794	75
80	1.49034	0.67099	0.01520	65.80231	0.01020	98.06771	36.84742	80
85	1.52797	0.65446	0.01447	69.10750	0.00947	105.59430	30.00646	85
90	1.56655	0.63834	0.01383	72.33130	0.00883	113.31094	41.14508	90
95	1.60611	0.62262	0.01325	75.47569	0.00825	121.22243	43.26333	95
100	1.64667	0.60729	0.01273	78.54264	0.00773	129.33370	45.36126	100

1% interest factors for discrete compounding periods

	Single Payment		Uniform Series					
	Compound Amount Factor	*Present Worth Factor*	*Capital Recovery Factor*	*Present Worth Factor*	*Sinking Fund Factor*	*Compound Amount Factor*	*Gradient Factor*	
N	*(F/P, 1, N)*	*(P/F, 1, N)*	*(A/P, 1, N)*	*(P/A, 1, N)*	*(A/F, 1, N)*	*(F/A, 1, N)*	*(A/G, 1, N)*	*N*
1	1.01000	0.99010	1.01000	0.99010	1.00000	1.00000	0.00000	1
2	1.02010	0.98030	0.50751	1.97040	0.49751	2.01000	0.49751	2
3	1.03030	0.97059	0.34002	2.94099	0.33002	2.03010	0.89337	3
4	1.04060	0.96098	0.25628	3.90197	0.24628	4.06040	1.48756	4
5	1.05101	0.95147	1.20604	4.85343	0.19604	5.10101	1.98010	5
6	1.06152	0.94206	0.17255	5.79548	0.16255	6.15202	2.47098	6
7	1.07214	0.93272	1.14863	6.72819	0.13863	7.21354	2.96020	7
8	1.08286	0.92348	0.13069	7.65168	0.12069	8.28567	3.44777	8
9	1.09369	0.91434	0.11674	8.56602	0.10674	9.36853	3.93367	9
10	1.10462	0.90529	0.10558	9.47130	0.09558	10.46221	4.41792	10
11	1.11567	0.89532	0.09645	10.36763	0.08645	11.56683	4.90052	11
12	1.12683	0.88745	0.08885	11.25508	0.07885	12.68250	5.38145	12
13	1.13809	0.87866	0.08241	12.13374	0.07241	13.80933	5.86073	13
14	1.14947	0.86996	0.07690	13.00370	0.09650	14.94742	6.33836	14
15	1.16097	0.86135	0.07212	13.86505	0.06212	16.09690	6.81433	15
16	1.17258	0.85282	0.06794	14.71787	0.05794	17.25786	7.28865	16
17	1.18430	0.84438	0.06426	15.56225	0.05426	18.43044	7.76131	17
18	1.19615	0.83602	0.06098	16.39627	0.05098	19.61475	8.23231	18
19	1.20811	0.82774	0.05805	17.22601	0.04805	20.81090	8.70167	19
20	1.22019	0.81954	0.05542	18.04555	0.04542	22.01900	9.16937	20
21	1.23239	0.81143	0.05303	18.86698	0.04303	23.23919	9.63542	21
22	1.24472	0.80340	0.05096	19.66038	0.04086	24.47158	10.09882	22
23	1.25716	0.79544	0.04889	20.45582	0.03889	25.71630	10.56257	23
24	1.26973	0.78757	0.04707	21.24339	0.03707	26.97346	11.02367	24
25	1.28243	0.77977	0.04541	22.02316	0.03541	28.24320	11.48312	25
26	1.29626	0.77205	0.04387	22.79520	0.03387	29.52663	11.94092	26
27	1.30821	0.76440	0.04245	23.55961	0.03245	30.82089	12.39707	27
28	1.32129	0.75684	0.04112	24.31644	0.03112	32.12910	12.85158	28
29	1.33450	0.74934	0.03990	25.06579	0.02990	33.45039	13.30444	29
30	1.34785	0.74192	0.03875	25.80771	0.02875	34.78489	13.75566	30

(Contd...)

31	1.36133	0.73458	0.03768	26.54229	0.02768	36.13274	14.20523	31
32	1.37494	0.72730	0.03667	27.26959	0.02667	37.49407	14.65317	32
33	1.38869	0.72010	0.03573	27.98969	0.02573	38.86901	15.09946	33
34	1.40258	0.71297	0.03484	28.70267	0.02484	40.25770	15.64410	34
35	1.41660	0.70591	0.03400	29.40958	0.02400	41.66028	15.98711	35
40	1.48896	0.67165	0.03046	32.83469	0.02046	48.88637	18.17761	40
45	1.56481	0.63905	0.02771	36.09451	0.01771	56.48107	20.32730	45
50	1.64463	0.60804	0.02551	39.19612	0.01551	64.46318	22.43635	50
55	1.72852	0.57853	0.02373	42.14719	0.01373	72.85246	24.50495	55
60	1.81670	0.55045	0.02224	44.96504	0.01224	81.66967	26.53331	60
65	1.90937	0.52373	0.02100	47.62661	0.01100	90.93665	28.52167	65
70	1.00676	0.49831	0.01993	50.16851	0.00993	100.67634	30.47026	70
75	1.10913	0.47413	0.01902	52.58705	0.00902	110.91285	32.37934	75
80	2.21672	0.45112	0.01822	54.98821	0.00022	121.57152	34.24920	80
85	2.32979	0.42922	0.01752	57.07768	0.00752	132.97900	36.08013	85
90	2.44863	0.40839	0.01690	59.16088	0.00690	144.86327	37.87245	90
95	2.57354	0.38857	0.01636	61.14298	0.00636	157.35376	39.62648	95
100	2.70481	0.36971	0.01587	63.02888	0.00687	170.48138	41.34257	100

1½% interest factors for discrete compuding periods

	Single Payment		*Uniform Series*					
	Compound Amount Factor	*Present Worth Factor*	*Capital Recovery Factor*	*Present Worth Factor*	*Sinking Fund Factor*	*Compound Amount Factor*	*Gradient Factor*	
N	*(F/P, 1½, N)*	*(P/F, 1½, N)*	*(A/P, 1½, N)*	*(P/A, 1½, N)*	*(A/F, 1½, N)*	*(F/A, 1½, N)*	*(A/G, 1½, N)*	*N*
1	1.01500	0.98522	1.01500	0.98522	1.00000	1.00000	0.00000	1
2	1.03023	0.97066	0.51128	0.95588	1.49628	2.01500	0.49628	2
3	1.04568	0.95632	0.34338	2.91220	0.32838	3.04523	0.99007	3
4	1.06136	0.94218	1.25944	3.85438	0.24444	4.09090	1.48139	4
5	1.07728	0.92826	0.20909	4.78264	0.19409	5.15227	1.97023	5
6	1.09344	0.91454	1.17553	5.69719	0.16053	6.22955	2.45658	6
7	1.10984	0.90103	1.15156	0.59821	0.13656	7.32299	2.94046	7
8	1.12649	0.88771	1.13358	7.48593	0.11858	8.43284	3.42185	8
9.	1.14339	0.87459	1.11961	8.36052	0.10461	9.55933	3.90077	9
10	1.16054	0.86167	0.10843	9.22218	0.09343	10.70272	4.37721	10
11	1.17795	0.84893	0.09929	10.07112	0.08429	11.86326	4.85118	11
12	1.19662	0.83639	1.09168	10.90751	0.07668	13.04121	5.32267	12
13	1.21355	0.82403	1.08524	11.73153	0.07024	14.23683	5.79169	13
14	1.23176	0.81185	1.07972	12.54338	0.06472	15.45038	6.25824	14
15	1.25023	0.79985	1.07494	13.34323	0.05994	16.68214	6.72231	15
16	1.26899	0.78803	0.07077	14.13126	0.05577	17.93237	7.18392	16
17	1.28802	0.77639	0.06708	14.90765	0.05208	19.20136	7.64306	17
18	1.30734	0.76491	0.06381	15.67256	0.04881	20.48938	8.09973	18
19	1.32695	0.75361	0.06088	16.42617	0.04588	21.79672	8.55394	19
20	1.34686	0.74247	0.05825	17.16864	0.04325	23.12367	9.00569	20
21	1.36706	0.73150	0.05587	17.90014	0.04087	24.47052	9.45497	21
22	1.38756	0.72069	0.05370	18.62082	0.03870	25.83758	9.90180	22
23	1.40838	0.71004	0.05173	19.33086	0.03673	27.22514	10.34618	23
24	1.42950	0.69954	0.04992	20.03041	0.03492	28.63352	10.78810	24
25	1.45095	0.68921	0.04826	20.71961	0.03326	30.06302	11.22758	25
26	1.47271	0.67902	0.04673	21.39863	0.03173	31.51397	11.66460	26
27	1.49480	0.66899	0.04532	22.06762	0.03032	32.98668	12.09918	27
28	1.51722	0.65910	0.04400	22.72672	0.02900	34.48148	12.53132	28
29	1.53998	0.64936	0.04278	23.37608	0.02778	35.99870	12.96102	29
30	1.56308	0.63976	0.04164	24.01584	0.02664	37.53868	13.38829	30

(Contd...)

31	1.58653	0.63031	0.04057	24.64615	0.02557	39.10176	13.81312	31
32	1.61032	0.62099	0.03958	25.26714	0.02458	40.68829	14.23553	32
33	1.63448	0.61182	0.03864	25.87895	0.02364	42.29861	14.65550	33
34	1.65900	0.60277	0.03776	26.48173	0.02276	43.93309	15.07306	34
35	1.68388	0.59387	0.03693	27.07559	0.02193	45.59209	15.48820	35
40	1.81402	0.55126	0.03343	29.91585	0.01843	54.26789	17.52773	40
45	1.95421	0.51171	0.03072	32.55234	0.01572	63.61420	19.50739	45
50	2.10524	0.47500	0.02857	34.99969	0.01357	73.68283	21.42772	50
55	2.26794	0.44093	0.02683	37.27147	0.01183	84.52960	23.28936	55
60	2.44322	0.40930	0.02539	39.38027	0.01039	96.21465	25.09296	60
65	2.63204	0.37993	0.02419	41.33779	0.00919	108.80277	26.83925	65
70	2.83546	0.35268	0.02317	43.15487	0.00817	122.36375	28.52901	70
75	3.05459	0.32738	0.02230	44.84160	0.00730	136.97278	30.16306	75
80	3.29066	0.30389	0.02155	46.40732	0.00655	152.71085	31.74228	80
85	3.54498	0.28209	0.02089	47.86072	0.00589	169.66523	33.26756	85
90	3.81895	0.26185	0.02032	49.20985	0.00532	187.92990	34.73987	90
95	4.11409	0.24307	0.01982	50.46220	0.00482	207.60614	36.16018	95
100	4.43206	0.22563	0.01937	51.62470	0.00437	228.80304	37.52953	100

2% interest factors for discrete compounding periods

	Single Payment		Uniform Series					
	Compound Amount Factor	*Present Worth Factor*	*Capital Recovery Factor*	*Present Worth Factor*	*Sinking Fund Factor*	*Compound Amount Factor*	*Gradient Factor*	
N	*(F/P, 2, N)*	*(P/F, 2, N)*	*(A/P, 2, N)*	*(P/A, 2, N)*	*(A/F, 2, N)*	*(F/A, 2, N)*	*(A/G, 2, N)*	*N*
1	1.02000	0.98039	1.02000	0.98039	1.00000	1.00000	0.00000	1
2	1.04040	0.96117	0.51505	1.94156	0.49505	2.02000	0.49505	2
3	1.06121	0.94232	0.34675	2.88388	0.32675	3.06040	0.98680	3
4	1.08243	0.92385	0.26262	3.80773	0.24262	4.12161	1.47525	4
5	1.10408	0.90573	0.21216	4.71346	0.19216	5.20404	1.96040	5
6	1.12616	0.88797	0.17853	5.60143	0.15853	6.30812	2.44226	6
7	1.14869	0.87056	0.15451	6.47199	0.13451	7.43428	2.92082	7
8	1.17166	0.85349	0.13651	7.32548	0.11651	8.58297	3.39608	8
9	1.19509	0.83676	0.12252	8.16224	0.10252	9.75463	3.86805	9
10	1.21899	0.82035	0.11133	8.98259	0.09133	10.94972	4.33674	10
11	1.24337	0.80426	0.10218	9.78685	0.08218	12.16872	4.80213	11
12	1.26824	0.78849	0.09456	10.57534	0.07456	13.41209	5.26424	12
13	1.29361	0.77303	0.08812	11.34837	0.06812	14.68033	5.72307	13
14	1.31948	0.75788	0.08260	12.10625	0.06260	15.97394	6.17862	14
15	1.34587	0.74301	0.07783	12.84926	0.05783	17.29342	6.63090	15
16	1.37279	0.72845	0.07365	13.57771	0.05365	18.63929	7.07990	16
17	1.40024	0.71416	0.06997	14.29187	0.04997	20.01207	7.52564	17
18	1.42825	0.70016	0.06670	14.99203	0.04670	21.41231	7.96811	18
19	1.45681	0.68643	0.06378	15.67846	0.04378	22.84056	8.40732	19
20	1.48595	0.67297	0.06116	16.35143	0.04116	24.29737	8.84328	20
21	1.51567	0.65978	0.05878	17.01121	0.03878	25.78332	9.27599	21
22	1.54598	0.64684	0.05663	17.65805	0.03663	27.29898	9.70546	22
23	1.57690	0.63416	0.05467	18.29220	0.03467	28.84496	10.13169	23
24	1.60844	0.62172	0.05287	18.91393	0.03287	30.42186	10.55468	24
25	1.64061	0.60953	0.05122	19.52346	0.03122	32.03030	10.97445	25
26	1.67342	0.59758	0.04970	20.12104	0.02970	33.67091	11.39100	26
27	1.70689	0.58586	0.04829	20.70690	0.02829	35.34432	11.80433	27
28	1.74102	0.57437	0.04699	21.28127	0.02699	37.05121	12.21446	28
29	1.77584	0.56311	0.04578	21.84438	0.02578	38.79223	12.62138	29
30	1.81136	0.55207	0.04465	22.39646	0.02465	40.56808	13.02512	30

(Contd...)

31	1.84759	0.54125	0.04360	22.93770	0.02360	42.37944	13.42566	31
32	1.88454	0.53063	0.04261	23.46833	0.02261	44.22703	13.82303	32
33	1.92223	0.52023	0.04169	23.98856	0.02169	46.11157	14.21722	33
34	1.96068	0.51003	0.04082	24.49859	0.02082	48.03380	14.60826	34
35	1.99989	0.50003	0.04000	24.99862	0.02000	49.99448	14.99613	35
40	2.20804	0.45289	0.03656	27.35548	0.01656	60.40198	16.88850	40
45	2.43785	0.41020	0.03391	29.49016	0.01391	71.89271	18.70336	45
50	2.69159	0.37153	0.03182	31.42361	0.01182	84.57940	20.44198	50
55	2.97173	0.33650	0.03014	33.17479	0.01014	98.58653	22.10572	55
60	3.28103	0.30478	0.02877	34.76089	0.00877	114.05154	23.69610	60
65	3.62252	0.27605	0.02763	36.19747	0.00763	131.12816	35.21471	65
70	3.99956	0.25003	0.02667	37.49862	0.00667	149.97791	26.66323	70
75	4.41584	0.22646	0.02586	38.67711	0.00586	170.79177	28.04344	75
80	4.87544	0.20511	0.02516	39.74451	0.00516	193.77196	29.35718	80
85	5.38288	0.18577	0.02456	40.71129	0.00456	219.14394	30.60635	85
90	5.94313	0.16826	0.02405	41.58693	0.00405	247.15666	31.79292	90
95	6.56170	0.15240	0.02360	42.38002	0.00360	278.08496	32.91889	95
100	7.24465	0.13803	0.02320	43.09835	0.00320	312.23231	33.98628	100

2½% interest factors for discrete compounding periods

	Single Payment		Uniform Series					
	Compound Amount Factor	*Present Worth Factor*	*Capital Recovery Factor*	*Present Worth Factor*	*Sinking Fund Factor*	*Compound Amount Factor*	*Gradient Factor*	
N	*(F/P, 2½,N)*	*(P/F, 2½,N)*	*(A/P, 2½,N)*	*(P/A, 2½,N)*	*(A/F, 2½,N)*	*(F/A, 2½,N)*	*(A/G, 2½,N)*	*N*
1	1.02500	0.97561	1.02500	0.97561	1.00000	1.00000	0.00000	1
2	1.05063	0.95181	0.51883	1.92742	0.49383	2.02500	0.49383	2
3	1.07689	0.92860	0.35014	2.85602	0.32514	3.07563	0.98354	3
4	1.10381	0.90595	0.26582	3.76197	0.24082	4.15252	1.46914	4
5	1.13141	0.88385	0.21525	4.64583	0.19025	5.25633	1.95063	5
6	1.15969	0.86230	0.18155	5.50813	0.15655	6.38774	2.42801	6
7	1.18869	0.84127	0.15750	6.34939	0.13250	7.54743	2.90128	7
8	1.21840	0.82075	0.13947	7.17014	0.11447	8.73612	3.37045	8
9	1.24886	0.80073	0.12546	7.97087	0.10046	9.95452	3.83552	9
10	1.28008	0.78120	0.11426	8.75206	0.08926	11.20338	4.29649	10
11	1.31209	0.76214	0.10511	9.51421	0.08011	12.48347	4.75338	11
12	1.34489	0.74356	0.09749	10.25776	0.07249	13.79555	5.20618	12
13	1.37851	0.72542	0.09105	10.98318	0.06605	15.14044	5.65490	13
14	1.41297	0.70773	0.08554	11.69091	0.06054	16.51895	6.09955	14
15	1.44630	0.69047	0.08077	12.38138	0.05577	17.93193	6.54013	15
16	1.48451	0.67362	0.07660	13.05500	0.05160	19.38022	6.97665	16
17	1.52162	0.65720	0.07293	13.71220	0.04793	20.86473	7.40912	17
18	1.55966	0.64117	0.06967	14.35336	0.04467	22.38635	7.83754	18
19	1.59865	0.62553	0.06676	14.97889	0.04176	23.94601	8.26193	19
20	1.63862	0.61027	0.06415	15.58916	0.03915	25.54466	8.68230	20
21	1.67958	0.59539	0.06179	16.18455	0.03679	27.18327	9.09865	21
22	1.72157	0.58086	0.05965	16.76541	0.03465	28.86286	9.51099	22
23	1.76461	0.56670	0.05770	17.33211	0.03270	30.58443	9.91933	23
24	1.80873	0.55288	0.05591	17.88499	0.03091	32.34904	10.32369	24
25	1.85394	0.53939	0.05428	18.42438	0.02928	34.15776	10.72408	25
26	1.90029	0.52623	0.05277	18.95061	0.02777	36.01171	11.12050	26
27	1.94780	0.51340	0.05138	19.46401	0.02638	37.91200	11.51298	27
28	1.99650	0.50088	0.05009	19.96489	0.02509	39.85980	11.90152	28
29	2.04641	0.48866	0.04889	20.45355	0.02389	41.85530	12.28613	29
30	2.09757	0.47674	0.04778	20.93029	0.02278	43.90270	12.66683	30

(Contd...)

31	2.15001	0.46511	0.04674	21.39541	0.02174	46.00027	13.04364	31
32	2.20376	0.45377	0.04577	21.84918	0.02077	48.15028	13.41656	32
33	2.25885	0.44270	0.04486	22.29188	0.01986	50.35403	13.78562	33
34	2.31532	0.43191	0.04401	22.72379	0.01901	52.61289	14.15082	34
35	2.37321	0.42137	0.04321	23.14516	0.01821	54.92821	14.51218	35
40	2.68506	0.37243	0.03984	25.10278	0.01484	67.40255	16.26203	40
45	3.03790	0.32917	0.03727	26.83302	0.01227	81.51613	17.91848	45
50	3.43711	0.29094	0.03526	28.36231	0.01026	97.48435	19.48389	50
55	3.88877	0.25715	0.03365	29.71398	0.00865	115.55092	20.96077	55
60	4.39979	0.22728	0.03235	30.90866	0.00735	135.99159	22.35185	60
65	4.97796	0.20089	0.03128	31.96458	0.00628	159.11833	23.65996	65
70	5.63210	0.17755	0.03040	32.89786	0.00540	185.28411	24.88807	70
75	6.37221	0.15693	0.02965	33.72274	0.00465	214.88830	26.03926	75
80	7.20957	0.13870	0.02903	34.45182	0.00403	248.38271	27.11666	80
85	8.15696	0.12259	0.02849	35.09621	0.00349	286.27857	28.12346	85
90	9.22886	0.10836	0.02804	35.66577	0.00304	329.15425	29.06288	90
95	10.44160	0.09577	0.02765	36.16917	0.00265	377.66415	29.93815	95
100	11.81372	0.08465	0.02731	36.61411	0.00231	432.54865	30.75249	100

3% interest factors for discrete compounding periods

	Single Payment		Uniform Series					
	Compound Amount Factor	Present Worth Factor	Capital Recovery Factor	Present Worth Factor	Sinking Fund Factor	Compound Amount Factor	Gradient Factor	
N	(F/P, 3, N)	(P/F, 3, N)	(A/P, 3, N)	(P/A, 3, N)	(A/F, 3, N)	(F/A, 3, N)	(A/G, 3, N)	N
1	1.03000	0.97087	1.03000	0.97087	1.00000	1.00000	0.00000	1
2	1.06090	0.94260	0.52261	1.91347	0.49261	2.03000	0.49261	2
3	1.09273	0.91514	0.35353	2.82861	0.32353	3.09090	0.98030	3
4	1.12551	0.88849	0.26903	3.71710	0.23903	4.18363	1.46306	4
5	1.15927	0.86261	0.21835	4.57971	0.18835	5.30914	1.94090	5
6	1.19405	0.83748	0.18460	5.41719	0.15460	6.46841	2.41383	6
7	1.22987	0.81309	0.16051	6.23028	0.13051	7.66246	2.88185	7
8	1.26677	0.78941	0.14246	7.01969	0.11246	8.89234	3.34496	8
9	1.30477	0.76642	0.12843	7.78611	0.09843	10.15911	3.80318	9
10	1.34392	0.74409	0.11723	8.53020	0.08723	11.46388	4.25650	10
11	1.38423	0.72242	0.10808	9.25262	0.07808	12.80780	4.70494	11
12	1.42576	0.70138	0.10046	9.95400	0.07046	14.19203	5.14850	12
13	1.46853	0.68095	0.09403	10.63496	0.06403	15.61779	5.58720	13
14	1.51259	0.66112	0.08853	11.29607	0.05853	17.08632	6.02104	14
15	1.55797	0.64186	0.08377	11.93794	0.05377	18.59891	6.45004	15
16	1.60471	0.62317	0.07961	12.56110	0.04961	20.15688	6.87421	16
17	1.65285	0.60502	0.07595	13.16612	0.04595	21.76159	7.29357	17
18	1.70243	0.58739	0.07271	13.75351	0.04271	23.41444	7.70812	18
19	1.75351	0.57029	0.06981	14.32380	0.03981	25.11687	8.11788	19
20	1.80611	0.55368	0.06722	14.87747	0.03722	26.87037	8.52286	20
21	1.86029	0.53755	0.06487	15.41502	0.03487	28.67649	8.92309	21
22	1.91610	0.52189	0.06275	15.93692	0.03275	30.53678	9.31858	22
23	1.97359	0.50669	0.06081	16.44361	0.03081	32.45288	9.70934	23
24	2.03279	0.49193	0.05905	16.93554	0.02905	34.42647	10.09540	24
25	2.09378	0.47761	0.05743	17.41315	0.02743	36.45926	10.47677	25
26	2.15659	0.46369	0.05594	17.87684	0.02594	38.55304	10.85348	26
27	2.22129	0.45019	0.05456	18.32703	0.02456	40.70963	11.22554	27
28	2.28793	0.43708	0.05329	18.76411	0.02329	42.93092	11.59298	28
29	2.35657	0.42435	0.05211	19.18845	0.02211	45.21885	11.95582	29
30	2.42726	0.41199	0.05102	19.60044	0.02102	47.57542	12.31407	30

(Contd...)

31	2.50008	0.39999	0.05000	20.00043	0.02000	50.00268	12.66777	31
32	2.57508	0.38834	0.04905	20.38877	0.01905	52.50276	13.01694	32
33	2.65234	0.37703	0.04816	20.76579	0.01816	55.07784	13.36160	33
34	2.73191	0.36604	0.04732	21.13184	0.01732	57.73018	13.70177	34
35	2.81386	0.35538	0.04654	21.48722	0.01654	60.46208	14.03749	35
40	3.26204	0.30656	0.04326	23.11477	0.01326	75.40126	15.65016	40
45	3.78160	0.26444	0.04079	24.51871	0.01079	92.71986	17.15557	45
50	4.38391	0.22811	0.03887	25.72976	0.00887	112.79687	18.55751	50
55	5.08215	0.19677	0.03735	26.77443	0.00735	136.07162	19.86004	55
60	5.89160	0.16973	0.03613	27.67556	0.00613	163.05344	21.06742	60
65	6.82998	0.14641	0.03515	28.45289	0.00515	194.33276	22.18407	65
70	7.91782	0.12630	0.03434	29.12342	0.00434	230.59406	23.21454	70
75	9.17893	0.10895	0.03367	29.70183	0.00367	272.63086	24.16342	75
80	10.64089	0.09398	0.03311	30.20076	0.00311	321.36302	25.03534	80
85	12.33571	0.08107	0.03265	30.63115	0.00265	377.85695	25.83490	85
90	14.30047	0.06993	0.03226	31.00241	0.00226	443.34890	26.56665	90
95	16.57816	0.06032	0.03193	31.32266	0.00193	519.27203	27.23505	95
100	19.21863	0.05203	0.03165	31.59891	0.00165	607.28773	27.84445	100

4% interest factors for discrete compounding periods

	Single Payment		Uniform Series					
	Compound Amount Factor	*Present Worth Factor*	*Capital Recovery Factor*	*Present Worth Factor*	*Sinking Fund Factor*	*Compound Amount Factor*	*Gradient Factor*	
N	*(F/P, 4, N)*	*(P/F, 4, N)*	*(A/P, 4, N)*	*(P/A, 4, N)*	*(A/F, 4, N)*	*(F/A, 4, N)*	*(A/G, 4, N)*	*N*
1	1.04000	0.96154	1.04000	0.96154	1.00000	1.00000	0.00000	1
2	1.08160	0.92456	0.53020	1.88609	0.49020	2.04000	0.49020	2
3	1.12486	0.88900	0.36035	2.77509	0.32035	3.12160	0.97386	3
4	1.16986	0.85480	0.27549	3.62990	0.23549	4.24646	1.45100	4
5	1.21665	0.82193	0.22463	4.45182	0.18463	5.41632	1.92161	5
6	1.26532	0.79031	0.19076	5.24214	0.15076	6.63298	2.38571	6
7	1.31593	0.75992	0.16661	6.00205	0.12661	7.89829	2.84332	7
8	1.36857	0.73069	0.14853	6.73274	0.10853	9.21423	3.29443	8
9	1.42331	0.70259	0.13449	7.43533	0.09449	10.58280	3.73908	9
10	1.48024	0.67556	0.12329	8.11090	0.08329	12.00611	4.17726	10
11	1.53945	0.64958	0.11415	8.76048	0.07415	13.48635	4.60901	11
12	1.60103	0.62460	0.10655	9.38507	0.06655	15.02581	5.03435	12
13	1.66507	0.60057	0.10014	9.98565	0.06014	16.62684	5.45329	13
14	1.73168	0.57748	0.09467	10.56312	0.05467	18.29191	5.86586	14
15	1.80094	0.55526	0.08994	11.11839	0.04994	20.02359	6.27209	15
16	1.87298	0.53391	0.08582	11.65230	0.04582	21.82453	6.67200	16
17	1.94790	0.51337	0.08220	12.16567	0.04220	23.69751	7.06563	17
18	2.02582	0.49363	0.07899	12.65930	0.03899	25.64541	7.45300	18
19	2.10685	0.47464	0.07614	13.13394	0.03614	27.67123	7.83416	19
20	2.19112	0.45639	0.07358	13.59033	0.03358	29.77808	8.20912	20
21	2.27877	0.43883	0.07128	14.02916	0.03128	31.96920	8.57794	21
22	2.36992	0.42196	0.06920	14.45112	0.02920	34.24797	8.94065	22
23	2.46472	0.40573	0.06731	14.85684	0.02731	36.61789	9.29729	23
24	2.56330	0.39012	0.06559	15.24696	0.02559	39.08260	9.64790	24
25	2.66584	0.37512	0.06401	15.62208	0.02401	41.64591	9.99252	25
26	2.77247	0.36069	0.06257	15.98277	0.02257	44.31174	10.33120	26
27	2.88337	0.34682	0.06124	16.32959	0.02124	47.08421	10.66399	27
28	2.99870	0.33348	0.06001	16.66306	0.02001	49.96758	10.99092	28
29	3.11865	0.32065	0.05888	16.98371	0.01888	52.96629	11.31205	29
30	3.24340	0.30832	0.05783	17.29203	0.01783	56.08494	11.62743	30

(Contd...)

31	3.37313	0.29646	0.05686	17.58849	0.01686	59.32834	11.93710	31
32	3.50806	0.28506	0.05595	17.87355	0.01595	62.70147	12.24113	32
33	3.64838	0.27409	0.05510	18.14765	0.01510	66.20953	12.53956	33
34	3.79432	0.26355	0.05431	18.41120	0.01431	69.85791	12.83244	34
35	3.94609	0.25342	0.05358	18.66461	0.01358	73.65222	13.11984	35
40	4.80102	0.20829	0.05052	19.79277	0.01052	95.02552	14.47651	40
45	5.84118	0.17120	0.04826	20.72004	0.00826	121.02939	15.70474	45
50	7.10668	0.14071	0.04655	21.48218	0.00655	152.66708	16.81225	50
55	8.64637	0.11566	0.04523	22.10861	0.00523	191.15917	17.80704	55
60	10.51963	0.09506	0.04420	22.62349	0.00420	237.99069	18.69723	60
65	12.79874	0.07813	0.04339	23.04668	0.00339	294.96838	19.49093	65
70	15.57162	0.06422	0.04275	23.39451	0.00275	364.29046	20.19614	70
75	18.94525	0.05278	0.04223	23.68041	0.00223	448.63137	20.82062	75
80	23.04980	0.04338	0.04181	23.91539	0.00181	551.24498	21.37185	80
85	28.04380	0.03566	0.04148	24.10853	0.00148	676.09012	21.85693	85
90	34.11933	0.02931	0.04121	24.26728	0.00121	827.98333	22.28255	90
95	41.51139	0.02409	0.04099	24.39776	0.00099	1012.78465	22.65498	95
100	50.50495	0.01980	0.04081	24.50500	0.00081	1237.62370	22.98000	100

5% interest factors for discrete compounding periods

	Single Payment		Uniform Series					
	Compound Amount Factor	*Present Worth Factor*	*Capital Recovery Factor*	*Present Worth Factor*	*Sinking Fund Factor*	*Compound Amount Factor*	*Gradient Factor*	
N	*(F/P, 5, N)*	*(P/F, 5, N)*	*(A/P, 5, N)*	*(P/A, 5, N)*	*(A/F, 5, N)*	*(F/A, 5, N)*	*(A/G, 5, N)*	*N*
1	1.05000	0.95238	1.05000	0.95238	1.00000	1.00000	0.00000	1
2	1.10250	0.90703	0.53780	1.85941	0.48780	2.05000	0.48780	2
3	1.15763	0.86384	0.36721	2.72325	0.31721	3.15250	0.96749	3
4	1.21551	0.82270	0.28201	3.54595	0.23201	4.31013	1.43905	4
5	1.27628	0.78353	0.23097	4.32948	0.18097	5.52563	1.90252	5
6	1.34010	0.74622	0.19702	5.07569	0.14702	6.80191	2.35790	6
7	1.40710	0.71068	0.17282	5.78637	0.12282	8.14201	2.80523	7
8	1.47746	0.67684	0.15472	6.46321	0.10472	9.54911	3.24451	8
9	1.55133	0.64461	0.14069	7.10782	0.09069	11.02656	3.67579	9
10	1.62889	0.61391	0.12950	7.72173	0.07950	12.57789	4.09909	10
11	1.71034	0.58468	0.12039	8.30641	0.07039	14.20679	4.51444	11
12	1.79586	0.55684	0.11283	8.86325	0.06283	15.91713	4.92190	12
13	1.88565	0.53032	0.10646	9.39357	0.05646	17.71298	5.32150	13
14	1.97993	0.50607	0.10102	9.89864	0.05102	19.59863	5.71329	14
15	2.07893	0.48102	0.09634	10.37966	0.04634	21.57856	6.09731	15
16	2.18287	0.45811	0.09227	10.83777	0.04227	23.65749	6.47363	16
17	2.29202	0.43630	0.08870	11.27407	0.03870	25.84037	6.84229	17
18	2.40662	0.41552	0.08555	11.68959	0.03555	28.13238	7.20336	18
19	2.52695	0.39573	0.08275	12.08532	0.03275	30.53900	7.55690	19
20	2.65330	0.37689	0.08024	12.46221	0.03024	33.06595	7.90297	20
21	2.78596	0.35894	0.07800	12.82115	0.02800	35.71925	8.24164	21
22	2.92526	0.34185	0.07597	13.16300	0.02597	38.50521	8.57298	22
23	3.07152	0.32557	0.07414	13.48857	0.02414	41.43048	8.89706	23
24	3.22510	0.31007	0.07247	13.79864	0.02247	44.50200	9.21397	24
25	3.38635	0.29530	0.07095	14.09394	0.02095	47.72710	9.52377	25
26	3.55567	0.28124	0.06956	14.37519	0.01956	51.11345	9.82655	26
27	3.73346	0.26785	0.06829	14.64303	0.01829	54.66913	10.12240	27
28	3.92013	0.25509	0.06712	14.89813	0.01712	58.40258	10.41138	28
29	4.11614	0.24295	0.06605	15.14107	0.01605	62.32271	10.69360	29
30	4.32194	0.23138	0.06505	15.37245	0.01505	66.43885	10.96914	30

(Contd...)

31	4.53804	0.22036	0.06413	15.59281	0.01413	70.76079	11.23809	31
32	4.76494	0.20987	0.06328	15.80268	0.01328	75.29883	11.50053	32
33	5.00319	0.19987	0.06249	16.00255	0.01249	80.06377	11.75657	33
34	5.25335	0.19035	0.06176	16.19290	0.01176	85.06696	12.00630	34
35	5.51602	0.18129	0.06107	16.37419	0.01107	90.32031	12.24980	35
40	7.03999	0.14205	0.05828	17.15909	0.00828	120.79977	13.37747	40
45	8.98501	0.11130	0.05626	17.77407	0.00626	159.70016	14.36444	45
50	11.46740	0.08720	0.05478	18.25593	0.00478	209.34800	15.22326	50
55	14.63563	0.06833	0.05367	18.63347	0.00367	272.71262	15.96645	55
60	18.67919	0.05354	0.05283	18.92929	0.00283	353.58372	16.60618	60
65	23.83990	0.04195	0.05219	19.16107	0.00219	456.79801	17.15410	65
70	30.42643	0.03287	0.05170	19.34268	0.00170	588.52851	17.62119	70
75	38.83269	0.02575	0.05132	19.48497	0.00132	756.65372	18.01759	75
80	49.56144	0.02018	0.05103	19.59646	0.00103	971.22882	18.35260	80
85	63.25435	0.01581	0.05080	19.68382	0.00080	1245.08707	18.63463	85
90	80.73037	0.01239	0.05063	19.75226	0.00063	1594.60730	18.87120	90
95	103.03468	0.00971	0.05049	19.80589	0.00049	2040.69353	19.06894	95
100	131.50126	0.00760	0.05038	19.84791	0.00038	2610.02516	19.23372	100

6% interest factors for discrete compounding periods

	Single Payment		*Uniform Series*					
	Compound Amount Factor	*Present Worth Factor*	*Capital Recovery Factor*	*Present Worth Factor*	*Sinking Fund Factor*	*Compound Amount Factor*	*Gradient Factor*	
N	*(F/P, 6, N)*	*(P/F, 6, N)*	*(A/P, 6, N)*	*(P/A, 6, N)*	*(A/F, 6, N)*	*(F/A, 6, N)*	*(A/G, 6, N)*	*N*
1	1.06000	0.94340	1.06000	0.94340	1.00000	1.00000	0.00000	1
2	1.12360	0.89000	0.54544	1.83339	0.48544	2.06000	0.48544	2
3	1.19102	0.83962	0.37411	2.67301	0.31411	3.18360	0.96118	3
4	1.26248	0.79209	0.28859	3.46511	0.22859	4.37462	1.42723	4
5	1.33823	0.74726	0.23740	4.21236	0.17740	5.63709	1.88363	5
6	1.41852	0.70496	0.20336	4.91732	0.14336	6.97532	2.33040	6
7	1.50363	0.66506	0.17914	5.58238	0.11914	8.39384	2.76758	7
8	1.59385	0.62741	0.16104	6.20979	0.10104	9.89747	3.19521	8
9	1.68948	0.59190	0.14702	6.80169	0.08702	11.49132	3.61333	9
10	1.79085	0.55839	0.13587	7.36009	0.07587	13.18079	4.02201	10
11	1.89830	0.52679	0.12679	7.88687	0.06679	14.97164	4.42129	11
12	2.01220	0.49697	0.11928	8.38384	0.05928	16.86994	4.81126	12
13	2.13293	0.46884	0.11296	8.85268	0.05296	18.88214	5.19198	13
14	2.26090	0.44230	0.10758	9.29498	0.04758	21.01507	5.56352	14
15	2.39656	0.41727	0.10296	9.71225	0.04296	23.27597	5.92598	15
16	2.54035	0.39365	0.09895	10.10590	0.03895	25.67253	6.27943	16
17	2.69277	0.37136	0.09544	10.47726	0.03544	28.21288	6.62397	17
18	2.85434	0.35034	0.09236	10.82760	0.03236	30.90565	6.95970	18
19	3.02560	0.33051	0.08962	11.15812	0.02962	33.75999	7.28673	19
20	3.20714	0.31180	0.08718	11.46992	0.02718	36.78559	7.60515	20
21	3.39956	0.29416	0.08500	11.76408	0.02500	39.99273	7.91508	21
22	3.60354	0.27751	0.08305	12.04158	0.02305	43.39229	8.21662	22
23	3.81975	0.26180	0.08128	12.30338	0.02128	46.99583	8.50991	23
24	4.04893	0.24698	0.07968	12.55036	0.01968	50.81558	8.79506	24
25	4.29187	0.23300	0.07823	12.78336	0.01823	54.86451	9.07220	25
26	4.54938	0.21981	0.07690	13.00317	0.01690	59.15638	9.34145	26
27	4.82235	0.20737	0.07570	13.21053	0.01570	63.70577	9.60294	27
28	5.11169	0.19563	0.07459	13.40616	0.01459	68.52811	9.85681	28
29	5.41839	0.18456	0.07358	13.59072	0.01358	73.63980	10.10319	29
30	5.74349	0.17411	0.07265	13.76483	0.01265	79.05819	10.34221	30

(Contd...)

31	6.08810	0.16425	0.07179	13.92909	0.01179	84.80168	10.57402	31
32	6.45339	0.15496	0.07100	14.08404	0.01100	90.88978	10.79875	32
33	6.84059	0.14619	0.07027	14.23023	0.01027	97.34316	11.01655	33
34	7.25103	0.13791	0.06960	14.36814	0.00960	104.18375	11.22756	34
35	7.68609	0.13011	0.06897	14.49825	0.00897	111.43478	11.43192	35
40	10.28572	0.09722	0.06640	15.04630	0.00646	154.76197	12.35898	40
45	13.76461	0.07265	0.06470	15.45583	0.00470	212.74351	13.14129	45
50	18.42015	0.05429	0.06344	15.76186	0.00344	290.33590	13.79643	50
55	24.65032	0.04057	0.06254	15.99054	0.00254	394.17203	14.34112	55
60	32.98769	0.03031	0.06188	16.16143	0.00188	533.12818	14.79095	60
65	44.14497	0.02265	0.06139	16.28912	0.00139	719.08286	15.16012	65
70	59.07593	0.01693	0.06103	16.38454	0.00103	967.93217	15.46135	70
75	79.05692	0.01265	0.06077	16.45585	0.00077	1300.94868	15.70583	75
80	105.79599	0.00945	0.06057	16.50913	0.00057	1746.59989	15.90328	80
85	141.57890	0.00706	0.06043	16.54895	0.00043	2342.98174	16.06202	85
90	189.46451	0.00528	0.06032	16.57870	0.00032	3141.07519	16.18912	90
95	253.54625	0.00394	0.06024	16.60093	0.00024	4209.10425	16.29050	95
100	339.30208	0.00295	0.06018	16.61755	0.00018	5638.36806	16.37107	100

7% interest factors for discrete compounding periods

	Single Payment		*Uniform Series*					
	Compound Amount Factor	*Present Worth Factor*	*Capital Recovery Factor*	*Present Worth Factor*	*Sinking Fund Factor*	*Compound Amount Factor*	*Gradient Factor*	
N	*(F/P, 7, N)*	*(P/F, 7, N)*	*(A/P, 7, N)*	*(P/A, 7, N)*	*(A/F, 7, N)*	*(F/A, 7, N)*	*(A/G, 7, N)*	*N*
1	1.07000	0.93458	1.07000	0.93458	1.00000	1.00000	0.00000	1
2	1.14490	0.87344	0.55309	1.80802	0.48309	2.07000	0.48309	2
3	1.22504	0.81630	0.38105	2.62432	0.31105	3.21490	0.95493	3
4	1.31080	0.76290	0.29523	3.38721	0.22523	4.43994	1.41554	4
5	1.40255	0.71299	0.24389	4.10020	0.17389	5.75074	1.86495	5
6	1.50073	0.66634	0.20980	4.76654	0.13980	7.15329	2.30322	6
7	1.60578	0.62275	0.18555	5.38929	0.11555	8.65402	2.73039	7
8	1.71819	0.58201	0.16747	5.97130	0.09747	10.25980	3.14654	8
9	1.83846	0.54393	0.15349	6.51523	0.08349	11.97799	3.55174	9
10	1.96715	0.50835	0.14238	7.02358	0.07238	13.81645	3.94607	10
11	2.10485	0.47509	0.13336	7.49867	0.06336	15.78360	4.32963	11
12	2.25219	0.44401	0.12590	7.94269	0.05590	17.88845	4.70252	12
13	2.40985	0.41496	0.11965	8.35765	0.04965	20.14064	5.06484	13
14	2.57853	0.38782	0.11434	8.74547	0.04434	22.55049	5.41673	14
15	2.75903	0.36245	0.10979	9.10791	0.03979	25.12902	5.75829	15
16	2.95216	0.33873	0.10586	9.44665	0.03586	27.88805	6.08968	16
17	3.15882	0.31657	0.10243	9.76322	0.03243	30.84022	6.41102	17
18	3.37993	0.29586	0.09941	10.05909	0.02941	33.99903	6.72247	18
19	3.61653	0.27651	0.09675	10.33560	0.02675	37.37896	7.02418	19
20	3.86968	0.25842	0.09439	10.59401	0.02439	40.99549	7.31631	20
21	4.14056	0.24151	0.09229	10.83553	0.02229	44.86518	7.59901	21
22	4.43040	0.22571	0.09041	11.06124	0.02041	49.00574	7.87247	22
23	4.74053	0.21095	0.08871	11.27219	0.01871	53.43614	8.13685	23
24	5.07237	0.19715	0.08719	11.46933	0.01719	58.17667	8.39234	24
25	5.42743	0.18425	0.08581	11.65358	0.01581	63.24904	8.63910	25
26	5.80735	0.17220	0.08456	11.82578	0.01456	68.67647	8.87733	26
27	6.21387	0.16093	0.08343	11.98671	0.01343	74.48382	9.10722	27
28	6.64884	0.15040	0.08239	12.13711	0.01239	80.69769	9.32894	28
29	7.11426	0.14056	0.08145	12.27767	0.01145	87.34653	9.54270	29
30	7.61226	0.13137	0.08059	12.40904	0.01059	94.46079	9.74868	30

(Contd...)

31	8.14511	0.12277	0.07980	12.53181	0.00980	102.07304	9.94708	31
32	8.71527	0.11474	0.07907	12.64656	0.00907	110.21815	10.13810	32
33	9.32534	0.10723	0.07841	12.75379	0.00841	118.93343	10.32191	33
34	9.97811	0.10022	0.07780	12.85401	0.00780	128.25876	10.49873	34
35	10.67658	0.09366	0.07723	12.94767	0.00723	138.23688	10.66873	35
40	14.97446	0.06678	0.07501	13.33171	0.00501	199.63511	11.42335	40
45	21.00245	0.04761	0.07350	13.60552	0.00350	285.74931	12.03599	45
50	29.45703	0.03395	0.07246	13.80075	0.00246	406.52893	12.52868	50
55	41.31500	0.02420	0.07174	13.93994	0.00174	575.92859	12.92146	55
60	57.94643	0.01726	0.07123	14.03918	0.00123	813.52038	13.23209	60
65	81.27286	0.01230	0.07087	14.10994	0.00087	1146.75516	13.47598	65
70	113.98939	0.00877	0.07062	14.16039	0.00062	1614.13417	13.66619	70
75	159.87602	0.00625	0.07044	14.19636	0.00044	2269.65742	13.81365	75
80	224.23439	0.00446	0.07031	14.22201	0.00031	3189.06268	13.92735	80
85	314.50033	0.00318	0.07022	14.24029	0.00022	4478.57612	14.01458	85
90	441.10298	0.00227	0.07016	14.25333	0.00016	6287.18543	14.08122	90
95	618.66975	0.00162	0.07011	14.26262	0.00011	8823.85354	14.13191	95
100	867.71633	0.00115	0.07008	14.26925	0.00008	12381.66179	14.17034	100.

8% interest factors for discrete compounding periods

	Single Payment		*Uniform Series*					
	Compound Amount Factor	*Present Worth Factor*	*Capital Recovery Factor*	*Present Worth Factor*	*Sinking Fund Factor*	*Compound Amount Factor*	*Gradient Factor*	
N	*(F/P, 8, N)*	*(P/F, 8, N)*	*(A/P, 8, N)*	*(P/A, 8, N)*	*(A/F, 8, N)*	*(F/A, 8, N)*	*(A/G, 8, N)*	*N*
1	1.08000	0.92593	1.08000	0.92593	1.00000	1.00000	0.00000	1
2	1.16640	0.85734	0.56077	1.78326	0.48077	2.08000	0.48077	2
3	1.25971	0.79383	0.38803	2.57710	0.30803	3.24640	0.94874	3
4	1.36049	0.73503	0.30192	3.31213	0.22192	4.50611	1.40396	4
5	1.46933	0.68058	0.25046	3.99271	0.17046	5.86660	1.84647	5
6	1.58687	0.63017	0.21632	4.62288	0.13632	7.33593	2.27635	6
7	1.71382	0.58349	0.19207	5.20637	0.11207	8.92280	2.69366	7
8	1.85093	0.54027	0.17401	5.74664	0.09401	10.63663	3.09852	8
9	1.99900	0.50025	0.16008	6.24689	0.08008	12.48756	3.49103	9
10	2.15892	0.46319	0.14903	6.71008	0.06903	14.48656	3.87131	10
11	2.33164	0.42888	0.14008	7.13896	0.06008	16.64549	4.23950	11
12	2.51817	0.39711	0.13270	7.53608	0.05270	18.97713	4.59575	12
13	2.71962	0.36770	0.12652	7.90378	0.04652	21.49530	4.94021	13
14	2.93719	0.34046	0.12130	8.24424	0.04130	24.21492	5.27305	14
15	3.17217	0.31524	0.11683	8.55948	0.03683	27.15211	5.59446	15
16	3.42594	0.29189	0.11298	8.85137	0.03298	30.32428	5.90463	16
17	3.70002	0.27027	0.10963	9.12164	0.02963	33.75023	6.20375	17
18	3.99602	0.25025	0.10670	9.37189	0.02670	37.45024	6.49203	18
19	4.31570	0.23171	0.10413	9.60360	0.02413	41.44626	6.76969	19
20	4.66096	0.21455	0.10185	9.81815	0.02185	45.76196	7.03695	20
21	5.03383	0.19866	0.09983	10.01680	0.01983	50.42292	7.29403	21
22	5.43654	0.18394	0.09803	10.20074	0.01803	55.45676	7.54118	22
23	5.87146	0.17032	0.09642	10.37106	0.01642	60.89330	7.77863	23
24	6.34118	0.15770	0.09498	10.52876	0.01498	66.76476	8.00661	24
25	6.84848	0.14602	0.09368	10.67478	0.01368	73.10594	8.22538	25
26	7.39635	0.13520	0.09251	10.80998	0.01251	79.95442	8.43518	26
27	7.98806	0.12519	0.09145	10.93516	0.01145	87.35077	8.63627	27
28	8.62711	0.11591	0.09049	11.05108	0.01049	95.33883	8.82888	28
29	9.31727	0.10733	0.08962	11.15841	0.00962	103.96594	9.01328	29
30	10.06266	0.09938	0.08883	11.25778	0.00883	113.28321	9.18971	30

(Contd...)

31	10.86767	0.09202	0.08811	11.34980	0.00811	123.34587	9.35843	31
32	11.73708	0.08520	0.08745	11.43500	0.00745	134.21354	9.51967	32
33	12.67605	0.07889	0.08685	11.51389	0.00685	145.95062	9.67370	33
34	13.69013	0.07305	0.08630	11.58693	0.00630	158.62667	9.82075	34
35	14.78534	0.06763	0.08580	11.65457	0.00580	172.31680	9.96107	35
40	21.72452	0.04603	0.08386	11.92461	0.00386	259.05652	10.56992	40
45	31.92045	0.03133	0.08259	12.10840	0.00259	386.50562	11.04465	45
50	46.90161	0.02132	0.08174	12.23348	0.00174	573.77016	11.41071	50
55	68.91386	0.01451	0.08118	12.31861	0.00118	848.92320	11.69015	55
60	101.25706	0.00988	0.08080	12.37655	0.00080	1253.21330	11.90154	60
65	148.77985	0.00672	0.08054	12.41598	0.00054	1847.24808	12.06016	65
70	218.60641	0.00457	0.08037	12.44282	0.00037	2720.08007	12.17832	70
75	321.20453	0.00311	0.08025	12.46108	0.00025	4002.55662	12.26577	75
80	471.95483	0.00212	0.08017	12.47351	0.00017	5886.93543	12.33013	80
85	693.45649	0.00144	0.08012	12.48197	0.00012	8655.70611	12.37725	85
90	1018.91509	0.00098	0.08008	12.48773	0.00008	12723.93862	12.41158	90
95	1497.12055	0.00067	0.08005	12.49165	0.00005	18701.50686	12.43650	95
100	2199.76126	0.00045	0.08004	12.49432	0.00004	27484.51570	12.45452	100

9% interest factors for discrete compounding periods

	Single Payment		*Uniform Series*					
	Compound Amount Factor	*Present Worth Factor*	*Capital Recovery Factor*	*Present Worth Factor*	*Sinking Fund Factor*	*Compound Amount Factor*	*Gradient Factor*	
N	*(F/P, 9, N)*	*(P/F, 9, N)*	*(A/P, 9, N)*	*(P/A, 9, N)*	*(A/F, 9, N)*	*(F/A, 9, N)*	*(A/G, 9, N)*	*N*
1	1.09000	0.91743	1.09000	0.91743	1.00000	1.00000	0.00000	1
2	1.18810	0.84168	0.56847	1.75911	0.47847	2.09000	0.47847	2
3	1.29503	0.77218	0.39505	2.53129	0.30505	3.27810	0.94262	3
4	1.41158	0.70843	0.30867	3.23972	0.21867	4.57313	1.39250	4
5	1.53862	0.64993	0.25709	3.88965	0.16709	5.98471	1.82820	5
6	1.67710	0.59627	0.22292	4.48592	0.13292	7.52333	2.24979	6
7	1.82804	0.54703	0.19869	5.03295	0.10869	9.20043	2.65740	7
8	1.99256	0.50187	0.18067	5.53482	0.09067	11.02847	3.05117	8
9	2.17189	0.46043	0.16680	5.99525	0.07680	13.02104	3.43123	9
10	2.36736	0.42241	0.15582	6.41766	0.06582	15.19293	3.79777	10
11	2.58043	0.38753	0.14695	6.80519	0.05695	17.56029	4.15096	11
12	2.81266	0.35553	0.13965	7.16073	0.04965	20.14072	4.49102	12
13	3.06580	0.32618	0.13357	7.48690	0.04357	22.95338	4.81816	13
14	3.34173	0.29925	0.12843	7.78615	0.03843	26.01919	5.13262	14
15	3.64248	0.27454	0.12406	8.06069	0.03406	29.36092	5.43463	15
16	3.97031	0.25187	0.12030	8.31256	0.03030	33.00340	5.72446	16
17	4.32763	0.23107	0.11705	8.54363	0.02705	36.97370	6.00238	17
18	4.71712	0.21199	0.11421	8.75563	0.02421	41.30134	6.26865	18
19	5.14166	0.19449	0.11173	8.95011	0.02173	46.01846	6.52358	19
20	5.60441	0.17843	0.10955	9.12855	0.01955	51.16012	6.76745	20
21	6.10881	0.16370	0.10762	9.29224	0.01762	56.76453	7.00056	21
22	6.65860	0.15018	0.10590	9.44243	0.01590	62.87334	7.22322	22
23	7.25787	0.13778	0.10438	9.58021	0.01438	69.53194	7.43574	23
24	7.91108	0.12640	0.10302	9.70661	0.01302	76.78981	7.63843	24
25	8.62308	0.11597	0.10181	9.82258	0.01181	84.70090	7.83160	25
26	9.39916	0.10639	0.10072	9.92897	0.01072	93.32398	8.01556	26
27	10.24508	0.09761	0.09973	10.02658	0.00973	102.72313	8.19064	27
28	11.16714	0.08955	0.09885	10.11613	0.00885	112.96822	8.35714	28
29	12.17218	0.08215	0.09806	10.19828	0.00806	124.13536	8.51538	29
30	13.26768	0.07537	0.09734	10.27365	0.00734	136.30754	8.66566	30

(Contd...)

31	14.46177	0.06915	0.09669	10.34280	0.00669	149.57522	8.80829	31
32	15.76333	0.06344	0.09610	10.40624	0.00610	164.03699	8.94358	32
33	17.18203	0.05820	0.09556	10.46444	0.00556	179.80032	9.07181	33
34	18.72841	0.05339	0.09508	10.51784	0.00508	196.98234	9.19329	34
35	20.41397	0.04899	0.09464	10.56682	0.00464	215.71075	9.30829	35
40	31.40942	0.03184	0.09296	10.75736	0.00296	337.88245	9.79573	40
45	48.32729	0.02069	0.09190	10.88120	0.00190	525.85873	10.16029	45
50	74.35752	0.01345	0.09123	10.96168	0.00123	815.08356	10.42952	50
55	114.40826	0.00874	0.09079	11.01399	0.00079	1260.09180	10.62614	55
60	176.03129	0.00568	0.09051	11.04799	0.00051	1944.79213	10.76832	60
65	270.84596	0.00369	0.09033	11.07009	0.00033	2998.28847	10.87023	65
70	416.73009	0.00240	0.09022	11.08445	0.00022	4619.22318	10.94273	70
75	841.19089	0.00156	0.09014	11.09378	0.00014	7113.23215	10.99396	75
80	986.55167	0.00101	0.09009	11.09985	0.00009	10950.57409	11.02994	80
85	1517.93203	0.00066	0.09006	11.10379	0.00006	16854.80033	11.05508	85
90	2335.52658	0.00043	0.09004	11.10635	0.00004	25939.18425	11.07256	90
95	3593.49715	0.00028	0.09003	11.10802	0.00003	39916.63496	11.08467	95
100	5529.04079	0.00018	0.09002	11.10910	0.00002	61422.67546	11.09302	100

10% interest factors for discrete compounding periods

	Single Payment		Uniform Series					
	Compound Amount Factor	Present Worth Factor	Capital Recovery Factor	Present Worth Factor	Sinking Fund Factor	Compound Amount Factor	Gradient Factor	
N	(F/P, 10, N)	(P/F, 10, N)	(A/P, 10, N)	(P/A, 10, N)	(A/F, 10, N)	(F/A, 10, N)	(A/G, 10, N)	N
1	1.10000	0.90909	1.10000	0.90909	1.00000	1.00000	0.00000	1
2	1.21000	0.82645	0.57619	1.73554	0.47619	2.10000	0.47619	2
3	1.33100	0.75131	0.40211	2.48685	0.30211	3.31000	0.93656	3
4	1.46410	0.68301	0.31547	3.16987	0.21547	4.64100	1.38117	4
5	1.61051	0.62092	0.26380	3.79079	0.16380	6.10510	1.81013	5
6	1.77156	0.56447	0.22961	4.35526	0.12961	7.71561	2.22356	6
7	1.94872	0.51316	0.20541	4.86842	0.10541	9.48717	2.62162	7
8	2.14359	0.46651	0.18744	5.33493	0.08744	11.43589	3.00448	8
9	2.35795	0.42410	0.17364	5.75902	0.07364	13.57948	3.37235	9
10	2.59374	0.38554	0.16275	6.14457	0.06275	15.93742	3.72546	10
11	2.85312	0.35049	0.15396	6.49506	0.05396	18.53117	4.06405	11
12	3.13843	0.31863	0.14676	6.81369	0.04676	21.38428	4.38840	12
13	3.45227	0.28966	0.14078	7.10336	0.04078	24.52271	4.69879	13
14	3.79750	0.26333	0.13575	7.36669	0.03575	27.97498	4.99553	14
15	4.17725	0.23939	0.13147	7.60608	0.03147	31.77248	5.27893	15
16	4.59497	0.21763	0.12782	7.82371	0.02782	35.94973	5.54934	16
17	5.05447	0.19784	0.12466	8.02155	0.02466	40.54470	5.80710	17
18	5.55992	0.17986	0.12193	8.20141	0.02193	45.59917	6.05256	18
19	6.11591	0.16351	0.11955	8.36492	0.01955	51.15909	6.28610	19
20	6.72750	0.14864	0.11746	8.51356	0.01746	57.27500	6.50808	20
21	7.40025	0.13513	0.11562	8.64869	0.01562	64.00250	6.71888	21
22	8.14027	0.12285	0.11401	8.77154	0.01401	71.40275	6.91889	22
23	8.95430	0.11168	0.11257	8.88322	0.01257	79.54302	7.10848	23
24	9.84973	0.10153	0.11130	8.98474	0.01130	88.49733	7.28805	24
25	10.83471	0.09230	0.11017	9.07704	0.01017	98.34706	7.45798	25
26	11.91818	0.08391	0.10916	9.16095	0.00916	109.18177	7.61865	26
27	13.10999	0.07628	0.10826	9.23722	0.00826	121.09994	7.77044	27
28	14.42099	0.06934	0.10745	9.30657	0.00745	134.20994	7.91372	28
29	15.86309	0.06304	0.10673	9.36961	0.00673	148.63093	8.04886	29
30	17.44940	0.05731	0.10608	9.42691	0.00608	164.49402	8.17623	30

(Contd...)

31	19.19434	0.05210	0.10550	9.47901	0.00550	181.94342	8.29617	31
32	21.11378	0.04736	0.10497	9.52636	0.00497	201.13777	8.40905	32
33	23.22515	0.04306	0.10450	9.56943	0.00450	222.25154	8.51520	33
34	25.54767	0.03914	0.10407	9.60857	0.00407	245.47670	8.61494	34
35	28.10244	0.03558	0.10389	9.64416	0.00369	271.02437	8.70860	35
40	45.25926	0.02209	0.10226	9.77905	0.00226	442.59256	9.09623	40
45	72.89048	0.01372	0.10139	9.86281	0.00139	718.90484	9.37405	45
50	117.39085	0.00852	0.10086	9.91481	0.00086	1163.90853	9.57041	50
55	189.05914	0.00529	0.10053	9.94711	0.00053	1880.59142	9.70754	55
60	304.48164	0.00328	0.10033	9.96716	0.00033	3034.81640	9.80229	60
65	490.37073	0.00204	0.10020	9.97961	0.00020	4893.70725	9.86718	65
70	789.74696	0.00127	0.10013	9.98734	0.00013	7887.46957	9.91125	70
75	1271.89537	0.00079	0.10008	9.99214	0.00008	12708.95371	9.94099	75
80	2048.40021	0.00049	0.10005	9.99512	0.00005	20474.00215	9.96093	80
85	3298.96903	0.00030	0.10003	9.99897	0.00003	32979.69030	9.97423	85
90	5313.02261	0.00019	0.10002	9.99812	0.00002	53120.22612	9.98306	90
95	8556.67605	0.00012	0.10001	9.99883	0.00001	85556.76047	9.98890	95
100	13780.61234	0.00007	0.10001	9.99927	0.00001	137796.12340	9.99274	100

11% interest factors for discrete compounding periods

	Single Payment		Uniform Series					
	Compound Amount Factor	*Present Worth Factor*	*Capital Recovery Factor*	*Present Worth Factor*	*Sinking Fund Factor*	*Compound Amount Factor*	*Gradient Factor*	
N	*(F/P, 11, N)*	*(P/F, 11, N)*	*(A/P, 11, N)*	*(P/A, 11, N)*	*(A/F, 11, N)*	*(F/A, 11, N)*	*(A/G, 11, N)*	*N*
1	1.11000	0.90090	1.11000	0.90090	1.00000	1.00000	0.00000	1
2	1.23210	0.81162	0.58393	1.71252	0.47393	2.11000	0.47393	2
3	1.36763	0.73119	0.40921	2.44371	0.29921	3.34210	0.93055	3
4	1.51807	0.65873	0.32233	3.10245	0.21233	4.70973	1.36995	4
5	1.68506	0.59345	0.27057	3.69590	0.16057	6.22780	1.79226	5
6	1.87041	0.53464	0.23638	4.23054	0.12638	7.91286	2.19764	6
7	2.07616	0.48166	0.21222	4.71220	0.10222	9.78327	2.58630	7
8	2.30454	0.43393	0.19432	5.14612	0.08432	11.85943	2.95847	8
9	2.55804	0.39092	0.18060	5.53705	0.07060	14.16397	3.31441	9
10	2.83942	0.36218	0.16980	5.88923	0.05980	16.72201	3.65442	10
11	3.15176	0.31728	0.16112	6.20662	0.06112	19.56143	3.97881	11
12	3.49845	0.28584	0.15403	6.49236	0.04403	22.71319	4.28793	12
13	3.88328	0.25751	0.14815	6.74987	0.03815	26.21164	4.58216	13
14	4.31044	0.23199	0.14323	6.98187	0.03323	30.09492	4.86187	14
15	4.78459	0.20900	0.13907	7.19087	0.02907	34.40536	5.12747	15
16	5.31089	0.18829	0.13552	7.37916	0.02552	39.18995	5.37938	16
17	5.89509	0.16963	0.13247	7.54879	0.02247	44.50084	5.61804	17
18	6.54355	0.15282	0.12984	7.70162	0.01984	50.39594	5.84389	18
19	7.26334	0.13768	0.12756	7.83929	0.01756	56.93949	6.05739	19
20	8.06231	0.12403	0.12558	7.96333	0.01558	64.20283	6.25898	20
21	8.94917	0.11174	0.12384	8.07507	0.01384	72.26514	6.44912	21
22	9.93357	0.10067	0.12231	8.17574	0.01231	81.21431	6.62829	22
23	11.02627	0.09069	0.12097	8.26643	0.01097	91.14788	6.79693	23
24	12.23916	0.08170	0.11979	8.34814	0.00979	102.17415	6.95552	24
25	13.58546	0.07361	0.11874	8.42174	0.00874	114.41331	7.10449	25
26	15.07986	0.06631	0.11781	8.48806	0.00781	127.99877	7.24430	26
27	16.73865	0.05974	0.11699	8.54780	0.00699	143.07864	7.37539	27
28	18.57990	0.05382	0.11626	8.60162	0.00626	159.81729	7.49818	28
29	20.62369	0.04849	0.11561	8.65011	0.00561	178.39719	7.61310	29
30	22.89230	0.04368	0.11502	8.69379	0.00502	199.02088	7.72056	30

(Contd...)

31	25.41045	0.03935	0.11451	8.73315	0.00451	221.91317	7.82096	31
32	28.20560	0.03545	0.11404	8.76860	0.00404	247.32362	7.91468	32
33	31.30821	0.03194	0.11363	8.80054	0.00363	275.52922	8.00210	33
34	34.75212	0.02878	0.11326	8.82932	0.00326	306.83744	8.08356	34
35	38.57485	0.02592	0.11293	8.85524	0.00293	341.58955	8.15944	35
40	65.00087	0.01538	0.11172	8.95105	0.00172	581.82807	8.46592	40
45	109.53024	0.00913	0.11101	9.00791	0.00101	986.63856	8.67628	45
50	184.56483	0.00542	0.11060	9.04165	0.00060	1668.77115	8.81853	50

12% interest factors for discrete compounding periods

	Single Payment		Uniform Series					
	Compound Amount Factor	*Present Worth Factor*	*Capital Recovery Factor*	*Present Worth Factor*	*Sinking Fund Factor*	*Compound Amount Factor*	*Gradient Factor*	
N	*(F/P, 12, N)*	*(P/F, 12, N)*	*(A/P, 12, N)*	*(P/A, 12, N)*	*(A/F, 12, N)*	*(F/A, 12, N)*	*(A/G, 12, N)*	*N*
1	1.12000	0.89286	1.12000	0.89286	1.00000	1.00000	0.00000	1
2	1.25440	0.79719	0.59170	1.69005	0.47170	2.12000	0.47170	2
3	1.40493	0.71178	0.41635	2.40183	0.29635	3.37440	0.92461	3
4	1.57352	0.63552	0.32923	3.03735	0.20923	4.77933	1.35885	4
5	1.76234	0.56743	0.27741	3.60478	0.15741	6.35285	1.77459	5
6	1.97382	0.50663	0.24323	4.11141	0.12323	8.11519	2.17205	6
7	2.21068	0.45235	0.21912	4.56376	0.09912	10.08901	2.55147	7
8	2.47596	0.40388	0.20130	4.96764	0.08130	12.29969	2.91314	8
9	2.77308	0.36061	0.18768	5.32825	0.06768	14.77566	3.25742	9
10	3.10585	0.32197	0.17698	5.65022	0.05698	17.54874	3.58465	10
11	3.47855	0.28748	0.16842	5.93770	0.04842	20.65458	3.89525	11
12	3.89598	0.25668	0.16144	6.19437	0.04144	24.13313	4.18965	12
13	4.36349	0.22917	0.15568	6.42355	0.03568	28.02911	4.46830	13
14	4.88711	0.20462	0.15087	6.62817	0.03087	32.39260	4.73169	14
15	5.47357	0.18270	0.14682	6.81086	0.02682	37.27971	4.98030	15
16	6.13039	0.16312	0.14339	6.97399	0.02339	42.75328	5.21466	16
17	6.86604	0.14564	0.14046	7.11963	0.02046	48.88367	5.43530	17
18	7.68997	0.13004	0.13794	7.24967	0.01794	55.74971	5.64274	18
19	8.61276	0.11611	0.13576	7.36578	0.01576	63.43968	5.83752	19
20	9.64629	0.10367	0.13388	7.46944	0.01388	72.05244	6.02020	20
21	10.80385	0.09256	0.13224	7.56200	0.01224	81.69874	6.19132	21
22	12.10031	0.08264	0.13081	7.64465	0.01081	92.50258	6.35141	22
23	13.55235	0.07379	0.12956	7.71843	0.00956	104.60289	6.50101	23
24	15.17863	0.06588	0.12846	7.78432	0.00846	118.15524	6.64064	24
25	17.00006	0.05882	0.12750	7.84314	0.00750	133.33387	6.77084	25
26	19.04007	0.05252	0.12665	7.89566	0.00665	150.33393	6.89210	26
27	21.32488	0.04689	0.12590	7.94255	0.00590	169.37401	7.00491	27
28	23.88387	0.04187	0.12524	7.98442	0.00524	190.69889	7.10976	28
29	26.74993	0.03738	0.12466	8.02181	0.00466	214.58275	7.20712	29
30	29.95992	0.03338	0.12414	8.05518	0.00414	241.33268	7.29742	30

(Contd...)

31	33.55511	0.02980	0.12369	8.08499	0.00369	271.29261	7.38110	31
32	37.58173	0.02661	0.12328	8.11159	0.00328	304.84772	7.45858	32
33	42.09153	0.02376	0.12292	8.13535	0.00292	342.42945	7.53025	33
34	47.14252	0.02121	0.12260	8.15656	0.00260	384.52098	7.59649	34
35	52.79962	0.01894	0.12232	8.17550	0.00232	431.66350	7.65765	35
40	93.05097	0.01075	0.12130	8.24378	0.00130	767.09142	7.89879	40
45	163.98760	0.00610	0.12074	8.28252	0.00074	1358.23003	8.05724	45
50	289.00219	0.00346	0.12042	8.30450	0.00042	2400.01825	8.15972	50

13% interest factors for discrete compounding periods

	Single Payment		Uniform Series					
	Compound Amount Factor	*Present Worth Factor*	*Capital Recovery Factor*	*Present Worth Factor*	*Sinking Fund Factor*	*Compound Amount Factor*	*Gradient Factor*	
N	*(F/P, 13,N)*	*(P/F, 13,N)*	*(A/P, 13,N)*	*(P/A, 13,N)*	*(A/F, 13,N)*	*(F/A, 13,N)*	*(A/G,13,N)*	*N*
1	1.13000	0.88496	1.13000	0.88496	1.00000	1.00000	0.00000	1
2	1.27690	0.78315	0.59948	1.66810	0.46948	2.13000	0.46948	2
3	1.44290	0.69305	0.42352	2.36115	0.29352	3.40690	0.91872	3
4	1.63047	0.61332	0.33619	2.97447	0.20619	4.84980	1.34787	4
5	1.84244	0.54276	0.28431	3.51723	0.15431	6.48027	1.75713	5
6	2.08195	0.48032	0.25015	3.99755	0.12015	8.32271	2.14677	6
7	2.35261	0.42506	0.22611	4.42261	0.09611	10.40466	2.51711	7
8	2.65844	0.37616	0.20839	4.79877	0.07839	12.75726	2.86851	8
9	3.00404	0.33288	0.19487	5.13166	0.06487	15.41571	3.20138	9
10	3.39457	0.29459	0.18429	5.42624	0.05429	18.41975	3.51619	10
11	3.83586	0.26070	0.17584	5.68694	0.04584	21.81432	3.81342	11
12	4.33452	0.23071	0.16899	5.91765	0.03899	25.65018	4.09359	12
13	4.89801	0.20416	0.16335	6.12181	0.03335	29.98470	4.35727	13
14	5.53475	0.18068	0.15867	6.30249	0.02867	34.88271	4.60504	14
15	6.25427	0.15989	0.15474	6.46238	0.02474	40.41746	4.83749	15
16	7.06733	0.14150	0.15143	6.60388	0.02143	46.67173	5.05523	16
17	7.98608	0.12522	0.14861	6.72909	0.01861	53.73906	5.25890	17
18	9.02427	0.11081	0.14620	6.83991	0.01620	61.72514	5.44911	18
19	10.19742	0.09806	0.14413	6.93797	0.01413	70.74941	5.62651	19
20	11.52309	0.08678	0.14235	7.02475	0.01235	80.94683	5.79172	20
21	13.02109	0.07680	0.14081	7.10155	0.01081	92.46992	5.94538	21
22	14.71383	0.06796	0.13948	7.16951	0.00948	105.49101	6.08809	22
23	16.62663	0.06014	0.13832	7.22966	0.00832	120.20484	6.22046	23
24	18.78809	0.05323	0.13731	7.28288	0.00731	136.83147	6.34309	24
25	21.23054	0.04710	0.13643	7.32998	0.00643	155.61956	6.45655	25
26	23.99051	0.04168	0.13565	7.37167	0.00565	176.85010	6.56141	26
27	27.10928	0.03689	0.13498	7.40856	0.00498	200.84061	6.65819	27
28	30.63349	0.03264	0.13439	7.44120	0.00439	227.94989	6.74743	28
29	34.61584	0.02889	0.13387	7.47009	0.00387	258.58338	6.82962	29
30	39.11590	0.02557	0.13341	7.49565	0.00341	293.19922	6.90523	30

(Contd...)

31	44.20096	0.02262	0.13301	7.51828	0.00301	332.31511	6.97473	31
32	49.94709	0.02002	0.13266	7.53830	0.00266	376.51608	7.03854	32
33	56.44021	0.01772	0.13234	7.55602	0.00234	426.46317	7.09707	33
34	63.77744	0.01568	0.13207	7.57170	0.00207	482.90338	7.15071	34
35	72.06851	0.01388	0.13183	7.58557	0.00183	546.68082	7.19983	35
40	132.78155	0.00753	0.13099	7.63438	0.00099	1013.70424	7.38878	40
45	244.64140	0.00409	0.13053	7.66086	0.00053	1874.16463	7.50761	45
50	450.73593	0.00222	0.13029	7.67524	0.00029	3459.50712	7.58113	50

14% interest factors for discrete compounding periods

	Single Payment		*Uniform Series*					
	Compound Amount Factor	*Present Worth Factor*	*Capital Recovery Factor*	*Present Worth Factor*	*Sinking Fund Factor*	*Compound Amount Factor*	*Gradient Factor*	
N	*(F/P, 14,N)*	*(P/F, 14,N)*	*(A/P, 14,N)*	*(P/A, 14,N)*	*(A/F, 14,N)*	*(F/A,14,N)*	*(A/G,14,N)*	*N*
1	1.14000	0.87719	1.14000	0.87719	1.00000	1.00000	0.00000	1
2	1.29960	0.76947	0.60729	1.64666	0.46729	2.14000	0.46729	2
3	1.48154	0.67497	0.43073	2.32163	0.29073	3.43960	0.91290	3
4	1.68896	0.59208	0.34320	2.91371	0.20320	4.92114	1.33701	4
5	1.92541	0.51937	0.29128	3.43308	0.15128	6.61010	1.73987	5
6	2.19497	0.45559	0.25716	3.88867	0.11716	8.53552	2.12182	6
7	2.50227	0.39964	0.23319	4.28830	0.09319	10.73049	2.48324	7
8	2.85259	0.35056	0.21557	4.63886	0.07557	13.23276	2.82457	8
9	3.25195	0.30751	0.20217	4.94637	0.06217	16.08535	3.14632	9
10	3.70722	0.26974	0.19171	5.21612	0.05171	19.33730	3.44903	10
11	4.22623	0.23662	0.18339	5.45273	0.04339	23.04452	3.73331	11
12	4.81790	0.20756	0.17667	5.66029	0.03667	27.27075	3.99977	12
13	5.49241	0.18207	0.17116	5.84236	0.03116	32.08865	4.24909	13
14	6.26135	0.15971	0.16661	6.00207	0.02661	37.58107	4.48194	14
15	7.13794	0.14010	0.16281	6.14217	0.02281	43.84241	4.69904	15
16	8.13725	0.12289	0.15962	6.26506	0.01962	50.98035	4.90110	16
17	9.27646	0.10780	0.15692	6.37286	0.01692	59.11760	5.08884	17
18	10.57517	0.09456	0.15462	6.46742	0.01462	68.39407	5.26299	18
19	12.05569	0.08295	0.15266	6.55037	0.01266	78.96923	5.42429	19
20	13.74349	0.07276	0.15099	6.62313	0.01099	91.02493	5.57343	20
21	15.66758	0.06383	0.14954	6.68696	0.00954	104.76842	5.71113	21
22	17.86104	0.05599	0.14830	6.74294	0.00830	120.43600	5.83807	22
23	20.36158	0.04911	0.14723	6.79206	0.00723	138.29704	5.95494	23
24	23.21221	0.04308	0.14630	6.83514	0.00630	158.65862	6.06237	24
25	26.46192	0.03779	0.14550	6.87293	0.00550	181.87083	6.16100	25
26	30.16658	0.03315	0.14480	6.90608	0.00480	208.33274	6.25143	26
27	34.38991	0.02908	0.14419	6.93515	0.00419	238.49933	6.33423	27
28	39.20449	0.02551	0.14366	6.96066	0.00366	272.88923	6.40996	28
29	44.69312	0.02237	0.14320	6.98304	0.00320	312.09373	6.47914	29
30	50.95016	0.01963	0.14280	7.00266	0.00280	356.78685	6.54226	30

(Contd...)

31	58.08318	0.01722	0.14245	7.01988	0.00245	407.73701	6.59979	31
32	66.21483	0.01510	0.14215	7.03498	0.00215	465.82019	6.65217	32
33	75.48490	0.01325	0.14188	7.04823	0.00188	532.03501	6.69981	33
34	86.05279	0.01162	0.14165	7.05985	0.00165	607.51991	6.74311	34
35	98.10018	0.01019	0.14144	7.07005	0.00144	693.57270	6.78240	35
40	188.88351	0.00529	0.14075	7.10504	0.00075	1342.02510	6.92996	40
45	363.67907	0.00275	0.14039	7.12322	0.00039	2590.56480	7.01878	45
50	700.23299	0.00143	0.14020	7.13266	0.00020	4994.52135	7.07135	50

15% interest factors for discrete compounding periods

	Single Payment		Uniform Series					
	Compound Amount Factor	*Present Worth Factor*	*Capital Recovery Factor*	*Present Worth Factor*	*Sinking Fund Factor*	*Compound Amount Factor*	*Gradient Factor*	
N	*(F/P, 15,N)*	*(P/F, 15,N)*	*(A/P, 15,N)*	*(P/A, 15,N)*	*(A/F, 15,N)*	*(F/A, 15,N)*	*(A/G, 15,N)*	*N*
1	1.15000	0.86957	1.15000	0.86957	1.00000	1.00000	0.00000	1
2	1.32250	0.75614	0.61512	1.62571	0.46512	2.15000	0.46512	2
3	1.52088	0.65752	0.43798	2.28323	0.28798	3.47250	0.90713	3
4	1.74901	0.57175	0.35027	2.85498	0.20027	4.99338	1.32626	4
5	2.01136	0.49718	0.29832	3.35216	0.14832	6.74238	1.72281	5
6	2.31306	0.43233	0.26424	3.78448	0.11424	8.75374	2.09719	6
7	2.66002	0.37594	0.24036	4.16042	0.09036	11.06680	2.44985	7
8	3.05902	0.32690	0.22285	4.48732	0.07285	13.72682	2.78133	8
9	3.51788	0.28426	0.20957	4.77158	0.05957	16.78584	3.09223	9
10	4.04556	0.24718	0.19925	5.01877	0.04925	20.30372	3.38320	10
11	4.65239	0.21494	0.19107	5.23371	0.04107	24.34928	3.65494	11
12	5.35025	0.18691	0.18448	5.42062	0.03448	29.00167	3.90820	12
13	6.15279	0.16253	0.17911	5.58315	0.02911	34.35192	4.14376	13
14	7.07571	0.14133	0.17469	5.72448	0.02469	40.50471	4.36241	14
15	8.13706	0.12289	0.17102	5.84737	0.02102	47.58041	4.56496	15
16	9.35762	0.10686	0.16795	5.95423	0.01795	55.71747	4.75225	16
17	10.76126	0.09293	0.16537	6.04716	0.01537	65.07509	4.92509	17
18	12.37545	0.08081	0.16319	6.12797	0.01319	75.83636	5.08431	18
19	14.23177	0.07027	0.16134	6.19823	0.01134	88.21181	5.23073	19
20	16.36654	0.06110	0.15976	6.25933	0.00976	102.44358	5.36514	20
21	18.82152	0.05313	0.15842	6.31246	0.00842	118.81012	5.48832	21
22	21.64475	0.04620	0.15727	6.35866	0.00727	137.63164	5.60102	22
23	24.89146	0.04017	0.15628	6.39884	0.00628	159.27638	5.70398	23
24	28.62518	0.03493	0.15543	6.43377	0.00543	184.16784	5.79789	24
25	32.91895	0.03038	0.15470	6.46415	0.00470	212.79302	5.88343	25
26	37.85680	0.02642	0.15407	6.49056	0.00407	245.71197	5.96123	26
27	43.53531	0.02297	0.15353	6.51353	0.00353	283.56877	6.03190	27
28	50.06561	0.01997	0.15306	6.53351	0.00306	327.10408	6.09600	28
29	57.57545	0.01737	0.15265	6.55088	0.00265	377.16969	6.15408	29
30	66.21177	0.01510	0.15230	6.56598	0.00230	434.74515	6.20663	30

(Contd...)

31	76.14354	0.01313	0.15200	6.57911	0.00200	500.95692	6.25412	31
32	87.56507	0.01142	0.15173	6.59053	0.00173	577.10046	6.29700	32
33	100.69983	0.00993	0.15150	6.60046	0.00150	664.66552	6.33567	33
34	115.80480	0.00864	0.15131	6.60910	0.00131	765.36535	6.37061	34
35	133.17552	0.00751	0.15113	6.61661	0.00113	881.17016	6.40187	35
40	267.86355	0.00373	0.15056	6.64178	0.00056	1779.09031	6.51678	40
45	538.76927	0.00186	0.15028	6.65429	0.00028	3585.12846	6.58299	45
50	1083.65744	0.00092	0.15014	6.66051	0.00014	7217.71628	6.62048	50

20% interest factors for discrete compounding periods

	Single Payment		Uniform Series					
	Compound Amount Factor	*Present Worth Factor*	*Capital Recovery Factor*	*Present Worth Factor*	*Sinking Fund Factor*	*Compound Amount Factor*	*Gradient Factor*	
N	*(F/P, 20,N)*	*(P/F, 20,N)*	*(A/P, 20,N)*	*(P/A, 20,N)*	*(A/F, 20,N)*	*(F/A,20,N)*	*(A/G, 20, N)*	*N*
1	1.20000	0.83333	1.20000	0.83333	1.00000	1.00000	0.00000	1
2	1.44000	0.69444	0.65455	1.52778	0.45455	2.20000	0.45455	2
3	1.72800	0.57870	0.47473	2.10648	0.27473	3.64000	0.87912	3
4	2.07360	0.48225	0.38629	2.58873	0.18629	5.36800	1.27422	4
5	2.48832	0.40188	0.33438	2.99061	0.13438	7.44160	1.64051	5
6	2.98598	0.33490	0.30071	3.32551	0.10071	9.92992	1.97883	6
7	3.58318	0.27908	0.27742	3.60459	0.07742	12.91590	2.29016	7
8	4.29982	0.23257	0.26061	3.83716	0.06061	16.49908	2.57562	8
9	5.15978	0.19381	0.24808	4.03097	0.04808	20.79890	2.83642	9
10	6.19174	0.16151	0.23852	4.19247	0.03852	25.95868	3.07386	10
11	7.43008	0.13459	0.23110	4.32706	0.03110	32.15042	3.28929	11
12	8.91610	0.11216	0.22526	4.43922	0.02526	39.58050	3.48410	12
13	10.69932	0.09346	0.22062	4.53268	0.02062	48.49660	3.65970	13
14	12.83918	0.07789	0.21689	4.61057	0.01689	59.19592	3.81749	14
15	15.40702	0.06491	0.21388	4.67547	0.01388	72.03511	3.95884	15
16	18.48843	0.05409	0.21144	4.72956	0.01144	87.44213	4.08511	16
17	22.18611	0.04507	0.20944	4.77463	0.00944	105.93056	4.19759	17
18	26.62333	0.03756	0.20781	4.81219	0.00781	128.11667	4.29752	18
19	31.94800	0.03130	0.20646	4.84350	0.00646	154.74000	4.38607	19
20	38.33760	0.02608	0.20536	4.86958	0.00536	186.68800	4.46435	20
21	46.00512	0.02174	0.20444	4.89132	0.00444	225.02560	4.53339	21
22	55.20614	0.01811	0.20369	4.90943	0.00369	271.03072	4.59414	22
23	66.24737	0.01509	0.20307	4.92453	0.00307	326.23686	4.64750	23
24	79.49685	0.01258	0.20255	4.93710	0.00255	392.48424	4.69426	24
25	95.39622	0.01048	0.20212	4.94759	0.00212	471.98108	4.73516	25
26	114.47546	0.00874	0.20176	4.95632	0.00176	567.37730	4.77088	26
27	137.37055	0.00728	0.20147	4.96360	0.00147	681.85276	4.80201	27
28	164.84466	0.00607	0.20122	4.96967	0.00122	819.22331	4.82911	28
29	197.81359	0.00506	0.20102	4.97472	0.00102	984.06797	4.85265	29
30	237.37631	0.00421	0.20085	4.97894	0.00085	1181.88157	4.87308	30

(Contd...)

31	284.85158	0.00351	0.20070	4.98245	0.00070	1419.25788	4.89079	31
32	341.82189	0.00293	0.20059	4.98537	0.00059	1704.10946	4.90611	32
33	410.18627	0.00244	0.20049	4.98781	0.00049	2045.93135	4.91935	33
34	492.22352	0.00203	0.20041	4.98984	0.00041	2456.11762	4.93079	34
35	590.66823	0.00169	0.20034	4.99154	0.00034	2948.34115	4.94064	35
40	1469.77157	0.00068	0.20014	4.99660	0.00014	7343.85784	4.97277	40
45	3657.26199	0.00027	0.20005	4.99863	0.00005	18281.30994	4.98769	45
50	9100.43815	0.00011	0.20002	4.99945	0.00002	45497.19075	4.99451	50

25% interest factors for discrete compounding periods

	Single Payment		*Uniform Series*					
	Compound Amount Factor	*Present Worth Factor*	*Capital Recovery Factor*	*Present Worth Factor*	*Sinking Fund Factor*	*Compound Amount Factor*	*Gradient Factor*	
N	*(F/P, 25, N)*	*(P/F, 25, N)*	*(A/P, 25, N)*	*(P/A, 25, N)*	*(A/F, 25, N)*	*(F/A, 25, N)*	*(A/G, 25, N)*	*N*
1	1.25000	0.80000	1.25000	0.80000	1.00000	1.00000	0.00000	1
2	1.56250	0.64000	0.69444	1.44000	0.44444	2.25000	0.44444	2
3	1.95313	0.51200	0.51230	1.95200	0.26230	3.81250	0.85246	3
4	2.44141	0.40960	0.42344	2.86160	0.17344	5.76563	1.22493	4
5	3.05176	0.32768	0.37185	2.68928	0.12185	8.20703	1.56307	5
6	3.81470	0.26214	0.33882	2.95142	0.08882	11.25879	1.86833	6
7	4.76837	0.20972	0.31634	3.16114	0.06634	15.07349	2.14243	7
8	5.96046	0.16777	0.30040	3.32891	0.05040	19.84186	2.38725	8
9	7.45058	0.13422	0.28876	3.46313	0.03876	25.80232	2.60478	9
10	9.31323	0.10737	0.28007	3.57050	0.03007	33.25290	2.79710	10
11	11.64153	0.08590	0.27349	3.65640	0.02349	42.56613	2.96631	11
12	44.55192	0.06872	0.26845	3.72512	0.01845	54.20766	3.11452	12
13	18.18989	0.05498	0.26454	3.78010	0.01454	68.75958	3.24374	13
14	22.73737	0.04398	0.26150	3.82408	0.01150	86.94947	3.35595	14
15	28.42171	0.03518	0.25912	3.85926	0.00912	109.68684	3.45299	15
16	35.52714	0.02815	0.25724	3.88741	0.00724	138.10855	3.53660	16
17	44.40892	0.02252	0.25576	3.90993	0.00576	173.63568	3.60838	17
18	55.51115	0.01801	0.25459	3.92794	0.00459	218.04460	3.66979	18
19	69.38894	0.01441	0.25366	3.94235	0.00366	273.55576	3.72218	19
20	86.73617	0.01153	0.25292	3.95388	0.00292	342.94470	3.76673	20
21	108.42022	0.00922	0.25233	3.96311	0.00233	429.68087	3.80451	21
22	135.52527	0.00738	0.25186	3.97049	0.00186	538.10109	3.83646	22
23	169.40659	0.00590	0.25148	3.97639	0.00148	873.62636	3.86343	23
24	211.75824	0.00472	0.25119	3.98111	0.00119	843.03295	3.88613	24
25	264.69780	0.00378	0.25095	3.98489	0.00095	1054.79118	3.90519	25
26	330.87225	0.00302	0.25076	3.98791	0.00076	1319.48898	3.92118	26
27	413.59031	0.00242	0.25061	3.99033	0.00061	1650.36123	3.93456	27
28	516.98788	0.00193	0.25048	3.99226	0.00048	2063.95153	3.94574	28
29	646.23485	0.00155	0.25039	3.99381	0.00039	2580.93941	3.95506	29
30	807.79357	0.00124	0.25031	3.99505	0.00031	3227.17427	3.96282	30

(Contd...)

31	1009.74196	0.00099	0.25025	3.99604	0.00025	4034.96783	3.96927	31
32	1262.17745	0.00079	0.25020	3.99683	0.00020	5044.70979	3.97463	32
33	1577.72181	0.00063	0.25016	3.99746	0.00016	6306.88724	3.97907	33
34	1972.15226	0.00051	0.25013	3.99797	0.00013	7884.60905	3.98275	34
35	2465.19033	0.00041	0.25010	3.99838	0.00010	9856.76132	3.98580	35

30% interest factors for discrete compounding periods

	Single Payment		Uniform Series					
	Compound Amount Factor	Present Worth Factor	Capital Recovery Factor	Present Worth Factor	Sinking Fund Factor	Compound Amount Factor	Gradient Factor	
N	(F/P, 30, N)	(P/F, 30, N)	(A/P, 30, N)	(P/A, 30, N)	(A/F, 30, N)	(F/A, 30, N)	(A/G, 30, N)	N
1	1.30000	0.76923	1.30000	0.76923	1.00000	1.00000	0.00000	1
2	1.69000	0.59172	0.73478	1.36095	0.43478	2.30000	0.43478	2
3	2.19700	0.45517	0.55063	1.81611	0.25063	3.99000	0.82707	3
4	2.85610	0.35013	0.46163	2.16624	0.16163	6.18700	1.17828	4
5	3.71293	0.26933	0.41058	2.43557	0.11058	9.04310	1.49031	5
6	4.82681	0.20718	0.37839	2.64275	0.07839	12.75603	1.76545	6
7	6.27485	0.15937	0.35687	2.80211	0.05687	17.58284	2.00628	7
8	8.15731	0.12259	0.34192	2.92470	0.04192	23.85769	2.21559	8
9	10.60450	0.09430	0.33124	3.01900	0.03124	32.01500	2.39627	9
10	13.78585	0.07254	0.32346	3.09154	0.02346	42.61950	2.55122	10
11	17.92160	0.05580	0.31773	3.14734	0.01773	56.40535	2.68328	11
12	23.29809	0.04292	0.31345	3.19026	0.01345	74.32695	2.79517	12
13	30.28751	0.03302	0.31024	3.22328	0.01024	97.62504	2.88946	13
14	39.37376	0.02540	0.30782	3.24867	0.00782	127.91255	2.96850	14
15	51.18589	0.01954	0.30598	3.26821	0.00598	167.28631	3.03444	15
16	66.54166	0.01503	0.30458	3.28324	0.00458	218.47220	3.08921	16
17	86.50416	0.01156	0.30351	3.29480	0.00351	285.01386	3.13451	17
18	112.45641	0.00889	0.30269	3.30369	0.00269	371.51802	3.17183	18
19	146.19203	0.00684	0.30207	3.31053	0.00207	483.97343	3.20247	19
20	190.04984	0.00526	0.30159	3.31579	0.00159	630.16546	3.22754	20
21	247.08453	0.00405	0.30122	3.31984	0.00122	820.21510	3.24799	21
22	321.18389	0.00311	0.30094	3.32296	0.00094	1067.27963	3.26462	22
23	417.53905	0.00239	0.30072	3.32535	0.00072	1388.46351	3.27812	23
24	542.80077	0.00184	0.30055	3.32719	0.00055	1806.00257	3.28904	24
25	705.64100	0.00142	0.30043	3.32861	0.00043	2348.80334	3.29785	25
26	917.33330	0.00109	0.30033	3.32970	0.00033	3054.44434	3.30496	26
27	1192.53329	0.00084	0.30025	3.33054	0.00025	3971.77764	3.31067	27
28	1550.29328	0.00065	0.30019	3.33118	0.00019	5164.31093	3.31526	28
29	2015.38126	0.00050	0.30015	3.33168	0.00015	6714.60421	3.31894	29
30	2619.99664	0.00038	0.30011	3.33206	0.00011	8729.98548	3.32188	30

(Contd...)

31	3405.99434	0.00029	0.30009	3.33235	0.00009	11349.98112	3.32423	31
32	4427.79264	0.00023	0.30007	3.33258	0.00007	14755.97546	3.32610	32
33	5756.13043	0.00017	0.30005	3.33275	0.00005	19183.76810	3.32760	33
34	7482.96956	0.00013	0.30004	3.33289	0.00004	24939.89853	3.32879	34
35	9727.88043	0.00010	0.30003	3.33299	0.00003	32422.86808	3.32974	35

40% interest factors for discrete compounding periods

	Single Payment		Uniform Series					
	Compound Amount Factor	*Present Worth Factor*	*Capital Recovery Factor*	*Present Worth Factor*	*Sinking Fund Factor*	*Compound Amount Factor*	*Gradient Factor*	
N	*(F/P, 40, N)*	*(P/F, 40, N)*	*(A/P, 40, N)*	*(P/A, 40, N)*	*(A/F, 40, N)*	*(F/A, 40, N)*	*(A/G, 40, N)*	*N*
1	1.40000	0.71429	1.40000	0.71429	1.00000	1.00000	0.00000	1
2	1.96000	0.51020	0.81667	1.22449	0.41667	2.40000	0.41667	2
3	2.74400	0.36443	0.62936	1.58892	0.22936	4.35000	0.77982	3
4	3.84160	0.26031	0.54077	1.84923	0.14077	7.10400	1.09234	4
5	5.37824	0.18593	0.49136	2.03516	0.09136	10.94560	1.35799	5
6	7.52954	0.13281	0.46126	2.16797	0.06126	16.32384	1.58110	6
7	10.54135	0.09486	0.44192	2.26284	0.04192	23.85338	1.76635	7
8	14.75789	0.06776	0.42907	2.33060	0.02907	34.39473	1.91852	8
9	20.66105	0.04840	0.42034	2.37900	0.02034	49.15262	2.04224	9
10	28.92547	0.03457	0.41432	2.41357	0.01432	69.81366	2.14190	10
11	40.49565	0.02469	0.41013	2.43826	0.01013	98.73913	2.22149	11
12	56.69391	0.01764	0.40718	2.45590	0.00718	139.23478	2.28454	12
13	79.37148	0.01260	0.40510	2.46850	0.00610	195.92869	2.33412	13
14	111.12007	0.00900	0.40363	2.47750	0.00363	275.30017	2.37287	14
15	155.56810	0.00643	0.40259	2.48393	0.00259	386.42024	2.40296	15
16	217.79533	0.00459	0.40185	2.48852	0.00185	541.98833	2.42620	16
17	304.91347	0.00328	0.40132	2.49180	0.00132	759.78367	2.44406	17
18	426.87885	0.00234	0.40094	2.49414	0.00094	1064.69714	2.45773	18
19	597.63040	0.00167	0.40067	2.49582	0.00067	1491.57599	2.46815	19
20	836.68255	0.00120	0.40048	2.49701	0.00048	2089.20639	2.47607	20
21	1171.35558	0.00085	0.40034	2.49787	0.00034	2925.88894	2.48206	21
22	1639.89781	0.00061	0.40024	2.49848	0.00024	4097.24452	2.48658	22
23	2296.85693	0.00044	0.40017	2.49891	0.00017	5737.14232	2.48998	23
24	3214.19970	0.00031	0.40012	2.49922	0.00012	8032.99925	2.49253	24
25	4499.87958	0.00022	0.40009	2.49944	0.00009	11247.19895	2.49444	25

50% interest factors for discrete compounding periods

	Single Payment		*Uniform Series*					
	Compound Amount Factor	*Present Worth Factor*	*Capital Recovery Factor*	*Present Worth Factor*	*Sinking Fund Factor*	*Compound Amount Factor*	*Gradient Factor*	
N	*(F/P, 50, N)*	*(P/F, 50, N)*	*(A/P, 50, N)*	*(P/A, 50, N)*	*(A/F, 50, N)*	*(F/A, 50, N)*	*(A/G, 50, N)*	*N*
1	1.50000	0.66667	1.50000	0.66667	1.00000	1.00000	0.00000	1
2	2.25000	0.44444	0.90000	1.11111	0.40000	2.50000	0.40000	2
3	3.37500	0.29630	0.71053	1.40741	0.21053	4.75000	0.73684	3
4	5.06250	0.19753	0.62308	1.60494	0.12308	8.12500	1.01538	4
5	7.59375	0.13169	0.57583	1.73663	0.07583	13.18750	1.24171	5
6	11.39063	0.08779	0.54812	1.82442	0.04812	20.78125	1.42256	6
7	17.08594	0.05853	0.53108	1.88294	0.03108	32.17188	1.56484	7
8	25.62891	0.03902	0.52030	1.92196	0.02030	49.25781	1.67518	8
9	38.44336	0.02601	0.51335	1.94798	0.01335	74.88672	1.75964	9
10	57.66504	0.01734	0.50882	1.96532	0.00882	113.33008	1.82352	10
11	86.49756	0.01156	0.50585	1.97688	0.00585	170.99512	1.87134	11
12	129.74634	0.00771	0.50388	1.98459	0.00388	257.49268	1.90679	12
13	194.61951	0.00514	0.50258	1.98972	0.00258	387.23901	1.93286	13
14	291.92926	0.00343	0.50172	1.99315	0.00172	581.85852	1.95188	14
15	437.89389	0.00228	0.50114	1.99543	0.00114	873.78778	1.96567	15
16	656.84084	0.00152	0.50076	1.99696	0.00076	1311.68167	1.97560	16
17	985.26125	0.00101	0.50051	1.99797	0.00051	1968.52251	1.98273	17
18	1477.89188	0.00068	0.50034	1.99865	0.00034	2953.78376	1.98781	18
19	2216.83782	0.00045	0.50023	1.99910	0.00023	4431.67564	1.99143	19
20	3325.25673	0.00030	0.50015	1.99940	0.00015	6648.51346	1.99398	20
21	4987.88510	0.00020	0.50010	1.99960	0.00010	9973.77019	1.99579	21
22	7481.82764	0.00013	0.50007	1.99973	0.00007	14961.65529	1.99706	22
23	11222.74146	0.00009	0.50004	1.99982	0.00004	22443.48293	1.99795	23
24	16834.11220	0.00006	0.50003	1.99988	0.00003	33666.22439	1.99857	24
25	25251.16829	0.00004	0.50002	1.99992	0.00002	50500.33659	1.99901	25.

60% interest factors for discrete compounding periods

	Single Payment		Uniform Series					
	Compound Amount Factor	*Present Worth Factor*	*Capital Recovery Factor*	*Present Worth Factor*	*Sinking Fund Factor*	*Compound Amount Factor*	*Gradient Factor*	
N	*(F/P, 60, N)*	*(P/F, 60, N)*	*(A/P, 60, N)*	*(P/A, 60, N)*	*(A/F, 60, N)*	*(F/A, 60, N)*	*(A/G, 60, N)*	*N*
1	1.60000	0.62500	1.60000	0.62500	1.00000	1.00000	0.00000	1
2	2.56000	0.39063	0.98462	1.01563	0.38462	2.60000	0.38462	2
3	4.09600	0.24414	0.79380	1.25977	0.19380	5.16000	0.69767	3
4	6.55360	0.15259	0.70804	1.41235	0.10804	9.25600	0.94641	4
5	10.48576	0.09537	0.66325	1.50772	0.06325	15.80980	1.13956	5
6	16.77722	0.05960	0.63803	1.56733	0.03803	26.29536	1.28637	6
7	26.84355	0.03725	0.62322	1.60458	0.02322	43.07258	1.39581	7
8	42.94967	0.02328	0.61430	1.62786	0.01430	69.91612	1.47596	8
9	68.71948	0.01455	0.60886	1.64241	0.00886	112.86579	1.53377	9
10	109.95116	0.00909	0.60551	1.65151	0.00551	181.58527	1.57488	10
11	175.92186	0.00568	0.60343	1.65719	0.00343	291.53643	1.60378	11
12	281.47498	0.00355	0.60214	1.66075	0.00214	467.45829	1.62388	12
13	450.35996	0.00222	0.60134	1.66297	0.00134	748.93327	1.63774	13
14	720.57594	0.00139	0.60083	1.66435	0.00083	1199.29323	1.64721	14
15	1152.92150	0.00087	0.60052	1.66522	0.00052	1919.86917	1.65364	15
16	1844.67441	0.00054	0.60033	1.66576	0.00033	3072.79068	1.65799	16
17	2951.47905	0.00034	0.60020	1.66610	0.00020	4917.46509	1.69090	17
18	4722.36648	0.00021	0.60013	1.66631	0.00013	7868.94414	1.66285	18
19	7555.78637	0.00013	0.60008	1.66645	0.00008	12591.31062	1.66415	19
20	12089.25820	0.00008	0.60005	1.66653	0.00005	20147.09699	1.66501	20

70% interest factors for discrete compounding periods

	Single Payment		Uniform Series					
	Compound Amount Factor	*Present Worth Factor*	*Capital Recovery Factor*	*Present Worth Factor*	*Sinking Fund Factor*	*Compound Amount Factor*	*Gradient Factor*	
N	*(F/P, 70, N)*	*(P/F, 70, N)*	*(A/P, 70, N)*	*(P/A, 70, N)*	*(A/F, 70, N)*	*(F/A, 70, N)*	*(A/G, 70, N)*	*N*
1	1.70000	0.58824	1.70000	0.58824	1.00000	1.00000	0.00000	1
2	2.89000	0.34602	1.07037	0.93426	0.37037	2.70000	0.37037	2
3	4.91300	0.20354	0.87889	1.13780	0.17889	5.59000	0.66190	3
4	8.35210	0.11973	0.79521	1.25753	0.09521	10.50300	0.88451	4
5	14.19857	0.07043	0.75304	1.32796	0.05304	18.85510	1.04974	5
6	24.13757	0.04143	0.73025	1.36939	0.03025	33.05367	1.16925	6
7	41.03387	0.02437	0.71749	1.39376	0.01749	57.19124	1.25372	7
8	69.75757	0.01434	0.71018	1.40809	0.01018	98.22511	1.31222	8
9	118.58788	0.00843	0.70595	1.41652	0.00595	167.98268	1.35203	9
10	201.59939	0.00496	0.70349	1.42149	0.00349	286.57056	1.37872	10
11	342.71896	0.00292	0.70205	1.42440	0.00205	486.16995	1.39638	11
12	582.62224	0.00172	0.70120	1.42612	0.00120	830.88891	1.40794	12
13	990.45780	0.00101	0.70071	1.42713	0.00071	1413.51115	1.41543	13
14	1683.77827	0.00059	0.70042	1.42772	0.00042	2403.96995	1.42020	14
15	2862.42305	0.00035	0.70024	1.42807	0.00024	4087.74722	1.42333	15
16	4866.11919	0.00021	0.70014	1.42828	0.00014	6950.17027	1.42528	16
17	8272.40262	0.00012	0.70008	1.42840	0.00008	11816.28946	1.42652	17
18	14063.08445	0.00007	0.70005	1.42847	0.00005	20088.69207	1.42729	18
19	23907.24357	0.00004	0.70003	1.42851	0.00003	34151.77653	1.42778	19
20	40642.31407	0.00002	0.70002	1.42854	0.00002	58059.02009	1.42806	20

80% interest factors for discrete compounding periods

	Single Payment		Uniform Series					
	Compound Amount Factor	*Present Worth Factor*	*Capital Recovery Factor*	*Present Worth Factor*	*Sinking Fund Factor*	*Compound Amount Factor*	*Gradient Factor*	
N	*(F/P, 80, N)*	*(P/F, 80, N)*	*(A/P, 80, N)*	*(P/A, 80, N)*	*(A/F, 80, N)*	*(F/A, 80, N)*	*(A/G, 80, N)*	*N*
1	1.80000	0.55556	1.80000	0.55556	1.00000	1.00000	0.00000	1
2	3.24000	0.30864	1.15714	0.86420	0.35714	2.80000	0.35714	2
3	5.83200	0.17147	0.96556	1.03567	0.16556	6.04000	0.62914	3
4	10.49760	0.09526	0.88423	1.13093	0.08423	11.87200	0.82884	4
5	18.89568	0.05292	0.84470	1.18385	0.04470	22.36960	0.97060	5
6	34.01222	0.02940	0.82423	1.21325	0.02423	41.26528	1.06825	6
7	61.22200	0.01633	0.81328	1.22958	0.01328	75.27750	1.13376	7
8	110.19961	0.00907	0.80733	1.23866	0.00733	136.49951	1.17674	8
9	198.35929	0.00504	0.80405	1.24370	0.00405	246.69911	1.20440	9
10	357.04672	0.00280	0.80225	1.24650	0.00225	445.05840	1.22191	10
11	642.68410	0.00156	0.80125	1.24806	0.00125	802.10513	1.23286	11
12	1156.83138	0.00086	0.80069	1.24892	0.00069	1444.78923	1.23962	12
13	2082.29649	0.00048	0.80038	1.24940	0.00038	2601.62061	1.24375	13
14	3748.13368	0.00027	0.80021	1.24967	0.00021	4683.91709	1.24626	14
15	6746.64082	0.00015	0.80012	1.24981	0.00012	8432.05077	1.24778	15

90% interest factors for discrete compounding periods

	Single Payment		*Uniform Series*					
	Compound Amount Factor	*Present Worth Factor*	*Capital Recovery Factor*	*Present Worth Factor*	*Sinking Fund Factor*	*Compound Amount Factor*	*Gradient Factor*	
N	*(F/P, 90, N)*	*(P/F, 90, N)*	*(A/P, 90, N)*	*(P/A, 90, N)*	*(A/F, 90, N)*	*(F3/A, 90, N)*	*(A/G, 90, N)*	*N*
1	1.90000	0.52632	1.90000	0.52632	1.00000	1.00000	0.00000	1
2	3.61000	0.27701	1.24483	0.80332	0.34483	2.90000	0.34483	2
3	6.85900	0.14579	1.05361	0.94912	0.15361	6.51000	0.59908	3
4	13.03210	0.07673	0.97480	1.02585	0.07480	13.36900	0.77867	4
5	24.76099	0.04039	0.93788	1.06624	0.03788	26.40110	0.90068	5
6	47.04588	0.02126	0.91955	1.08749	0.01955	51.16209	0.98081	6
7	89.38717	0.01119	0.91018	1.09868	0.01018	98.20797	1.03191	7
8	169.83563	0.00589	0.90533	1.10457	0.00533	187.59514	1.06373	8
9	322.68770	0.00310	0.90280	1.10767	0.00280	357.43078	1.08313	9
10	613.10863	0.00163	0.90147	1.10930	0.00147	680.11847	1.09477	10

100% interest factors for discrete compounding periods

	Single Payment		*Uniform Series*					
	Compound Amount Factor	*Present Worth Factor*	*Capital Recovery Factor*	*Present Worth Factor*	*Sinking Fund Factor*	*Compound Amount Factor*	*Gradient Factor*	
N	*(F/P, 100, N)*	*(P/F, 100, N)*	*(A/P, 100, N)*	*(P/A, 100, N)*	*(A/F, 100, N)*	*(F/A, 100, N)*	*(A/G, 100, N)*	*N*
1	2.00000	0.50000	2.00000	0.50000	1.00000	1.00000	0.00000	1
2	4.00000	0.25000	1.33333	0.75000	0.33333	3.00000	0.33333	2
3	8.00000	0.12500	1.14286	0.87500	0.14286	7.00000	0.57143	3
4	16.00000	0.06250	1.06667	0.93750	0.06667	15.00000	0.73333	4
5	32.00000	0.03125	1.03226	0.96875	0.03226	31.00000	0.83871	5
6	64.00000	0.01563	1.01587	0.98438	0.01587	63.00000	0.90476	6
7	128.00000	0.00781	1.00787	0.99219	0.00787	127.00000	0.94488	7
8	256.00000	0.00391	1.00392	0.99609	0.00392	255.00000	0.96863	8
9	512.00000	0.00195	1.00196	0.99805	0.00196	511.00000	0.98239	9
10	1024.00000	0.00098	1.00098	0.99902	0.00098	1023.00000	0.99022	10